华章数学原版精品系列

数学分析原理

(英文版·原书第3版·典藏版)

Principles of Mathematical Analysis
(Third Edition)

[美] 沃尔特·鲁丁（Walter Rudin）著

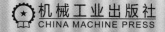

图书在版编目（CIP）数据

数学分析原理（英文版·原书第3版·典藏版）/（美）沃尔特·鲁丁（Walter Rudin）著 . —北京：机械工业出版社，2019.2（2023.8重印）

（华章数学原版精品系列）

书名原文：Principles of Mathematical Analysis, Third Edition

ISBN 978-7-111-61954-3

I. 数… II. 沃… III. 数学分析–英文 IV. O17

中国版本图书馆 CIP 数据核字（2019）第 024497 号

北京市版权局著作权合同登记　图字：01-2018-8342 号。

Walter Rudin: Principles of Mathematical Analysis, Third Edition (ISBN: 0-07-054235-X).

Copyright © 1964, 1976 by McGraw-Hill Education.

All Rights reserved. No part of this publication may be reproduced or transmitted in any form or by any means, electronic or mechanical, including without limitation photocopying, recording, taping, or any database, information or retrieval system, without the prior written permission of the publisher.

This authorized English reprint edition is jointly published by McGraw-Hill Education and China Machine Press. This edition is authorized for sale in the Chinese mainland (excluding Hong Kong SAR, Macao SAR and Taiwan).

Copyright © 2019 by McGraw-Hill Education and China Machine Press.

本书英文影印版由麦格劳-希尔教育出版公司和机械工业出版社合作出版．此版本仅限在中国大陆地区（不包括香港、澳门特别行政区及台湾地区）销售．未经许可之出口，视为违反著作权法，将受法律之制裁．

未经出版者预先书面许可，不得以任何方式复制或抄袭本书的任何部分．

本书封面贴有 McGraw-Hill 公司防伪标签，无标签者不得销售．

版权所有，侵权必究．

出版发行：机械工业出版社（北京市西城区百万庄大街 22 号　邮政编码：100037）	
责任编辑：刘立卿	责任校对：殷　虹
印　　刷：北京铭成印刷有限公司	版　　次：2023 年 8 月第 1 版第 7 次印刷
开　　本：186mm×240mm　1/16	印　　张：22
书　　号：ISBN 978-7-111-61954-3	定　　价：69.00 元

客服电话：（010）88361066　68326294

版权所有·侵权必究
封底无防伪标均为盗版

PREFACE

This book is intended to serve as a text for the course in analysis that is usually taken by advanced undergraduates or by first-year students who study mathematics.

The present edition covers essentially the same topics as the second one, with some additions, a few minor omissions, and considerable rearrangement. I hope that these changes will make the material more accessible amd more attractive to the students who take such a course.

Experience has convinced me that it is pedagogically unsound (though logically correct) to start off with the construction of the real numbers from the rational ones. At the beginning, most students simply fail to appreciate the need for doing this. Accordingly, the real number system is introduced as an ordered field with the least-upper-bound property, and a few interesting applications of this property are quickly made. However, Dedekind's construction is not omitted. It is now in an Appendix to Chapter 1, where it may be studied and enjoyed whenever the time seems ripe.

The material on functions of several variables is almost completely rewritten, with many details filled in, and with more examples and more motivation. The proof of the inverse function theorem—the key item in Chapter 9—is simplified by means of the fixed point theorem about contraction mappings. Differential forms are discussed in much greater detail. Several applications of Stokes' theorem are included.

As regards other changes, the chapter on the Riemann-Stieltjes integral has been trimmed a bit, a short do-it-yourself section on the gamma function has been added to Chapter 8, and there is a large number of new exercises, most of them with fairly detailed hints.

I have also included several references to articles appearing in the *American Mathematical Monthly* and in *Mathematics Magazine*, in the hope that students will develop the habit of looking into the journal literature. Most of these references were kindly supplied by R. B. Burckel.

Over the years, many people, students as well as teachers, have sent me corrections, criticisms, and other comments concerning the previous editions of this book. I have appreciated these, and I take this opportunity to express my sincere thanks to all who have written me.

WALTER RUDIN

CONTENTS

Preface

Chapter 1	**The Real and Complex Number Systems**	**1**
	Introduction	1
	Ordered Sets	3
	Fields	5
	The Real Field	8
	The Extended Real Number System	11
	The Complex Field	12
	Euclidean Spaces	16
	Appendix	17
	Exercises	21
Chapter 2	**Basic Topology**	**24**
	Finite, Countable, and Uncountable Sets	24
	Metric Spaces	30
	Compact Sets	36
	Perfect Sets	41

	Connected Sets	42
	Exercises	43

Chapter 3 Numerical Sequences and Series — 47

	Convergent Sequences	47
	Subsequences	51
	Cauchy Sequences	52
	Upper and Lower Limits	55
	Some Special Sequences	57
	Series	58
	Series of Nonnegative Terms	61
	The Number e	63
	The Root and Ratio Tests	65
	Power Series	69
	Summation by Parts	70
	Absolute Convergence	71
	Addition and Multiplication of Series	72
	Rearrangements	75
	Exercises	78

Chapter 4 Continuity — 83

	Limits of Functions	83
	Continuous Functions	85
	Continuity and Compactness	89
	Continuity and Connectedness	93
	Discontinuities	94
	Monotonic Functions	95
	Infinite Limits and Limits at Infinity	97
	Exercises	98

Chapter 5 Differentiation — 103

	The Derivative of a Real Function	103
	Mean Value Theorems	107
	The Continuity of Derivatives	108
	L'Hospital's Rule	109
	Derivatives of Higher Order	110
	Taylor's Theorem	110
	Differentiation of Vector-valued Functions	111
	Exercises	114

Chapter 6 The Riemann-Stieltjes Integral — 120
Definition and Existence of the Integral — 120
Properties of the Integral — 128
Integration and Differentiation — 133
Integration of Vector-valued Functions — 135
Rectifiable Curves — 136
Exercises — 138

Chapter 7 Sequences and Series of Functions — 143
Discussion of Main Problem — 143
Uniform Convergence — 147
Uniform Convergence and Continuity — 149
Uniform Convergence and Integration — 151
Uniform Convergence and Differentiation — 152
Equicontinuous Families of Functions — 154
The Stone-Weierstrass Theorem — 159
Exercises — 165

Chapter 8 Some Special Functions — 172
Power Series — 172
The Exponential and Logarithmic Functions — 178
The Trigonometric Functions — 182
The Algebraic Completeness of the Complex Field — 184
Fourier Series — 185
The Gamma Function — 192
Exercises — 196

Chapter 9 Functions of Several Variables — 204
Linear Transformations — 204
Differentiation — 211
The Contraction Principle — 220
The Inverse Function Theorem — 221
The Implicit Function Theorem — 223
The Rank Theorem — 228
Determinants — 231
Derivatives of Higher Order — 235
Differentiation of Integrals — 236
Exercises — 239

Chapter 10 Integration of Differential Forms — 245
Integration — 245

Primitive Mappings	248
Partitions of Unity	251
Change of Variables	252
Differential Forms	253
Simplexes and Chains	266
Stokes' Theorem	273
Closed Forms and Exact Forms	275
Vector Analysis	280
Exercises	288

Chapter 11   The Lebesgue Theory — **300**

Set Functions	300
Construction of the Lebesgue Measure	302
Measure Spaces	310
Measurable Functions	310
Simple Functions	313
Integration	314
Comparison with the Riemann Integral	322
Integration of Complex Functions	325
Functions of Class $\mathscr{L}^2$	325
Exercises	332

Bibliography   **335**

List of Special Symbols   **337**

Index   **339**

1
THE REAL AND COMPLEX NUMBER SYSTEMS

INTRODUCTION

A satisfactory discussion of the main concepts of analysis (such as convergence, continuity, differentiation, and integration) must be based on an accurately defined number concept. We shall not, however, enter into any discussion of the axioms that govern the arithmetic of the integers, but assume familiarity with the rational numbers (i.e., the numbers of the form m/n, where m and n are integers and $n \neq 0$).

The rational number system is inadequate for many purposes, both as a field and as an ordered set. (These terms will be defined in Secs. 1.6 and 1.12.) For instance, there is no rational p such that $p^2 = 2$. (We shall prove this presently.) This leads to the introduction of so-called "irrational numbers" which are often written as infinite decimal expansions and are considered to be "approximated" by the corresponding finite decimals. Thus the sequence

$$1, 1.4, 1.41, 1.414, 1.4142, \ldots$$

"tends to $\sqrt{2}$." But unless the irrational number $\sqrt{2}$ has been clearly defined, the question must arise: Just what is it that this sequence "tends to"?

This sort of question can be answered as soon as the so-called "real number system" is constructed.

1.1 Example We now show that the equation

(1) $$p^2 = 2$$

is not satisfied by any rational p. If there were such a p, we could write $p = m/n$ where m and n are integers that are not both even. Let us assume this is done. Then (1) implies

(2) $$m^2 = 2n^2,$$

This shows that m^2 is even. Hence m is even (if m were odd, m^2 would be odd), and so m^2 is divisible by 4. It follows that the right side of (2) is divisible by 4, so that n^2 is even, which implies that n is even.

The assumption that (1) holds thus leads to the conclusion that both m and n are even, contrary to our choice of m and n. Hence (1) is impossible for rational p.

We now examine this situation a little more closely. Let A be the set of all positive rationals p such that $p^2 < 2$ and let B consist of all positive rationals p such that $p^2 > 2$. We shall show that *A contains no largest number and B contains no smallest.*

More explicitly, for every p in A we can find a rational q in A such that $p < q$, and for every p in B we can find a rational q in B such that $q < p$.

To do this, we associate with each rational $p > 0$ the number

(3) $$q = p - \frac{p^2 - 2}{p + 2} = \frac{2p + 2}{p + 2}.$$

Then

(4) $$q^2 - 2 = \frac{2(p^2 - 2)}{(p + 2)^2}.$$

If p is in A then $p^2 - 2 < 0$, (3) shows that $q > p$, and (4) shows that $q^2 < 2$. Thus q is in A.

If p is in B then $p^2 - 2 > 0$, (3) shows that $0 < q < p$, and (4) shows that $q^2 > 2$. Thus q is in B.

1.2 Remark The purpose of the above discussion has been to show that the rational number system has certain gaps, in spite of the fact that between any two rationals there is another: If $r < s$ then $r < (r + s)/2 < s$. The real number system fills these gaps. This is the principal reason for the fundamental role which it plays in analysis.

In order to elucidate its structure, as well as that of the complex numbers, we start with a brief discussion of the general concepts of *ordered set* and *field*.

Here is some of the standard set-theoretic terminology that will be used throughout this book.

1.3 Definitions If A is any set (whose elements may be numbers or any other objects), we write $x \in A$ to indicate that x is a member (or an element) of A.

If x is not a member of A, we write: $x \notin A$.

The set which contains no element will be called the *empty set*. If a set has at least one element, it is called *nonempty*.

If A and B are sets, and if every element of A is an element of B, we say that A is a subset of B, and write $A \subset B$, or $B \supset A$. If, in addition, there is an element of B which is not in A, then A is said to be a *proper* subset of B. Note that $A \subset A$ for every set A.

If $A \subset B$ and $B \subset A$, we write $A = B$. Otherwise $A \neq B$.

1.4 Definition Throughout Chap. 1, the set of all rational numbers will be denoted by Q.

ORDERED SETS

1.5 Definition Let S be a set. An *order* on S is a relation, denoted by $<$, with the following two properties:

(i) If $x \in S$ and $y \in S$ then one and only one of the statements

$$x < y, \quad x = y, \quad y < x$$

is true.

(ii) If $x, y, z \in S$, if $x < y$ and $y < z$, then $x < z$.

The statement "$x < y$" may be read as "x is less than y" or "x is smaller than y" or "x precedes y".

It is often convenient to write $y > x$ in place of $x < y$.

The notation $x \leq y$ indicates that $x < y$ or $x = y$, without specifying which of these two is to hold. In other words, $x \leq y$ is the negation of $x > y$.

1.6 Definition An *ordered set* is a set S in which an order is defined.

For example, Q is an ordered set if $r < s$ is defined to mean that $s - r$ is a positive rational number.

1.7 Definition Suppose S is an ordered set, and $E \subset S$. If there exists a $\beta \in S$ such that $x \leq \beta$ for every $x \in E$, we say that E is *bounded above*, and call β an *upper bound* of E.

Lower bounds are defined in the same way (with $\geq$ in place of $\leq$).

1.8 Definition Suppose S is an ordered set, $E \subset S$, and E is bounded above. Suppose there exists an $\alpha \in S$ with the following properties:

(i) α is an upper bound of E.
(ii) If $\gamma < \alpha$ then γ is not an upper bound of E.

Then α is called the *least upper bound of E* [that there is at most one such α is clear from (ii)] or the *supremum of E*, and we write

$$\alpha = \sup E.$$

The *greatest lower bound*, or *infimum*, of a set E which is bounded below is defined in the same manner: The statement

$$\alpha = \inf E$$

means that α is a lower bound of E and that no β with $\beta > \alpha$ is a lower bound of E.

1.9 Examples

(a) Consider the sets A and B of Example 1.1 as subsets of the ordered set Q. The set A is bounded above. In fact, the upper bounds of A are exactly the members of B. Since B contains no smallest member, A has no least upper bound in Q.

Similarly, B is bounded below: The set of all lower bounds of B consists of A and of all $r \in Q$ with $r \leq 0$. Since A has no lasgest member, B has no greatest lower bound in Q.

(b) If $\alpha = \sup E$ exists, then α may or may not be a member of E. For instance, let E_1 be the set of all $r \in Q$ with $r < 0$. Let E_2 be the set of all $r \in Q$ with $r \leq 0$. Then

$$\sup E_1 = \sup E_2 = 0,$$

and $0 \notin E_1$, $0 \in E_2$.

(c) Let E consist of all numbers $1/n$, where $n = 1, 2, 3, \ldots$. Then $\sup E = 1$, which is in E, and $\inf E = 0$, which is not in E.

1.10 Definition An ordered set S is said to have the *least-upper-bound property* if the following is true:

If $E \subset S$, E is not empty, and E is bounded above, then $\sup E$ exists in S.

Example 1.9(a) shows that Q does not have the least-upper-bound property.

We shall now show that there is a close relation between greatest lower bounds and least upper bounds, and that every ordered set with the least-upper-bound property also has the greatest-lower-bound property.

1.11 Theorem *Suppose S is an ordered set with the least-upper-bound property, $B \subset S$, B is not empty, and B is bounded below. Let L be the set of all lower bounds of B. Then*
$$\alpha = \sup L$$
exists in S, and $\alpha = \inf B$.

In particular, inf B *exists in* S.

Proof Since B is bounded below, L is not empty. Since L consists of exactly those $y \in S$ which satisfy the inequality $y \leq x$ for every $x \in B$, we see that *every $x \in B$ is an upper bound of L*. Thus L is bounded above. Our hypothesis about S implies therefore that L has a supremum in S; call it α.

If $\gamma < \alpha$ then (see Definition 1.8) γ is not an upper bound of L, hence $\gamma \notin B$. It follows that $\alpha \leq x$ for every $x \in B$. Thus $\alpha \in L$.

If $\alpha < \beta$ then $\beta \notin L$, since α is an upper bound of L.

We have shown that $\alpha \in L$ but $\beta \notin L$ if $\beta > \alpha$. In other words, α is a lower bound of B, but β is not if $\beta > \alpha$. This means that $\alpha = \inf B$.

FIELDS

1.12 Definition A *field* is a set F with two operations, called *addition* and *multiplication*, which satisfy the following so-called "field axioms" (A), (M), and (D):

(A) Axioms for addition

(A1) If $x \in F$ and $y \in F$, then their sum $x + y$ is in F.
(A2) Addition is commutative: $x + y = y + x$ for all $x, y \in F$.
(A3) Addition is associative: $(x + y) + z = x + (y + z)$ for all $x, y, z \in F$.
(A4) F contains an element 0 such that $0 + x = x$ for every $x \in F$.
(A5) To every $x \in F$ corresponds an element $-x \in F$ such that
$$x + (-x) = 0.$$

(M) Axioms for multiplication

(M1) If $x \in F$ and $y \in F$, then their product xy is in F.
(M2) Multiplication is commutative: $xy = yx$ for all $x, y \in F$.
(M3) Multiplication is associative: $(xy)z = x(yz)$ for all $x, y, z \in F$.
(M4) F contains an element $1 \neq 0$ such that $1x = x$ for every $x \in F$.
(M5) If $x \in F$ and $x \neq 0$ then there exists an element $1/x \in F$ such that
$$x \cdot (1/x) = 1.$$

6 PRINCIPLES OF MATHEMATICAL ANALYSIS

(D) The distributive law
$$x(y + z) = xy + xz$$
holds for all $x, y, z \in F$.

1.13 Remarks

(a) One usually writes (in any field)
$$x - y, \frac{x}{y}, x + y + z, xyz, x^2, x^3, 2x, 3x, \ldots$$
in place of
$$x + (-y), x \cdot \left(\frac{1}{y}\right), (x + y) + z, (xy)z, xx, xxx, x + x, x + x + x, \ldots.$$

(b) The field axioms clearly hold in Q, the set of all rational numbers, if addition and multiplication have their customary meaning. Thus Q is a field.

(c) Although it is not our purpose to study fields (or any other algebraic structures) in detail, it is worthwhile to prove that some familiar properties of Q are consequences of the field axioms; once we do this, we will not need to do it again for the real numbers and for the complex numbers.

1.14 Proposition *The axioms for addition imply the following statements.*

(a) *If $x + y = x + z$ then $y = z$.*
(b) *If $x + y = x$ then $y = 0$.*
(c) *If $x + y = 0$ then $y = -x$.*
(d) $-(-x) = x$.

Statement (a) is a cancellation law. Note that (b) asserts the uniqueness of the element whose existence is assumed in (A4), and that (c) does the same for (A5).

Proof If $x + y = x + z$, the axioms (A) give
$$y = 0 + y = (-x + x) + y = -x + (x + y)$$
$$= -x + (x + z) = (-x + x) + z = 0 + z = z.$$

This proves (a). Take $z = 0$ in (a) to obtain (b). Take $z = -x$ in (a) to obtain (c).
Since $-x + x = 0$, (c) (with $-x$ in place of x) gives (d).

1.15 Proposition *The axioms for multiplication imply the following statements.*

(a) *If $x \neq 0$ and $xy = xz$ then $y = z$.*
(b) *If $x \neq 0$ and $xy = x$ then $y = 1$.*
(c) *If $x \neq 0$ and $xy = 1$ then $y = 1/x$.*
(d) *If $x \neq 0$ then $1/(1/x) = x$.*

The proof is so similar to that of Proposition 1.14 that we omit it.

1.16 Proposition *The field axioms imply the following statements, for any x, y, $z \in F$.*

(a) $0x = 0$.
(b) *If $x \neq 0$ and $y \neq 0$ then $xy \neq 0$.*
(c) $(-x)y = -(xy) = x(-y)$.
(d) $(-x)(-y) = xy$.

Proof $0x + 0x = (0 + 0)x = 0x$. Hence 1.14(b) implies that $0x = 0$, and (a) holds.

Next, assume $x \neq 0$, $y \neq 0$, but $xy = 0$. Then (a) gives

$$1 = \left(\frac{1}{y}\right)\left(\frac{1}{x}\right)xy = \left(\frac{1}{y}\right)\left(\frac{1}{x}\right)0 = 0,$$

a contradiction. Thus (b) holds.

The first equality in (c) comes from

$$(-x)y + xy = (-x + x)y = 0y = 0,$$

combined with 1.14(c); the other half of (c) is proved in the same way. Finally,

$$(-x)(-y) = -[x(-y)] = -[-(xy)] = xy$$

by (c) and 1.14(d).

1.17 Definition An *ordered field* is a *field F* which is also an *ordered set*, such that

(i) $x + y < x + z$ if $x, y, z \in F$ and $y < z$,
(ii) $xy > 0$ if $x \in F$, $y \in F$, $x > 0$, and $y > 0$.

If $x > 0$, we call x *positive*; if $x < 0$, x is *negative*.

For example, Q is an ordered field.

All the familiar rules for working with inequalities apply in every ordered field: Multiplication by positive [negative] quantities preserves [reverses] inequalities, no square is negative, etc. The following proposition lists some of these.

1.18 Proposition *The following statements are true in every ordered field.*

(a) If $x > 0$ then $-x < 0$, and vice versa.
(b) If $x > 0$ and $y < z$ then $xy < xz$.
(c) If $x < 0$ and $y < z$ then $xy > xz$.
(d) If $x \neq 0$ then $x^2 > 0$. In particular, $1 > 0$.
(e) If $0 < x < y$ then $0 < 1/y < 1/x$.

Proof
(a) If $x > 0$ then $0 = -x + x > -x + 0$, so that $-x < 0$. If $x < 0$ then $0 = -x + x < -x + 0$, so that $-x > 0$. This proves (a).
(b) Since $z > y$, we have $z - y > y - y = 0$, hence $x(z - y) > 0$, and therefore

$$xz = x(z - y) + xy > 0 + xy = xy.$$

(c) By (a), (b), and Proposition 1.16(c),

$$-[x(z - y)] = (-x)(z - y) > 0,$$

so that $x(z - y) < 0$, hence $xz < xy$.
(d) If $x > 0$, part (ii) of Definition 1.17 gives $x^2 > 0$. If $x < 0$, then $-x > 0$, hence $(-x)^2 > 0$. But $x^2 = (-x)^2$, by Proposition 1.16(d). Since $1 = 1^2$, $1 > 0$.
(e) If $y > 0$ and $v \leq 0$, then $yv \leq 0$. But $y \cdot (1/y) = 1 > 0$. Hence $1/y > 0$. Likewise, $1/x > 0$. If we multiply both sides of the inequality $x < y$ by the positive quantity $(1/x)(1/y)$, we obtain $1/y < 1/x$.

THE REAL FIELD

We now state the *existence theorem* which is the core of this chapter.

1.19 Theorem *There exists an ordered field R which has the least-upper-bound property.*

Moreover, R contains Q as a subfield.

The second statement means that $Q \subset R$ and that the operations of addition and multiplication in R, when applied to members of Q, coincide with the usual operations on rational numbers; also, the positive rational numbers are positive elements of R.

The members of R are called *real numbers*.

The proof of Theorem 1.19 is rather long and a bit tedious and is therefore presented in an Appendix to Chap. 1. The proof actually constructs R from Q.

The next theorem could be extracted from this construction with very little extra effort. However, we prefer to derive it from Theorem 1.19 since this provides a good illustration of what one can do with the least-upper-bound property.

1.20 Theorem

(a) *If $x \in R$, $y \in R$, and $x > 0$, then there is a positive integer n such that*
$$nx > y.$$
(b) *If $x \in R, y \in R$, and $x < y$, then there exists a $p \in Q$ such that $x < p < y$.*

Part (a) is usually referred to as the *archimedean property* of R. Part (b) may be stated by saying that Q is *dense* in R: Between any two real numbers there is a rational one.

Proof
(a) Let A be the set of all nx, where n runs through the positive integers. If (a) were false, then y would be an upper bound of A. But then A has a *least* upper bound in R. Put $\alpha = \sup A$. Since $x > 0$, $\alpha - x < \alpha$, and $\alpha - x$ is not an upper bound of A. Hence $\alpha - x < mx$ for some positive integer m. But then $\alpha < (m + 1)x \in A$, which is impossible, since α is an upper bound of A.
(b) Since $x < y$, we have $y - x > 0$, and (a) furnishes a positive integer n such that
$$n(y - x) > 1.$$
Apply (a) again, to obtain positive integers m_1 and m_2 such that $m_1 > nx$, $m_2 > -nx$. Then
$$-m_2 < nx < m_1.$$
Hence there is an integer m (with $-m_2 \leq m \leq m_1$) such that
$$m - 1 \leq nx < m.$$
If we combine these inequalities, we obtain
$$nx < m \leq 1 + nx < ny.$$
Since $n > 0$, it follows that
$$x < \frac{m}{n} < y.$$
This proves (b), with $p = m/n$.

We shall now prove the existence of nth roots of positive reals. This proof will show how the difficulty pointed out in the Introduction (irrationality of $\sqrt{2}$) can be handled in R.

1.21 Theorem *For every real $x > 0$ and every integer $n > 0$ there is one and only one positive real y such that $y^n = x$.*

This number y is written $\sqrt[n]{x}$ or $x^{1/n}$.

Proof That there is at most one such y is clear, since $0 < y_1 < y_2$ implies $y_1^n < y_2^n$.

Let E be the set consisting of all positive real numbers t such that $t^n < x$.

If $t = x/(1 + x)$ then $0 \leq t < 1$. Hence $t^n \leq t < x$. Thus $t \in E$, and E is not empty.

If $t > 1 + x$ then $t^n \geq t > x$, so that $t \notin E$. Thus $1 + x$ is an upper bound of E.

Hence Theorem 1.19 implies the existence of

$$y = \sup E.$$

To prove that $y^n = x$ we will show that each of the inequalities $y^n < x$ and $y^n > x$ leads to a contradiction.

The identity $b^n - a^n = (b - a)(b^{n-1} + b^{n-2}a + \cdots + a^{n-1})$ yields the inequality

$$b^n - a^n < (b - a)nb^{n-1}$$

when $0 < a < b$.

Assume $y^n < x$. Choose h so that $0 < h < 1$ and

$$h < \frac{x - y^n}{n(y + 1)^{n-1}}.$$

Put $a = y$, $b = y + h$. Then

$$(y + h)^n - y^n < hn(y + h)^{n-1} < hn(y + 1)^{n-1} < x - y^n.$$

Thus $(y + h)^n < x$, and $y + h \in E$. Since $y + h > y$, this contradicts the fact that y is an upper bound of E.

Assume $y^n > x$. Put

$$k = \frac{y^n - x}{ny^{n-1}}.$$

Then $0 < k < y$. If $t \geq y - k$, we conclude that

$$y^n - t^n \leq y^n - (y - k)^n < kny^{n-1} = y^n - x.$$

Thus $t^n > x$, and $t \notin E$. It follows that $y - k$ is an upper bound of E.

But $y - k < y$, which contradicts the fact that y is the *least* upper bound of E.

Hence $y^n = x$, and the proof is complete.

Corollary *If a and b are positive real numbers and n is a positive integer, then*
$$(ab)^{1/n} = a^{1/n} b^{1/n}.$$

Proof Put $\alpha = a^{1/n}$, $\beta = b^{1/n}$. Then
$$ab = \alpha^n \beta^n = (\alpha\beta)^n,$$
since multiplication is commutative. [Axiom (M2) in Definition 1.12.] The uniqueness assertion of Theorem 1.21 shows therefore that
$$(ab)^{1/n} = \alpha\beta = a^{1/n} b^{1/n}.$$

1.22 Decimals We conclude this section by pointing out the relation between real numbers and decimals.

Let $x > 0$ be real. Let n_0 be the largest integer such that $n_0 \leq x$. (Note that the existence of n_0 depends on the archimedean property of R.) Having chosen $n_0, n_1, \ldots, n_{k-1}$, let n_k be the largest integer such that
$$n_0 + \frac{n_1}{10} + \cdots + \frac{n_k}{10^k} \leq x.$$

Let E be the set of these numbers

(5) $$n_0 + \frac{n_1}{10} + \cdots + \frac{n_k}{10^k} \qquad (k = 0, 1, 2, \ldots).$$

Then $x = \sup E$. The decimal expansion of x is

(6) $$n_0 \cdot n_1 n_2 n_3 \cdots.$$

Conversely, for any infinite decimal (6) the set E of numbers (5) is bounded above, and (6) is the decimal expansion of $\sup E$.

Since we shall never use decimals, we do not enter into a detailed discussion.

THE EXTENDED REAL NUMBER SYSTEM

1.23 Definition The extended real number system consists of the real field R and two symbols, $+\infty$ and $-\infty$. We preserve the original order in R, and define
$$-\infty < x < +\infty$$
for every $x \in R$.

It is then clear that $+\infty$ is an upper bound of every subset of the extended real number system, and that every nonempty subset has a least upper bound. If, for example, E is a nonempty set of real numbers which is not bounded above in R, then $\sup E = +\infty$ in the extended real number system.

Exactly the same remarks apply to lower bounds.

The extended real number system does not form a field, but it is customary to make the following conventions:

(a) If x is real then
$$x + \infty = +\infty, \qquad x - \infty = -\infty, \qquad \frac{x}{+\infty} = \frac{x}{-\infty} = 0.$$

(b) If $x > 0$ then $x \cdot (+\infty) = +\infty$, $x \cdot (-\infty) = -\infty$.
(c) If $x < 0$ then $x \cdot (+\infty) = -\infty$, $x \cdot (-\infty) = +\infty$.

When it is desired to make the distinction between real numbers on the one hand and the symbols $+\infty$ and $-\infty$ on the other quite explicit, the former are called *finite*.

THE COMPLEX FIELD

1.24 Definition A *complex number* is an ordered pair (a, b) of real numbers. "Ordered" means that (a, b) and (b, a) are regarded as distinct if $a \neq b$.

Let $x = (a, b)$, $y = (c, d)$ be two complex numbers. We write $x = y$ if and only if $a = c$ and $b = d$. (Note that this definition is not entirely superfluous; think of equality of rational numbers, represented as quotients of integers.) We define
$$x + y = (a + c, b + d),$$
$$xy = (ac - bd, ad + bc).$$

1.25 Theorem *These definitions of addition and multiplication turn the set of all complex numbers into a field, with $(0, 0)$ and $(1, 0)$ in the role of 0 and 1.*

Proof We simply verify the field axioms, as listed in Definition 1.12. (Of course, we use the field structure of R.)

Let $x = (a, b)$, $y = (c, d)$, $z = (e, f)$.
(A1) is clear.
(A2) $x + y = (a + c, b + d) = (c + a, d + b) = y + x$.

(A3) $(x+y)+z = (a+c, b+d) + (e,f)$
$= (a+c+e, b+d+f)$
$= (a,b) + (c+e, d+f) = x + (y+z)$.
(A4) $x + 0 = (a,b) + (0,0) = (a,b) = x$.
(A5) Put $-x = (-a, -b)$. Then $x + (-x) = (0,0) = 0$.
(M1) is clear.
(M2) $xy = (ac - bd, ad + bc) = (ca - db, da + cb) = yx$.
(M3) $(xy)z = (ac - bd, ad + bc)(e, f)$
$= (ace - bde - adf - bcf, acf - bdf + ade + bce)$
$= (a,b)(ce - df, cf + de) = x(yz)$.
(M4) $1x = (1, 0)(a, b) = (a, b) = x$.
(M5) If $x \neq 0$ then $(a, b) \neq (0, 0)$, which means that at least one of the real numbers a, b is different from 0. Hence $a^2 + b^2 > 0$, by Proposition 1.18(d), and we can define

$$\frac{1}{x} = \left(\frac{a}{a^2+b^2}, \frac{-b}{a^2+b^2} \right).$$

Then

$$x \cdot \frac{1}{x} = (a, b) \left(\frac{a}{a^2+b^2}, \frac{-b}{a^2+b^2} \right) = (1, 0) = 1.$$

(D) $x(y+z) = (a,b)(c+e, d+f)$
$= (ac + ae - bd - bf, ad + af + bc + be)$
$= (ac - bd, ad + bc) + (ae - bf, af + be)$
$= xy + xz$.

1.26 Theorem *For any real numbers a and b we have*

$$(a, 0) + (b, 0) = (a + b, 0), \qquad (a, 0)(b, 0) = (ab, 0).$$

The proof is trivial.

Theorem 1.26 shows that the complex numbers of the form $(a, 0)$ have the same arithmetic properties as the corresponding real numbers a. We can therefore identify $(a, 0)$ with a. This identification gives us the real field as a subfield of the complex field.

The reader may have noticed that we have defined the complex numbers without any reference to the mysterious square root of -1. We now show that the notation (a, b) is equivalent to the more customary $a + bi$.

1.27 Definition $i = (0, 1)$.

1.28 Theorem $i^2 = -1$.

Proof $i^2 = (0, 1)(0, 1) = (-1, 0) = -1$.

1.29 Theorem *If a and b are real, then* $(a, b) = a + bi$.

Proof
$$a + bi = (a, 0) + (b, 0)(0, 1)$$
$$= (a, 0) + (0, b) = (a, b).$$

1.30 Definition If a, b are real and $z = a + bi$, then the complex number $\bar{z} = a - bi$ is called the *conjugate* of z. The numbers a and b are the *real part* and the *imaginary part* of z, respectively.

We shall occasionally write
$$a = \operatorname{Re}(z), \quad b = \operatorname{Im}(z).$$

1.31 Theorem *If z and w are complex, then*

(a) $\overline{z + w} = \bar{z} + \bar{w}$,
(b) $\overline{zw} = \bar{z} \cdot \bar{w}$,
(c) $z + \bar{z} = 2 \operatorname{Re}(z)$, $z - \bar{z} = 2i \operatorname{Im}(z)$,
(d) $z\bar{z}$ *is real and positive* (*except when* $z = 0$).

Proof (a), (b), and (c) are quite trivial. To prove (d), write $z = a + bi$, and note that $z\bar{z} = a^2 + b^2$.

1.32 Definition If z is a complex number, its *absolute value* $|z|$ is the non-negative square root of $z\bar{z}$; that is, $|z| = (z\bar{z})^{1/2}$.

The existence (and uniqueness) of $|z|$ follows from Theorem 1.21 and part (d) of Theorem 1.31.

Note that when x is real, then $\bar{x} = x$, hence $|x| = \sqrt{x^2}$. Thus $|x| = x$ if $x \geq 0$, $|x| = -x$ if $x < 0$.

1.33 Theorem *Let z and w be complex numbers. Then*

(a) $|z| > 0$ *unless* $z = 0$, $|0| = 0$,
(b) $|\bar{z}| = |z|$,
(c) $|zw| = |z||w|$,
(d) $|\operatorname{Re} z| \leq |z|$,
(e) $|z + w| \leq |z| + |w|$.

Proof (a) and (b) are trivial. Put $z = a + bi$, $w = c + di$, with a, b, c, d real. Then
$$|zw|^2 = (ac - bd)^2 + (ad + bc)^2 = (a^2 + b^2)(c^2 + d^2) = |z|^2|w|^2$$
or $|zw|^2 = (|z||w|)^2$. Now (c) follows from the uniqueness assertion of Theorem 1.21.

To prove (d), note that $a^2 \leq a^2 + b^2$, hence
$$|a| = \sqrt{a^2} \leq \sqrt{a^2 + b^2}.$$

To prove (e), note that $\bar{z}w$ is the conjugate of $z\bar{w}$, so that $z\bar{w} + \bar{z}w = 2\,\mathrm{Re}\,(z\bar{w})$. Hence
$$\begin{aligned}|z + w|^2 &= (z + w)(\bar{z} + \bar{w}) = z\bar{z} + z\bar{w} + \bar{z}w + w\bar{w} \\ &= |z|^2 + 2\,\mathrm{Re}\,(z\bar{w}) + |w|^2 \\ &\leq |z|^2 + 2|z\bar{w}| + |w|^2 \\ &= |z|^2 + 2|z||w| + |w|^2 = (|z| + |w|)^2.\end{aligned}$$

Now (e) follows by taking square roots.

1.34 Notation If $x_1, \ldots, x_n$ are complex numbers, we write
$$x_1 + x_2 + \cdots + x_n = \sum_{j=1}^{n} x_j.$$

We conclude this section with an important inequality, usually known as the *Schwarz inequality*.

1.35 Theorem *If $a_1, \ldots, a_n$ and $b_1, \ldots, b_n$ are complex numbers, then*
$$\left|\sum_{j=1}^{n} a_j \bar{b}_j\right|^2 \leq \sum_{j=1}^{n} |a_j|^2 \sum_{j=1}^{n} |b_j|^2.$$

Proof Put $A = \Sigma|a_j|^2$, $B = \Sigma|b_j|^2$, $C = \Sigma a_j \bar{b}_j$ (in all sums in this proof, j runs over the values $1, \ldots, n$). If $B = 0$, then $b_1 = \cdots = b_n = 0$, and the conclusion is trivial. Assume therefore that $B > 0$. By Theorem 1.31 we have
$$\begin{aligned}\sum |Ba_j - Cb_j|^2 &= \sum (Ba_j - Cb_j)(B\bar{a}_j - \overline{Cb_j}) \\ &= B^2 \sum |a_j|^2 - B\bar{C} \sum a_j \bar{b}_j - BC \sum \bar{a}_j b_j + |C|^2 \sum |b_j|^2 \\ &= B^2 A - B|C|^2 \\ &= B(AB - |C|^2).\end{aligned}$$

Since each term in the first sum is nonnegative, we see that
$$B(AB - |C|^2) \geq 0.$$
Since $B > 0$, it follows that $AB - |C|^2 \geq 0$. This is the desired inequality.

EUCLIDEAN SPACES

1.36 Definitions For each positive integer k, let R^k be the set of all ordered k-tuples
$$\mathbf{x} = (x_1, x_2, \ldots, x_k),$$
where $x_1, \ldots, x_k$ are real numbers, called the *coordinates* of $\mathbf{x}$. The elements of R^k are called points, or vectors, especially when $k > 1$. We shall denote vectors by boldfaced letters. If $\mathbf{y} = (y_1, \ldots, y_k)$ and if α is a real number, put
$$\mathbf{x} + \mathbf{y} = (x_1 + y_1, \ldots, x_k + y_k),$$
$$\alpha \mathbf{x} = (\alpha x_1, \ldots, \alpha x_k)$$
so that $\mathbf{x} + \mathbf{y} \in R^k$ and $\alpha \mathbf{x} \in R^k$. This defines addition of vectors, as well as multiplication of a vector by a real number (a scalar). These two operations satisfy the commutative, associative, and distributive laws (the proof is trivial, in view of the analogous laws for the real numbers) and make R^k into a *vector space over the real field*. The zero element of R^k (sometimes called the *origin* or the *null vector*) is the point $\mathbf{0}$, all of whose coordinates are 0.

We also define the so-called "inner product" (or scalar product) of $\mathbf{x}$ and $\mathbf{y}$ by
$$\mathbf{x} \cdot \mathbf{y} = \sum_{i=1}^{k} x_i y_i$$
and the *norm* of $\mathbf{x}$ by
$$|\mathbf{x}| = (\mathbf{x} \cdot \mathbf{x})^{1/2} = \left(\sum_{1}^{k} x_i^2 \right)^{1/2}.$$

The structure now defined (the vector space R^k with the above inner product and norm) is called euclidean k-space.

1.37 Theorem *Suppose* $\mathbf{x}, \mathbf{y}, \mathbf{z} \in R^k$, *and α is real. Then*

(a) $|\mathbf{x}| \geq 0$;
(b) $|\mathbf{x}| = 0$ *if and only if* $\mathbf{x} = \mathbf{0}$;
(c) $|\alpha \mathbf{x}| = |\alpha| |\mathbf{x}|$;
(d) $|\mathbf{x} \cdot \mathbf{y}| \leq |\mathbf{x}| |\mathbf{y}|$;
(e) $|\mathbf{x} + \mathbf{y}| \leq |\mathbf{x}| + |\mathbf{y}|$;
(f) $|\mathbf{x} - \mathbf{z}| \leq |\mathbf{x} - \mathbf{y}| + |\mathbf{y} - \mathbf{z}|$.

Proof (*a*), (*b*), and (*c*) are obvious, and (*d*) is an immediate consequence of the Schwarz inequality. By (*d*) we have

$$|\mathbf{x} + \mathbf{y}|^2 = (\mathbf{x} + \mathbf{y}) \cdot (\mathbf{x} + \mathbf{y})$$
$$= \mathbf{x} \cdot \mathbf{x} + 2\mathbf{x} \cdot \mathbf{y} + \mathbf{y} \cdot \mathbf{y}$$
$$\leq |\mathbf{x}|^2 + 2|\mathbf{x}||\mathbf{y}| + |\mathbf{y}|^2$$
$$= (|\mathbf{x}| + |\mathbf{y}|)^2,$$

so that (*e*) is proved. Finally, (*f*) follows from (*e*) if we replace $\mathbf{x}$ by $\mathbf{x} - \mathbf{y}$ and $\mathbf{y}$ by $\mathbf{y} - \mathbf{z}$.

1.38 Remarks Theorem 1.37 (*a*), (*b*), and (*f*) will allow us (see Chap. 2) to regard R^k as a metric space.

R^1 (the set of all real numbers) is usually called the line, or the real line. Likewise, R^2 is called the plane, or the complex plane (compare Definitions 1.24 and 1.36). In these two cases the norm is just the absolute value of the corresponding real or complex number.

APPENDIX

Theorem 1.19 will be proved in this appendix by constructing R from Q. We shall divide the construction into several steps.

Step 1 The members of R will be certain subsets of Q, called *cuts*. A cut is, by definition, any set $\alpha \subset Q$ with the following three properties.

(I) α is not empty, and $\alpha \neq Q$.
(II) If $p \in \alpha$, $q \in Q$, and $q < p$, then $q \in \alpha$.
(III) If $p \in \alpha$, then $p < r$ for some $r \in \alpha$.

The letters $p, q, r, \ldots$ will always denote rational numbers, and $\alpha, \beta, \gamma, \ldots$ will denote cuts.

Note that (III) simply says that α has no largest member; (II) implies two facts which will be used freely:

If $p \in \alpha$ and $q \notin \alpha$ then $p < q$.
If $r \notin \alpha$ and $r < s$ then $s \notin \alpha$.

Step 2 Define "$\alpha < \beta$" to mean: α is a proper subset of β.

Let us check that this meets the requirements of Definition 1.5.

If $\alpha < \beta$ and $\beta < \gamma$ it is clear that $\alpha < \gamma$. (A proper subset of a proper subset is a proper subset.) It is also clear that at most one of the three relations

$$\alpha < \beta, \quad \alpha = \beta, \quad \beta < \alpha$$

can hold for any pair α, β. To show that at least one holds, assume that the first two fail. Then α is not a subset of β. Hence there is a $p \in \alpha$ with $p \notin \beta$. If $q \in \beta$, it follows that $q < p$ (since $p \notin \beta$), hence $q \in \alpha$, by (II). Thus $\beta \subset \alpha$. Since $\beta \neq \alpha$, we conclude: $\beta < \alpha$.

Thus R is now an ordered set.

Step 3 *The ordered set R has the least-upper-bound property.*

To prove this, let A be a nonempty subset of R, and assume that $\beta \in R$ is an upper bound of A. Define γ to be the union of all $\alpha \in A$. In other words, $p \in \gamma$ if and only if $p \in \alpha$ for some $\alpha \in A$. We shall prove that $\gamma \in R$ and that $\gamma = \sup A$.

Since A is not empty, there exists an $\alpha_0 \in A$. This α_0 is not empty. Since $\alpha_0 \subset \gamma$, γ is not empty. Next, $\gamma \subset \beta$ (since $\alpha \subset \beta$ for every $\alpha \in A$), and therefore $\gamma \neq Q$. Thus γ satisfies property (I). To prove (II) and (III), pick $p \in \gamma$. Then $p \in \alpha_1$ for some $\alpha_1 \in A$. If $q < p$, then $q \in \alpha_1$, hence $q \in \gamma$; this proves (II). If $r \in \alpha_1$ is so chosen that $r > p$, we see that $r \in \gamma$ (since $\alpha_1 \subset \gamma$), and therefore γ satisfies (III).

Thus $\gamma \in R$.

It is clear that $\alpha \leq \gamma$ for every $\alpha \in A$.

Suppose $\delta < \gamma$. Then there is an $s \in \gamma$ and that $s \notin \delta$. Since $s \in \gamma$, $s \in \alpha$ for some $\alpha \in A$. Hence $\delta < \alpha$, and δ is not an upper bound of A.

This gives the desired result: $\gamma = \sup A$.

Step 4 If $\alpha \in R$ and $\beta \in R$ we define $\alpha + \beta$ to be the set of all sums $r + s$, where $r \in \alpha$ and $s \in \beta$.

We define 0^* to be the set of all negative rational numbers. It is clear that 0^* is a cut. *We verify that the axioms for addition* (see Definition 1.12) *hold in R, with 0^* playing the role of* 0.

(A1) We have to show that $\alpha + \beta$ is a cut. It is clear that $\alpha + \beta$ is a nonempty subset of Q. Take $r' \notin \alpha$, $s' \notin \beta$. Then $r' + s' > r + s$ for all choices of $r \in \alpha$, $s \in \beta$. Thus $r' + s' \notin \alpha + \beta$. It follows that $\alpha + \beta$ has property (I).

Pick $p \in \alpha + \beta$. Then $p = r + s$, with $r \in \alpha$, $s \in \beta$. If $q < p$, then $q - s < r$, so $q - s \in \alpha$, and $q = (q - s) + s \in \alpha + \beta$. Thus (II) holds. Choose $t \in \alpha$ so that $t > r$. Then $p < t + s$ and $t + s \in \alpha + \beta$. Thus (III) holds.

(A2) $\alpha + \beta$ is the set of all $r + s$, with $r \in \alpha$, $s \in \beta$. By the same definition, $\beta + \alpha$ is the set of all $s + r$. Since $r + s = s + r$ for all $r \in Q$, $s \in Q$, we have $\alpha + \beta = \beta + \alpha$.

(A3) As above, this follows from the associative law in Q.

(A4) If $r \in \alpha$ and $s \in 0^*$, then $r + s < r$, hence $r + s \in \alpha$. Thus $\alpha + 0^* \subset \alpha$. To obtain the opposite inclusion, pick $p \in \alpha$, and pick $r \in \alpha$, $r > p$. Then

$p - r \in 0^*$, and $p = r + (p - r) \in \alpha + 0^*$. Thus $\alpha \subset \alpha + 0^*$. We conclude that $\alpha + 0^* = \alpha$.

(A5) Fix $\alpha \in R$. Let β be the set of all p with the following property:

There exists $r > 0$ such that $-p - r \notin \alpha$.

In other words, some rational number smaller than $-p$ fails to be in α.

We show that $\beta \in R$ and that $\alpha + \beta = 0^*$.

If $s \notin \alpha$ and $p = -s - 1$, then $-p - 1 \notin \alpha$, hence $p \in \beta$. So β is not empty. If $q \in \alpha$, then $-q \notin \beta$. So $\beta \neq Q$. Hence β satisfies (I).

Pick $p \in \beta$, and pick $r > 0$, so that $-p - r \notin \alpha$. If $q < p$, then $-q - r > -p - r$, hence $-q - r \notin \alpha$. Thus $q \in \beta$, and (II) holds. Put $t = p + (r/2)$. Then $t > p$, and $-t - (r/2) = -p - r \notin \alpha$, so that $t \in \beta$. Hence β satisfies (III).

We have proved that $\beta \in R$.

If $r \in \alpha$ and $s \in \beta$, then $-s \notin \alpha$, hence $r < -s$, $r + s < 0$. Thus $\alpha + \beta \subset 0^*$.

To prove the opposite inclusion, pick $v \in 0^*$, put $w = -v/2$. Then $w > 0$, and there is an integer n such that $nw \in \alpha$ but $(n + 1)w \notin \alpha$. (Note that this depends on the fact that Q has the archimedean property!) Put $p = -(n + 2)w$. Then $p \in \beta$, since $-p - w \notin \alpha$, and

$$v = nw + p \in \alpha + \beta.$$

Thus $0^* \subset \alpha + \beta$.

We conclude that $\alpha + \beta = 0^*$.

This β will of course be denoted by $-\alpha$.

Step 5 Having proved that the addition defined in Step 4 satisfies Axioms (A) of Definition 1.12, it follows that Proposition 1.14 is valid in R, and we can prove one of the requirements of Definition 1.17:

If $\alpha, \beta, \gamma \in R$ and $\beta < \gamma$, then $\alpha + \beta < \alpha + \gamma$.

Indeed, it is obvious from the definition of $+$ in R that $\alpha + \beta \subset \alpha + \gamma$; if we had $\alpha + \beta = \alpha + \gamma$, the cancellation law (Proposition 1.14) would imply $\beta = \gamma$.

It also follows that $\alpha > 0^*$ if and only if $-\alpha < 0^*$.

Step 6 Multiplication is a little more bothersome than addition in the present context, since products of negative rationals are positive. For this reason we confine ourselves first to R^+, the set of all $\alpha \in R$ with $\alpha > 0^*$.

If $\alpha \in R^+$ and $\beta \in R^+$, we define $\alpha\beta$ to be the set of all p such that $p \leq rs$ for some choice of $r \in \alpha$, $s \in \beta$, $r > 0$, $s > 0$.

We define 1^* to be the set of all $q < 1$.

Then the axioms (M) and (D) of Definition 1.12 hold, with R^+ in place of F, and with 1^* in the role of 1.

The proofs are so similar to the ones given in detail in Step 4 that we omit them.

Note, in particular, that the second requirement of Definition 1.17 holds: If $\alpha > 0^*$ and $\beta > 0^*$ then $\alpha\beta > 0^*$.

Step 7 We complete the definition of multiplication by setting $\alpha 0^* = 0^*\alpha = 0^*$, and by setting

$$\alpha\beta = \begin{cases} (-\alpha)(-\beta) & \text{if } \alpha < 0^*, \beta < 0^*, \\ -[(-\alpha)\beta] & \text{if } \alpha < 0^*, \beta > 0^*, \\ -[\alpha \cdot (-\beta)] & \text{if } \alpha > 0^*, \beta < 0^*. \end{cases}$$

The products on the right were defined in Step 6.

Having proved (in Step 6) that the axioms (M) hold in R^+, it is now perfectly simple to prove them in R, by repeated application of the identity $\gamma = -(-\gamma)$ which is part of Proposition 1.14. (See Step 5.)

The proof of the distributive law

$$\alpha(\beta + \gamma) = \alpha\beta + \alpha\gamma$$

breaks into cases. For instance, suppose $\alpha > 0^*$, $\beta < 0^*$, $\beta + \gamma > 0^*$. Then $\gamma = (\beta + \gamma) + (-\beta)$, and (since we already know that the distributive law holds in R^+)

$$\alpha\gamma = \alpha(\beta + \gamma) + \alpha \cdot (-\beta).$$

But $\alpha \cdot (-\beta) = -(\alpha\beta)$. Thus

$$\alpha\beta + \alpha\gamma = \alpha(\beta + \gamma).$$

The other cases are handled in the same way.

We have now completed the proof that R is an ordered field with the least-upper-bound property.

Step 8 We associate with each $r \in Q$ the set r^* which consists of all $p \in Q$ such that $p < r$. It is clear that each r^* is a cut; that is, $r^* \in R$. These cuts satisfy the following relations:

(a) $r^* + s^* = (r + s)^*$,
(b) $r^* s^* = (rs)^*$,
(c) $r^* < s^*$ if and only if $r < s$.

To prove (a), choose $p \in r^* + s^*$. Then $p = u + v$, where $u < r$, $v < s$. Hence $p < r + s$, which says that $p \in (r + s)^*$.

Conversely, suppose $p \in (r+s)^*$. Then $p < r + s$. Choose t so that $2t = r + s - p$, put
$$r' = r - t, s' = s - t.$$
Then $r' \in r^*$, $s' \in s^*$, and $p = r' + s'$, so that $p \in r^* + s^*$.

This proves (a). The proof of (b) is similar.

If $r < s$ then $r \in s^*$, but $r \notin r^*$; hence $r^* < s^*$.

If $r^* < s^*$, then there is a $p \in s^*$ such that $p \notin r^*$. Hence $r \leq p < s$, so that $r < s$.

This proves (c).

Step 9 We saw in Step 8 that the replacement of the rational numbers r by the corresponding "rational cuts" $r^* \in R$ preserves sums, products, and order. This fact may be expressed by saying that the ordered field Q is *isomorphic* to the ordered field Q^* whose elements are the rational cuts. Of course, r^* is by no means the same as r, but the properties we are concerned with (arithmetic and order) are the same in the two fields.

It is this identification of Q with Q^* which allows us to regard Q as a subfield of R.

The second part of Theorem 1.19 is to be understood in terms of this identification. Note that the same phenomenon occurs when the real numbers are regarded as a subfield of the complex field, and it also occurs at a much more elementary level, when the integers are identified with a certain subset of Q.

It is a fact, which we will not prove here, that *any two ordered fields with the least-upper-bound property are isomorphic*. The first part of Theorem 1.19 therefore characterizes the real field R completely.

The books by Landau and Thurston cited in the Bibliography are entirely devoted to number systems. Chapter 1 of Knopp's book contains a more leisurely description of how R can be obtained from Q. Another construction, in which each real number is defined to be an equivalence class of Cauchy sequences of rational numbers (see Chap. 3), is carried out in Sec. 5 of the book by Hewitt and Stromberg.

The cuts in Q which we used here were invented by Dedekind. The construction of R from Q by means of Cauchy sequences is due to Cantor. Both Cantor and Dedekind published their constructions in 1872.

EXERCISES

Unless the contrary is explicitly stated, all numbers that are mentioned in these exercises are understood to be real.

1. If r is rational ($r \neq 0$) and x is irrational, prove that $r + x$ and rx are irrational.

2. Prove that there is no rational number whose square is 12.
3. Prove Proposition 1.15.
4. Let E be a nonempty subset of an ordered set; suppose α is a lower bound of E and β is an upper bound of E. Prove that $\alpha \leq \beta$.
5. Let A be a nonempty set of real numbers which is bounded below. Let $-A$ be the set of all numbers $-x$, where $x \in A$. Prove that
$$\inf A = -\sup(-A).$$
6. Fix $b > 1$.
   (a) If m, n, p, q are integers, $n > 0$, $q > 0$, and $r = m/n = p/q$, prove that
   $$(b^m)^{1/n} = (b^p)^{1/q}.$$
   Hence it makes sense to define $b^r = (b^m)^{1/n}$.
   (b) Prove that $b^{r+s} = b^r b^s$ if r and s are rational.
   (c) If x is real, define $B(x)$ to be the set of all numbers b^t, where t is rational and $t \leq x$. Prove that
   $$b^r = \sup B(r)$$
   when r is rational. Hence it makes sense to define
   $$b^x = \sup B(x)$$
   for every real x.
   (d) Prove that $b^{x+y} = b^x b^y$ for all real x and y.
7. Fix $b > 1$, $y > 0$, and prove that there is a unique real x such that $b^x = y$, by completing the following outline. (This x is called the *logarithm of y to the base b*.)
   (a) For any positive integer n, $b^n - 1 \geq n(b - 1)$.
   (b) Hence $b - 1 \geq n(b^{1/n} - 1)$.
   (c) If $t > 1$ and $n > (b-1)/(t-1)$, then $b^{1/n} < t$.
   (d) If w is such that $b^w < y$, then $b^{w+(1/n)} < y$ for sufficiently large n; to see this, apply part (c) with $t = y \cdot b^{-w}$.
   (e) If $b^w > y$, then $b^{w-(1/n)} > y$ for sufficiently large n.
   (f) Let A be the set of all w such that $b^w < y$, and show that $x = \sup A$ satisfies $b^x = y$.
   (g) Prove that this x is unique.
8. Prove that no order can be defined in the complex field that turns it into an ordered field. *Hint:* -1 is a square.
9. Suppose $z = a + bi$, $w = c + di$. Define $z < w$ if $a < c$, and also if $a = c$ but $b < d$. Prove that this turns the set of all complex numbers into an ordered set. (This type of order relation is called a *dictionary order*, or *lexicographic order*, for obvious reasons.) Does this ordered set have the least-upper-bound property?
10. Suppose $z = a + bi$, $w = u + iv$, and
$$a = \left(\frac{|w| + u}{2}\right)^{1/2}, \quad b = \left(\frac{|w| - u}{2}\right)^{1/2}.$$

Prove that $z^2 = w$ if $v \geq 0$ and that $(\bar{z})^2 = w$ if $v \leq 0$. Conclude that every complex number (with one exception!) has two complex square roots.

11. If z is a complex number, prove that there exists an $r \geq 0$ and a complex number w with $|w| = 1$ such that $z = rw$. Are w and r always uniquely determined by z?

12. If $z_1, \ldots, z_n$ are complex, prove that
$$|z_1 + z_2 + \cdots + z_n| \leq |z_1| + |z_2| + \cdots + |z_n|.$$

13. If x, y are complex, prove that
$$||x| - |y|| \leq |x - y|.$$

14. If z is a complex number such that $|z| = 1$, that is, such that $z\bar{z} = 1$, compute
$$|1 + z|^2 + |1 - z|^2.$$

15. Under what conditions does equality hold in the Schwarz inequality?

16. Suppose $k \geq 3$, $\mathbf{x}, \mathbf{y} \in R^k$, $|\mathbf{x} - \mathbf{y}| = d > 0$, and $r > 0$. Prove:
    (a) If $2r > d$, there are infinitely many $\mathbf{z} \in R^k$ such that
    $$|\mathbf{z} - \mathbf{x}| = |\mathbf{z} - \mathbf{y}| = r.$$
    (b) If $2r = d$, there is exactly one such $\mathbf{z}$.
    (c) If $2r < d$, there is no such $\mathbf{z}$.
    How must these statements be modified if k is 2 or 1?

17. Prove that
$$|\mathbf{x} + \mathbf{y}|^2 + |\mathbf{x} - \mathbf{y}|^2 = 2|\mathbf{x}|^2 + 2|\mathbf{y}|^2$$
if $\mathbf{x} \in R^k$ and $\mathbf{y} \in R^k$. Interpret this geometrically, as a statement about parallelograms.

18. If $k \geq 2$ and $\mathbf{x} \in R^k$, prove that there exists $\mathbf{y} \in R^k$ such that $\mathbf{y} \neq \mathbf{0}$ but $\mathbf{x} \cdot \mathbf{y} = 0$. Is this also true if $k = 1$?

19. Suppose $\mathbf{a} \in R^k$, $\mathbf{b} \in R^k$. Find $\mathbf{c} \in R^k$ and $r > 0$ such that
$$|\mathbf{x} - \mathbf{a}| = 2|\mathbf{x} - \mathbf{b}|$$
if and only if $|\mathbf{x} - \mathbf{c}| = r$.
(*Solution:* $3\mathbf{c} = 4\mathbf{b} - \mathbf{a}$, $3r = 2|\mathbf{b} - \mathbf{a}|$.)

20. With reference to the Appendix, suppose that property (III) were omitted from the definition of a cut. Keep the same definitions of order and addition. Show that the resulting ordered set has the least-upper-bound property, that addition satisfies axioms (A1) to (A4) (with a slightly different zero-element!) but that (A5) fails.

2
BASIC TOPOLOGY

FINITE, COUNTABLE, AND UNCOUNTABLE SETS

We begin this section with a definition of the function concept.

2.1 Definition Consider two sets A and B, whose elements may be any objects whatsoever, and suppose that with each element x of A there is associated, in some manner, an element of B, which we denote by $f(x)$. Then f is said to be a *function* from A to B (or a *mapping* of A into B). The set A is called the *domain* of f (we also say f is defined on A), and the elements $f(x)$ are called the *values* of f. The set of all values of f is called the *range* of f.

2.2 Definition Let A and B be two sets and let f be a mapping of A into B. If $E \subset A$, $f(E)$ is defined to be the set of all elements $f(x)$, for $x \in E$. We call $f(E)$ the *image* of E under f. In this notation, $f(A)$ is the range of f. It is clear that $f(A) \subset B$. If $f(A) = B$, we say that f maps A onto B. (Note that, according to this usage, *onto* is more specific than *into*.)

If $E \subset B$, $f^{-1}(E)$ denotes the set of all $x \in A$ such that $f(x) \in E$. We call $f^{-1}(E)$ the *inverse image* of E under f. If $y \in B$, $f^{-1}(y)$ is the set of all $x \in A$

such that $f(x) = y$. If, for each $y \in B, f^{-1}(y)$ consists of at most one element of A, then f is said to be a 1-1 (*one-to-one*) mapping of A into B. This may also be expressed as follows: f is a 1-1 mapping of A into B provided that $f(x_1) \neq f(x_2)$ whenever $x_1 \neq x_2$, $x_1 \in A$, $x_2 \in A$.

(The notation $x_1 \neq x_2$ means that x_1 and x_2 are distinct elements; otherwise we write $x_1 = x_2$.)

2.3 Definition If there exists a 1-1 mapping of A *onto* B, we say that A and B can be put in 1-1 *correspondence*, or that A and B have the same *cardinal number*, or, briefly, that A and B are *equivalent*, and we write $A \sim B$. This relation clearly has the following properties:

> It is reflexive: $A \sim A$.
> It is symmetric: If $A \sim B$, then $B \sim A$.
> It is transitive: If $A \sim B$ and $B \sim C$, then $A \sim C$.

Any relation with these three properties is called an *equivalence relation*.

2.4 Definition For any positive integer n, let J_n be the set whose elements are the integers $1, 2, \ldots, n$; let J be the set consisting of all positive integers. For any set A, we say:

(a) A is *finite* if $A \sim J_n$ for some n (the empty set is also considered to be finite).
(b) A is *infinite* if A is not finite.
(c) A is *countable* if $A \sim J$.
(d) A is *uncountable* if A is neither finite nor countable.
(e) A is *at most countable* if A is finite or countable.

Countable sets are sometimes called *enumerable*, or *denumerable*.

For two finite sets A and B, we evidently have $A \sim B$ if and only if A and B contain the same number of elements. For infinite sets, however, the idea of "having the same number of elements" becomes quite vague, whereas the notion of 1-1 correspondence retains its clarity.

2.5 Example Let A be the set of all integers. Then A is countable. For, consider the following arrangement of the sets A and J:

$$A: \quad 0, 1, -1, 2, -2, 3, -3, \ldots$$
$$J: \quad 1, 2, 3, 4, 5, 6, 7, \ldots$$

We can, in this example, even give an explicit formula for a function f from J to A which sets up a 1-1 correspondence:

$$f(n) = \begin{cases} \dfrac{n}{2} & (n \text{ even}), \\ -\dfrac{n-1}{2} & (n \text{ odd}). \end{cases}$$

2.6 Remark A finite set cannot be equivalent to one of its proper subsets. That this is, however, possible for infinite sets, is shown by Example 2.5, in which J is a proper subset of A.

In fact, we could replace Definition 2.4(*b*) by the statement: A is infinite if A is equivalent to one of its proper subsets.

2.7 Definition By a *sequence*, we mean a function f defined on the set J of all positive integers. If $f(n) = x_n$, for $n \in J$, it is customary to denote the sequence f by the symbol $\{x_n\}$, or sometimes by $x_1, x_2, x_3, \ldots$. The values of f, that is, the elements x_n, are called the *terms* of the sequence. If A is a set and if $x_n \in A$ for all $n \in J$, then $\{x_n\}$ is said to be a *sequence in A*, or a *sequence of elements of A*.

Note that the terms $x_1, x_2, x_3, \ldots$ of a sequence need not be distinct.

Since every countable set is the range of a 1-1 function defined on J, we may regard every countable set as the range of a sequence of distinct terms. Speaking more loosely, we may say that the elements of any countable set can be "arranged in a sequence."

Sometimes it is convenient to replace J in this definition by the set of all nonnegative integers, i.e., to start with 0 rather than with 1.

2.8 Theorem *Every infinite subset of a countable set A is countable.*

Proof Suppose $E \subset A$, and E is infinite. Arrange the elements x of A in a sequence $\{x_n\}$ of distinct elements. Construct a sequence $\{n_k\}$ as follows:

Let n_1 be the smallest positive integer such that $x_{n_1} \in E$. Having chosen $n_1, \ldots, n_{k-1}$ ($k = 2, 3, 4, \ldots$), let n_k be the smallest integer greater than n_{k-1} such that $x_{n_k} \in E$.

Putting $f(k) = x_{n_k}$ ($k = 1, 2, 3, \ldots$), we obtain a 1-1 correspondence between E and J.

The theorem shows that, roughly speaking, countable sets represent the "smallest" infinity: No uncountable set can be a subset of a countable set.

2.9 Definition Let A and Ω be sets, and suppose that with each element α of A there is associated a subset of Ω which we denote by E_α.

The set whose elements are the sets E_α will be denoted by $\{E_\alpha\}$. Instead of speaking of sets of sets, we shall sometimes speak of a collection of sets, or a family of sets.

The *union* of the sets E_α is defined to be the set S such that $x \in S$ if and only if $x \in E_\alpha$ for at least one $\alpha \in A$. We use the notation

(1) $$S = \bigcup_{\alpha \in A} E_\alpha.$$

If A consists of the integers $1, 2, \ldots, n$, one usually writes

(2) $$S = \bigcup_{m=1}^{n} E_m$$

or

(3) $$S = E_1 \cup E_2 \cup \cdots \cup E_n.$$

If A is the set of all positive integers, the usual notation is

(4) $$S = \bigcup_{m=1}^{\infty} E_m.$$

The symbol ∞ in (4) merely indicates that the union of a *countable* collection of sets is taken, and should not be confused with the symbols $+\infty$, $-\infty$, introduced in Definition 1.23.

The *intersection* of the sets E_α is defined to be the set P such that $x \in P$ if and only if $x \in E_\alpha$ for every $\alpha \in A$. We use the notation

(5) $$P = \bigcap_{\alpha \in A} E_\alpha,$$

or

(6) $$P = \bigcap_{m=1}^{n} E_m = E_1 \cap E_2 \cap \cdots \cap E_n,$$

or

(7) $$P = \bigcap_{m=1}^{\infty} E_m,$$

as for unions. If $A \cap B$ is not empty, we say that A and B *intersect*; otherwise they are *disjoint*.

2.10 Examples

(a) Suppose E_1 consists of $1, 2, 3$ and E_2 consists of $2, 3, 4$. Then $E_1 \cup E_2$ consists of $1, 2, 3, 4$, whereas $E_1 \cap E_2$ consists of $2, 3$.

(b) Let A be the set of real numbers x such that $0 < x \leq 1$. For every $x \in A$, let E_x be the set of real numbers y such that $0 < y < x$. Then

(i) $\quad E_x \subset E_z$ if and only if $0 < x \leq z \leq 1$;
(ii) $\quad \bigcup_{x \in A} E_x = E_1$;
(iii) $\quad \bigcap_{x \in A} E_x$ is empty;

(i) and (ii) are clear. To prove (iii), we note that for every $y > 0$, $y \notin E_x$ if $x < y$. Hence $y \notin \bigcap_{x \in A} E_x$.

2.11 Remarks Many properties of unions and intersections are quite similar to those of sums and products; in fact, the words sum and product were sometimes used in this connection, and the symbols Σ and Π were written in place of $\bigcup$ and $\bigcap$.

The commutative and associative laws are trivial:

(8) $\qquad A \cup B = B \cup A; \qquad A \cap B = B \cap A.$

(9) $\qquad (A \cup B) \cup C = A \cup (B \cup C); \qquad (A \cap B) \cap C = A \cap (B \cap C).$

Thus the omission of parentheses in (3) and (6) is justified.

The distributive law also holds:

(10) $\qquad A \cap (B \cup C) = (A \cap B) \cup (A \cap C).$

To prove this, let the left and right members of (10) be denoted by E and F, respectively.

Suppose $x \in E$. Then $x \in A$ and $x \in B \cup C$, that is, $x \in B$ or $x \in C$ (possibly both). Hence $x \in A \cap B$ or $x \in A \cap C$, so that $x \in F$. Thus $E \subset F$.

Next, suppose $x \in F$. Then $x \in A \cap B$ or $x \in A \cap C$. That is, $x \in A$, and $x \in B \cup C$. Hence $x \in A \cap (B \cup C)$, so that $F \subset E$.

It follows that $E = F$.

We list a few more relations which are easily verified:

(11) $\qquad A \subset A \cup B,$

(12) $\qquad A \cap B \subset A.$

If 0 denotes the empty set, then

(13) $\qquad A \cup 0 = A, \qquad A \cap 0 = 0.$

If $A \subset B$, then

(14) $\qquad A \cup B = B, \qquad A \cap B = A.$

2.12 Theorem *Let $\{E_n\}$, $n = 1, 2, 3, \ldots$, be a sequence of countable sets, and put*

(15) $$S = \bigcup_{n=1}^{\infty} E_n.$$

Then S is countable.

Proof Let every set E_n be arranged in a sequence $\{x_{nk}\}$, $k = 1, 2, 3, \ldots$, and consider the infinite array

(16)
$$\begin{array}{cccccc}
x_{11} & x_{12} & x_{13} & x_{14} & \cdots \\
x_{21} & x_{22} & x_{23} & x_{24} & \cdots \\
x_{31} & x_{32} & x_{33} & x_{34} & \cdots \\
x_{41} & x_{42} & x_{43} & x_{44} & \cdots \\
\end{array}$$

in which the elements of E_n form the nth row. The array contains all elements of S. As indicated by the arrows, these elements can be arranged in a sequence

(17) $$x_{11}; x_{21}, x_{12}; x_{31}, x_{22}, x_{13}; x_{41}, x_{32}, x_{23}, x_{14}; \ldots$$

If any two of the sets E_n have elements in common, these will appear more than once in (17). Hence there is a subset T of the set of all positive integers such that $S \sim T$, which shows that S is at most countable (Theorem 2.8). Since $E_1 \subset S$, and E_1 is infinite, S is infinite, and thus countable.

Corollary *Suppose A is at most countable, and, for every $\alpha \in A$, B_α is at most countable. Put*

$$T = \bigcup_{\alpha \in A} B_\alpha.$$

Then T is at most countable.

For T is equivalent to a subset of (15).

2.13 Theorem *Let A be a countable set, and let B_n be the set of all n-tuples $(a_1, \ldots, a_n)$, where $a_k \in A$ ($k = 1, \ldots, n$), and the elements $a_1, \ldots, a_n$ need not be distinct. Then B_n is countable.*

Proof That B_1 is countable is evident, since $B_1 = A$. Suppose B_{n-1} is countable ($n = 2, 3, 4, \ldots$). The elements of B_n are of the form

(18) $$(b, a) \quad (b \in B_{n-1}, a \in A).$$

For every fixed b, the set of pairs (b, a) is equivalent to A, and hence countable. Thus B_n is the union of a countable set of countable sets. By Theorem 2.12, B_n is countable.

The theorem follows by induction.

Corollary *The set of all rational numbers is countable.*

Proof We apply Theorem 2.13, with $n = 2$, noting that every rational r is of the form b/a, where a and b are integers. The set of pairs (a, b), and therefore the set of fractions b/a, is countable.

In fact, even the set of all algebraic numbers is countable (see Exercise 2).

That not all infinite sets are, however, countable, is shown by the next theorem.

2.14 Theorem *Let A be the set of all sequences whose elements are the digits 0 and 1. This set A is uncountable.*

The elements of A are sequences like $1, 0, 0, 1, 0, 1, 1, 1, \ldots$.

Proof Let E be a countable subset of A, and let E consist of the sequences $s_1, s_2, s_3, \ldots$. We construct a sequence s as follows. If the nth digit in s_n is 1, we let the nth digit of s be 0, and vice versa. Then the sequence s differs from every member of E in at least one place; hence $s \notin E$. But clearly $s \in A$, so that E is a proper subset of A.

We have shown that every countable subset of A is a proper subset of A. It follows that A is uncountable (for otherwise A would be a proper subset of A, which is absurd).

The idea of the above proof was first used by Cantor, and is called Cantor's diagonal process; for, if the sequences $s_1, s_2, s_3, \ldots$ are placed in an array like (16), it is the elements on the diagonal which are involved in the construction of the new sequence.

Readers who are familiar with the binary representation of the real numbers (base 2 instead of 10) will notice that Theorem 2.14 implies that the set of all real numbers is uncountable. We shall give a second proof of this fact in Theorem 2.43.

METRIC SPACES

2.15 Definition A set X, whose elements we shall call *points*, is said to be a *metric space* if with any two points p and q of X there is associated a real number $d(p, q)$, called the *distance* from p to q, such that

(a) $d(p, q) > 0$ if $p \neq q$; $d(p, p) = 0$;
(b) $d(p, q) = d(q, p)$;
(c) $d(p, q) \leq d(p, r) + d(r, q)$, for any $r \in X$.

Any function with these three properties is called a *distance function*, or a *metric*.

2.16 Examples The most important examples of metric spaces, from our standpoint, are the euclidean spaces R^k, especially R^1 (the real line) and R^2 (the complex plane); the distance in R^k is defined by

$$(19) \qquad d(\mathbf{x}, \mathbf{y}) = |\mathbf{x} - \mathbf{y}| \qquad (\mathbf{x}, \mathbf{y} \in R^k).$$

By Theorem 1.37, the conditions of Definition 2.15 are satisfied by (19).

It is important to observe that every subset Y of a metric space X is a metric space in its own right, with the same distance function. For it is clear that if conditions (a) to (c) of Definition 2.15 hold for $p, q, r \in X$, they also hold if we restrict p, q, r to lie in Y.

Thus every subset of a euclidean space is a metric space. Other examples are the spaces $\mathscr{C}(K)$ and $\mathscr{L}^2(\mu)$, which are discussed in Chaps. 7 and 11, respectively.

2.17 Definition By the *segment* (a, b) we mean the set of all real numbers x such that $a < x < b$.

By the *interval* $[a, b]$ we mean the set of all real numbers x such that $a \leq x \leq b$.

Occasionally we shall also encounter "half-open intervals" $[a, b)$ and $(a, b]$; the first consists of all x such that $a \leq x < b$, the second of all x such that $a < x \leq b$.

If $a_i < b_i$ for $i = 1, \ldots, k$, the set of all points $\mathbf{x} = (x_1, \ldots, x_k)$ in R^k whose coordinates satisfy the inequalities $a_i \leq x_i \leq b_i$ $(1 \leq i \leq k)$ is called a *k-cell*. Thus a 1-cell is an interval, a 2-cell is a rectangle, etc.

If $\mathbf{x} \in R^k$ and $r > 0$, the *open* (or *closed*) *ball* B with center at $\mathbf{x}$ and radius r is defined to be the set of all $\mathbf{y} \in R^k$ such that $|\mathbf{y} - \mathbf{x}| < r$ (or $|\mathbf{y} - \mathbf{x}| \leq r$).

We call a set $E \subset R^k$ *convex* if

$$\lambda \mathbf{x} + (1 - \lambda) \mathbf{y} \in E$$

whenever $\mathbf{x} \in E$, $\mathbf{y} \in E$, and $0 < \lambda < 1$.

For example, *balls are convex*. For if $|\mathbf{y} - \mathbf{x}| < r$, $|\mathbf{z} - \mathbf{x}| < r$, and $0 < \lambda < 1$, we have

$$\begin{aligned}|\lambda \mathbf{y} + (1 - \lambda)\mathbf{z} - \mathbf{x}| &= |\lambda(\mathbf{y} - \mathbf{x}) + (1 - \lambda)(\mathbf{z} - \mathbf{x})| \\ &\leq \lambda |\mathbf{y} - \mathbf{x}| + (1 - \lambda)|\mathbf{z} - \mathbf{x}| < \lambda r + (1 - \lambda)r \\ &= r.\end{aligned}$$

The same proof applies to closed balls. It is also easy to see that k-cells are convex.

2.18 Definition Let X be a metric space. All points and sets mentioned below are understood to be elements and subsets of X.

(a) A *neighborhood* of p is a set $N_r(p)$ consisting of all q such that $d(p, q) < r$, for some $r > 0$. The number r is called the *radius* of $N_r(p)$.

(b) A point p is a *limit point* of the set E if *every* neighborhood of p contains a point $q \neq p$ such that $q \in E$.

(c) If $p \in E$ and p is not a limit point of E, then p is called an *isolated point* of E.

(d) E is *closed* if every limit point of E is a point of E.

(e) A point p is an *interior* point of E if there is a neighborhood N of p such that $N \subset E$.

(f) E is *open* if every point of E is an interior point of E.

(g) The *complement* of E (denoted by E^c) is the set of all points $p \in X$ such that $p \notin E$.

(h) E is *perfect* if E is closed and if every point of E is a limit point of E.

(i) E is *bounded* if there is a real number M and a point $q \in X$ such that $d(p, q) < M$ for all $p \in E$.

(j) E is *dense in X* if every point of X is a limit point of E, or a point of E (or both).

Let us note that in R^1 neighborhoods are segments, whereas in R^2 neighborhoods are interiors of circles.

2.19 Theorem *Every neighborhood is an open set.*

Proof Consider a neighborhood $E = N_r(p)$, and let q be any point of E. Then there is a positive real number h such that

$$d(p, q) = r - h.$$

For all points s such that $d(q, s) < h$, we have then

$$d(p, s) \leq d(p, q) + d(q, s) < r - h + h = r,$$

so that $s \in E$. Thus q is an interior point of E.

2.20 Theorem *If p is a limit point of a set E, then every neighborhood of p contains infinitely many points of E.*

Proof Suppose there is a neighborhood N of p which contains only a finite number of points of E. Let $q_1, \ldots, q_n$ be those points of $N \cap E$, which are distinct from p, and put

$$r = \min_{1 \leq m \leq n} d(p, q_m)$$

[we use this notation to denote the smallest of the numbers $d(p, q_1), \ldots, d(p, q_n)$]. The minimum of a finite set of positive numbers is clearly positive, so that $r > 0$.

The neighborhood $N_r(p)$ contains no point q of E such that $q \neq p$, so that p is not a limit point of E. This contradiction establishes the theorem.

Corollary *A finite point set has no limit points.*

2.21 Examples Let us consider the following subsets of R^2:

(a) The set of all complex z such that $|z| < 1$.
(b) The set of all complex z such that $|z| \leq 1$.
(c) A nonempty finite set.
(d) The set of all integers.
(e) The set consisting of the numbers $1/n$ ($n = 1, 2, 3, \ldots$). Let us note that this set E has a limit point (namely, $z = 0$) but that no point of E is a limit point of E; we wish to stress the difference between having a limit point and containing one.
(f) The set of all complex numbers (that is, R^2).
(g) The segment (a, b).

Let us note that (d), (e), (g) can be regarded also as subsets of R^1. Some properties of these sets are tabulated below:

	Closed	Open	Perfect	Bounded
(a)	No	Yes	No	Yes
(b)	Yes	No	Yes	Yes
(c)	Yes	No	No	Yes
(d)	Yes	No	No	No
(e)	No	No	No	Yes
(f)	Yes	Yes	Yes	No
(g)	No		No	Yes

In (g), we left the second entry blank. The reason is that the segment (a, b) is not open if we regard it as a subset of R^2, but it is an open subset of R^1.

2.22 Theorem *Let $\{E_\alpha\}$ be a (finite or infinite) collection of sets E_α. Then*

(20) $$\left(\bigcup_\alpha E_\alpha\right)^c = \bigcap_\alpha (E_\alpha^c).$$

Proof Let A and B be the left and right members of (20). If $x \in A$, then $x \notin \bigcup_\alpha E_\alpha$, hence $x \notin E_\alpha$ for any α, hence $x \in E_\alpha^c$ for every α, so that $x \in \bigcap E_\alpha^c$. Thus $A \subset B$.

Conversely, if $x \in B$, then $x \in E_\alpha^c$ for every α, hence $x \notin E_\alpha$ for any α, hence $x \notin \bigcup_\alpha E_\alpha$, so that $x \in (\bigcup_\alpha E_\alpha)^c$. Thus $B \subset A$.

It follows that $A = B$.

2.23 Theorem *A set E is open if and only if its complement is closed.*

Proof First, suppose E^c is closed. Choose $x \in E$. Then $x \notin E^c$, and x is not a limit point of E^c. Hence there exists a neighborhood N of x such that $E^c \cap N$ is empty, that is, $N \subset E$. Thus x is an interior point of E, and E is open.

Next, suppose E is open. Let x be a limit point of E^c. Then every neighborhood of x contains a point of E^c, so that x is not an interior point of E. Since E is open, this means that $x \in E^c$. It follows that E^c is closed.

Corollary *A set F is closed if and only if its complement is open.*

2.24 Theorem

(a) For any collection $\{G_\alpha\}$ of open sets, $\bigcup_\alpha G_\alpha$ is open.
(b) For any collection $\{F_\alpha\}$ of closed sets, $\bigcap_\alpha F_\alpha$ is closed.
(c) For any finite collection $G_1, \ldots, G_n$ of open sets, $\bigcap_{i=1}^n G_i$ is open.
(d) For any finite collection $F_1, \ldots, F_n$ of closed sets, $\bigcup_{i=1}^n F_i$ is closed.

Proof Put $G = \bigcup_\alpha G_\alpha$. If $x \in G$, then $x \in G_\alpha$ for some α. Since x is an interior point of G_α, x is also an interior point of G, and G is open. This proves (a).

By Theorem 2.22,

$$(21) \qquad \left(\bigcap_\alpha F_\alpha\right)^c = \bigcup_\alpha (F_\alpha^c),$$

and F_α^c is open, by Theorem 2.23. Hence (a) implies that (21) is open so that $\bigcap_\alpha F_\alpha$ is closed.

Next, put $H = \bigcap_{i=1}^n G_i$. For any $x \in H$, there exist neighborhoods N_i of x, with radii r_i, such that $N_i \subset G_i$ ($i = 1, \ldots, n$). Put

$$r = \min(r_1, \ldots, r_n),$$

and let N be the neighborhood of x of radius r. Then $N \subset G_i$ for $i = 1, \ldots, n$, so that $N \subset H$, and H is open.

By taking complements, (d) follows from (c):

$$\left(\bigcup_{i=1}^n F_i\right)^c = \bigcap_{i=1}^n (F_i^c).$$

2.25 Examples In parts (c) and (d) of the preceding theorem, the finiteness of the collections is essential. For let G_n be the segment $\left(-\frac{1}{n}, \frac{1}{n}\right)$ $(n = 1, 2, 3, \ldots)$. Then G_n is an open subset of R^1. Put $G = \bigcap_{n=1}^{\infty} G_n$. Then G consists of a single point (namely, $x = 0$) and is therefore not an open subset of R^1.

Thus the intersection of an infinite collection of open sets need not be open. Similarly, the union of an infinite collection of closed sets need not be closed.

2.26 Definition If X is a metric space, if $E \subset X$, and if E' denotes the set of all limit points of E in X, then the *closure* of E is the set $\bar{E} = E \cup E'$.

2.27 Theorem *If X is a metric space and $E \subset X$, then*

(a) $\bar{E}$ *is closed,*
(b) $E = \bar{E}$ *if and only if E is closed,*
(c) $\bar{E} \subset F$ *for every closed set $F \subset X$ such that $E \subset F$.*

By (a) and (c), $\bar{E}$ is the *smallest* closed subset of X that contains E.

Proof
(a) If $p \in X$ and $p \notin \bar{E}$ then p is neither a point of E nor a limit point of E. Hence p has a neighborhood which does not intersect E. The complement of $\bar{E}$ is therefore open. Hence $\bar{E}$ is closed.
(b) If $E = \bar{E}$, (a) implies that E is closed. If E is closed, then $E' \subset E$ [by Definitions 2.18(d) and 2.26], hence $\bar{E} = E$.
(c) If F is closed and $F \supset E$, then $F \supset F'$, hence $F \supset E'$. Thus $F \supset \bar{E}$.

2.28 Theorem *Let E be a nonempty set of real numbers which is bounded above. Let $y = \sup E$. Then $y \in \bar{E}$. Hence $y \in E$ if E is closed.*

Compare this with the examples in Sec. 1.9.

Proof If $y \in E$ then $y \in \bar{E}$. Assume $y \notin E$. For every $h > 0$ there exists then a point $x \in E$ such that $y - h < x < y$, for otherwise $y - h$ would be an upper bound of E. Thus y is a limit point of E. Hence $y \in \bar{E}$.

2.29 Remark Suppose $E \subset Y \subset X$, where X is a metric space. To say that E is an open subset of X means that to each point $p \in E$ there is associated a positive number r such that the conditions $d(p, q) < r$, $q \in X$ imply that $q \in E$. But we have already observed (Sec. 2.16) that Y is also a metric space, so that our definitions may equally well be made within Y. To be quite explicit, let us say that E is *open relative to* Y if to each $p \in E$ there is associated an $r > 0$ such that $q \in E$ whenever $d(p, q) < r$ and $q \in Y$. Example 2.21(g) showed that a set may be open relative to Y without being an open subset of X. However, there is a simple relation between these concepts, which we now state.

2.30 Theorem *Suppose $Y \subset X$. A subset E of Y is open relative to Y if and only if $E = Y \cap G$ for some open subset G of X.*

Proof Suppose E is open relative to Y. To each $p \in E$ there is a positive number r_p such that the conditions $d(p, q) < r_p$, $q \in Y$ imply that $q \in E$. Let V_p be the set of all $q \in X$ such that $d(p, q) < r_p$, and define

$$G = \bigcup_{p \in E} V_p.$$

Then G is an open subset of X, by Theorems 2.19 and 2.24.

Since $p \in V_p$ for all $p \in E$, it is clear that $E \subset G \cap Y$.

By our choice of V_p, we have $V_p \cap Y \subset E$ for every $p \in E$, so that $G \cap Y \subset E$. Thus $E = G \cap Y$, and one half of the theorem is proved.

Conversely, if G is open in X and $E = G \cap Y$, every $p \in E$ has a neighborhood $V_p \subset G$. Then $V_p \cap Y \subset E$, so that E is open relative to Y.

COMPACT SETS

2.31 Definition By an *open cover* of a set E in a metric space X we mean a collection $\{G_\alpha\}$ of open subsets of X such that $E \subset \bigcup_\alpha G_\alpha$.

2.32 Definition A subset K of a metric space X is said to be *compact* if every open cover of K contains a *finite* subcover.

More explicitly, the requirement is that if $\{G_\alpha\}$ is an open cover of K, then there are finitely many indices $\alpha_1, \ldots, \alpha_n$ such that

$$K \subset G_{\alpha_1} \cup \cdots \cup G_{\alpha_n}.$$

The notion of compactness is of great importance in analysis, especially in connection with continuity (Chap. 4).

It is clear that every finite set is compact. The existence of a large class of infinite compact sets in R^k will follow from Theorem 2.41.

We observed earlier (in Sec. 2.29) that if $E \subset Y \subset X$, then E may be open relative to Y without being open relative to X. The property of being open thus depends on the space in which E is embedded. The same is true of the property of being closed.

Compactness, however, behaves better, as we shall now see. To formulate the next theorem, let us say, temporarily, that K is compact relative to X if the requirements of Definition 2.32 are met.

2.33 Theorem *Suppose $K \subset Y \subset X$. Then K is compact relative to X if and only if K is compact relative to Y.*

By virtue of this theorem we are able, in many situations, to regard compact sets as metric spaces in their own right, without paying any attention to any embedding space. In particular, although it makes little sense to talk of *open* spaces, or of *closed* spaces (every metric space X is an open subset of itself, and is a closed subset of itself), it does make sense to talk of *compact* metric spaces.

Proof Suppose K is compact relative to X, and let $\{V_\alpha\}$ be a collection of sets, open relative to Y, such that $K \subset \bigcup_\alpha V_\alpha$. By theorem 2.30, there are sets G_α, open relative to X, such that $V_\alpha = Y \cap G_\alpha$, for all α; and since K is compact relative to X, we have

(22) $$K \subset G_{\alpha_1} \cup \cdots \cup G_{\alpha_n}$$

for some choice of finitely many indices $\alpha_1, \ldots, \alpha_n$. Since $K \subset Y$, (22) implies

(23) $$K \subset V_{\alpha_1} \cup \cdots \cup V_{\alpha_n}.$$

This proves that K is compact relative to Y.

Conversely, suppose K is compact relative to Y, let $\{G_\alpha\}$ be a collection of open subsets of X which covers K, and put $V_\alpha = Y \cap G_\alpha$. Then (23) will hold for some choice of $\alpha_1, \ldots, \alpha_n$; and since $V_\alpha \subset G_\alpha$, (23) implies (22).

This completes the proof.

2.34 Theorem *Compact subsets of metric spaces are closed.*

Proof Let K be a compact subset of a metric space X. We shall prove that the complement of K is an open subset of X.

Suppose $p \in X$, $p \notin K$. If $q \in K$, let V_q and W_q be neighborhoods of p and q, respectively, of radius less than $\tfrac{1}{2}d(p,q)$ [see Definition 2.18(a)]. Since K is compact, there are finitely many points $q_1, \ldots, q_n$ in K such that

$$K \subset W_{q_1} \cup \cdots \cup W_{q_n} = W.$$

If $V = V_{q_1} \cap \cdots \cap V_{q_n}$, then V is a neighborhood of p which does not intersect W. Hence $V \subset K^c$, so that p is an interior point of K^c. The theorem follows.

2.35 Theorem *Closed subsets of compact sets are compact.*

Proof Suppose $F \subset K \subset X$, F is closed (relative to X), and K is compact. Let $\{V_\alpha\}$ be an open cover of F. If F^c is adjoined to $\{V_\alpha\}$, we obtain an open cover Ω of K. Since K is compact, there is a finite subcollection Φ of Ω which covers K, and hence F. If F^c is a member of Φ, we may remove it from Φ and still retain an open cover of F. We have thus shown that a finite subcollection of $\{V_\alpha\}$ covers F.

Corollary *If F is closed and K is compact, then $F \cap K$ is compact.*

Proof Theorems 2.24(b) and 2.34 show that $F \cap K$ is closed; since $F \cap K \subset K$, Theorem 2.35 shows that $F \cap K$ is compact.

2.36 Theorem *If $\{K_\alpha\}$ is a collection of compact subsets of a metric space X such that the intersection of every finite subcollection of $\{K_\alpha\}$ is nonempty, then $\bigcap K_\alpha$ is nonempty.*

Proof Fix a member K_1 of $\{K_\alpha\}$ and put $G_\alpha = K_\alpha^c$. Assume that no point of K_1 belongs to every K_α. Then the sets G_α form an open cover of K_1; and since K_1 is compact, there are finitely many indices $\alpha_1, \ldots, \alpha_n$ such that $K_1 \subset G_{\alpha_1} \cup \cdots \cup G_{\alpha_n}$. But this means that

$$K_1 \cap K_{\alpha_1} \cap \cdots \cap K_{\alpha_n}$$

is empty, in contradiction to our hypothesis.

Corollary *If $\{K_n\}$ is a sequence of nonempty compact sets such that $K_n \supset K_{n+1}$ ($n = 1, 2, 3, \ldots$), then $\bigcap_1^\infty K_n$ is not empty.*

2.37 Theorem *If E is an infinite subset of a compact set K, then E has a limit point in K.*

Proof If no point of K were a limit point of E, then each $q \in K$ would have a neighborhood V_q which contains at most one point of E (namely, q, if $q \in E$). It is clear that no finite subcollection of $\{V_q\}$ can cover E; and the same is true of K, since $E \subset K$. This contradicts the compactness of K.

2.38 Theorem *If $\{I_n\}$ is a sequence of intervals in R^1, such that $I_n \supset I_{n+1}$ ($n = 1, 2, 3, \ldots$), then $\bigcap_1^\infty I_n$ is not empty.*

Proof If $I_n = [a_n, b_n]$, let E be the set of all a_n. Then E is nonempty and bounded above (by b_1). Let x be the sup of E. If m and n are positive integers, then

$$a_n \le a_{m+n} \le b_{m+n} \le b_m,$$

so that $x \le b_m$ for each m. Since it is obvious that $a_m \le x$, we see that $x \in I_m$ for $m = 1, 2, 3, \ldots$.

2.39 Theorem *Let k be a positive integer. If $\{I_n\}$ is a sequence of k-cells such that $I_n \supset I_{n+1}$ ($n = 1, 2, 3, \ldots$), then $\bigcap_1^\infty I_n$ is not empty.*

Proof Let I_n consist of all points $\mathbf{x} = (x_1, \ldots, x_k)$ such that

$$a_{n,j} \leq x_j \leq b_{n,j} \quad (1 \leq j \leq k; n = 1, 2, 3, \ldots),$$

and put $I_{n,j} = [a_{n,j}, b_{n,j}]$. For each j, the sequence $\{I_{n,j}\}$ satisfies the hypotheses of Theorem 2.38. Hence there are real numbers x_j^* ($1 \leq j \leq k$) such that

$$a_{n,j} \leq x_j^* \leq b_{n,j} \quad (1 \leq j \leq k; n = 1, 2, 3, \ldots).$$

Setting $\mathbf{x}^* = (x_1^*, \ldots, x_k^*)$, we see that $\mathbf{x}^* \in I_n$ for $n = 1, 2, 3, \ldots$. The theorem follows.

2.40 Theorem *Every k-cell is compact.*

Proof Let I be a k-cell, consisting of all points $\mathbf{x} = (x_1, \ldots, x_k)$ such that $a_j \leq x_j \leq b_j$ ($1 \leq j \leq k$). Put

$$\delta = \left\{ \sum_1^k (b_j - a_j)^2 \right\}^{1/2}.$$

Then $|\mathbf{x} - \mathbf{y}| \leq \delta$, if $\mathbf{x} \in I$, $\mathbf{y} \in I$.

Suppose, to get a contradiction, that there exists an open cover $\{G_\alpha\}$ of I which contains no finite subcover of I. Put $c_j = (a_j + b_j)/2$. The intervals $[a_j, c_j]$ and $[c_j, b_j]$ then determine 2^k k-cells Q_i whose union is I. At least one of these sets Q_i, call it I_1, cannot be covered by any finite subcollection of $\{G_\alpha\}$ (otherwise I could be so covered). We next subdivide I_1 and continue the process. We obtain a sequence $\{I_n\}$ with the following properties:

(a) $I \supset I_1 \supset I_2 \supset I_3 \supset \cdots$;
(b) I_n is not covered by any finite subcollection of $\{G_\alpha\}$;
(c) if $\mathbf{x} \in I_n$ and $\mathbf{y} \in I_n$, then $|\mathbf{x} - \mathbf{y}| \leq 2^{-n}\delta$.

By (a) and Theorem 2.39, there is a point $\mathbf{x}^*$ which lies in every I_n. For some α, $\mathbf{x}^* \in G_\alpha$. Since G_α is open, there exists $r > 0$ such that $|\mathbf{y} - \mathbf{x}^*| < r$ implies that $\mathbf{y} \in G_\alpha$. If n is so large that $2^{-n}\delta < r$ (there is such an n, for otherwise $2^n \leq \delta/r$ for all positive integers n, which is absurd since R is archimedean), then (c) implies that $I_n \subset G_\alpha$, which contradicts (b).

This completes the proof.

The equivalence of (a) and (b) in the next theorem is known as the Heine-Borel theorem.

2.41 Theorem *If a set E in R^k has one of the following three properties, then it has the other two:*

(a) *E is closed and bounded.*
(b) *E is compact.*
(c) *Every infinite subset of E has a limit point in E.*

Proof If (a) holds, then $E \subset I$ for some k-cell I, and (b) follows from Theorems 2.40 and 2.35. Theorem 2.37 shows that (b) implies (c). It remains to be shown that (c) implies (a).

If E is not bounded, then E contains points $\mathbf{x}_n$ with
$$|\mathbf{x}_n| > n \qquad (n = 1, 2, 3, \ldots).$$

The set S consisting of these points $\mathbf{x}_n$ is infinite and clearly has no limit point in R^k, hence has none in E. Thus (c) implies that E is bounded.

If E is not closed, then there is a point $\mathbf{x}_0 \in R^k$ which is a limit point of E but not a point of E. For $n = 1, 2, 3, \ldots$, there are points $\mathbf{x}_n \in E$ such that $|\mathbf{x}_n - \mathbf{x}_0| < 1/n$. Let S be the set of these points $\mathbf{x}_n$. Then S is infinite (otherwise $|\mathbf{x}_n - \mathbf{x}_0|$ would have a constant positive value, for infinitely many n), S has $\mathbf{x}_0$ as a limit point, and S has no other limit point in R^k. For if $\mathbf{y} \in R^k$, $\mathbf{y} \neq \mathbf{x}_0$, then

$$|\mathbf{x}_n - \mathbf{y}| \geq |\mathbf{x}_0 - \mathbf{y}| - |\mathbf{x}_n - \mathbf{x}_0|$$
$$\geq |\mathbf{x}_0 - \mathbf{y}| - \frac{1}{n} \geq \frac{1}{2}|\mathbf{x}_0 - \mathbf{y}|$$

for all but finitely many n; this shows that $\mathbf{y}$ is not a limit point of S (Theorem 2.20).

Thus S has no limit point in E; hence E must be closed if (c) holds.

We should remark, at this point, that (b) and (c) are equivalent in any metric space (Exercise 26) but that (a) does not, in general, imply (b) and (c). Examples are furnished by Exercise 16 and by the space $\mathscr{L}^2$, which is discussed in Chap. 11.

2.42 Theorem (Weierstrass) *Every bounded infinite subset of R^k has a limit point in R^k.*

Proof Being bounded, the set E in question is a subset of a k-cell $I \subset R^k$. By Theorem 2.40, I is compact, and so E has a limit point in I, by Theorem 2.37.

PERFECT SETS

2.43 Theorem *Let P be a nonempty perfect set in R^k. Then P is uncountable.*

Proof Since P has limit points, P must be infinite. Suppose P is countable, and denote the points of P by $\mathbf{x}_1, \mathbf{x}_2, \mathbf{x}_3, \ldots$. We shall construct a sequence $\{V_n\}$ of neighborhoods, as follows.

Let V_1 be any neighborhood of $\mathbf{x}_1$. If V_1 consists of all $\mathbf{y} \in R^k$ such that $|\mathbf{y} - \mathbf{x}_1| < r$, the closure $\bar{V}_1$ of V_1 is the set of all $\mathbf{y} \in R^k$ such that $|\mathbf{y} - \mathbf{x}_1| \le r$.

Suppose V_n has been constructed, so that $V_n \cap P$ is not empty. Since every point of P is a limit point of P, there is a neighborhood V_{n+1} such that (i) $\bar{V}_{n+1} \subset V_n$, (ii) $\mathbf{x}_n \notin \bar{V}_{n+1}$, (iii) $V_{n+1} \cap P$ is not empty. By (iii), V_{n+1} satisfies our induction hypothesis, and the construction can proceed.

Put $K_n = \bar{V}_n \cap P$. Since $\bar{V}_n$ is closed and bounded, $\bar{V}_n$ is compact. Since $\mathbf{x}_n \notin K_{n+1}$, no point of P lies in $\bigcap_1^\infty K_n$. Since $K_n \subset P$, this implies that $\bigcap_1^\infty K_n$ is empty. But each K_n is nonempty, by (iii), and $K_n \supset K_{n+1}$, by (i); this contradicts the Corollary to Theorem 2.36.

Corollary *Every interval $[a, b]$ $(a < b)$ is uncountable. In particular, the set of all real numbers is uncountable.*

2.44 The Cantor set The set which we are now going to construct shows that there exist perfect sets in R^1 which contain no segment.

Let E_0 be the interval $[0, 1]$. Remove the segment $(\frac{1}{3}, \frac{2}{3})$, and let E_1 be the union of the intervals

$$[0, \tfrac{1}{3}] \quad [\tfrac{2}{3}, 1].$$

Remove the middle thirds of these intervals, and let E_2 be the union of the intervals

$$[0, \tfrac{1}{9}], \, [\tfrac{2}{9}, \tfrac{3}{9}], \, [\tfrac{6}{9}, \tfrac{7}{9}], \, [\tfrac{8}{9}, 1].$$

Continuing in this way, we obtain a sequence of compact sets E_n, such that

(a) $E_1 \supset E_2 \supset E_3 \supset \cdots$;
(b) E_n is the union of 2^n intervals, each of length 3^{-n}.

The set

$$P = \bigcap_{n=1}^\infty E_n$$

is called the *Cantor set*. P is clearly compact, and Theorem 2.36 shows that P is not empty.

No segment of the form

$$\left(\frac{3k+1}{3^m}, \frac{3k+2}{3^m}\right), \tag{24}$$

where k and m are positive integers, has a point in common with P. Since every segment (α, β) contains a segment of the form (24), if

$$3^{-m} < \frac{\beta - \alpha}{6},$$

P contains no segment.

To show that P is perfect, it is enough to show that P contains no isolated point. Let $x \in P$, and let S be any segment containing x. Let I_n be that interval of E_n which contains x. Choose n large enough, so that $I_n \subset S$. Let x_n be an endpoint of I_n, such that $x_n \neq x$.

It follows from the construction of P that $x_n \in P$. Hence x is a limit point of P, and P is perfect.

One of the most interesting properties of the Cantor set is that it provides us with an example of an uncountable set of measure zero (the concept of measure will be discussed in Chap. 11).

CONNECTED SETS

2.45 Definition Two subsets A and B of a metric space X are said to be *separated* if both $A \cap \bar{B}$ and $\bar{A} \cap B$ are empty, i.e., if no point of A lies in the closure of B and no point of B lies in the closure of A.

A set $E \subset X$ is said to be *connected* if E is *not* a union of two nonempty separated sets.

2.46 Remark Separated sets are of course disjoint, but disjoint sets need not be separated. For example, the interval $[0, 1]$ and the segment $(1, 2)$ are *not* separated, since 1 is a limit point of $(1, 2)$. However, the segments $(0, 1)$ and $(1, 2)$ *are* separated.

The connected subsets of the line have a particularly simple structure:

2.47 Theorem *A subset E of the real line R^1 is connected if and only if it has the following property: If $x \in E$, $y \in E$, and $x < z < y$, then $z \in E$.*

Proof If there exist $x \in E$, $y \in E$, and some $z \in (x, y)$ such that $z \notin E$, then $E = A_z \cup B_z$ where

$$A_z = E \cap (-\infty, z), \qquad B_z = E \cap (z, \infty).$$

Since $x \in A_z$ and $y \in B_z$, A and B are nonempty. Since $A_z \subset (-\infty, z)$ and $B_z \subset (z, \infty)$, they are separated. Hence E is not connected.

To prove the converse, suppose E is not connected. Then there are nonempty separated sets A and B such that $A \cup B = E$. Pick $x \in A$, $y \in B$, and assume (without loss of generality) that $x < y$. Define

$$z = \sup (A \cap [x, y]).$$

By Theorem 2.28, $z \in \bar{A}$; hence $z \notin B$. In particular, $x \leq z < y$.
If $z \notin A$, it follows that $x < z < y$ and $z \notin E$.
If $z \in A$, then $z \notin \bar{B}$, hence there exists z_1 such that $z < z_1 < y$ and $z_1 \notin B$. Then $x < z_1 < y$ and $z_1 \notin E$.

EXERCISES

1. Prove that the empty set is a subset of every set.
2. A complex number z is said to be *algebraic* if there are integers $a_0, \ldots, a_n$, not all zero, such that
$$a_0 z^n + a_1 z^{n-1} + \cdots + a_{n-1} z + a_n = 0.$$
Prove that the set of all algebraic numbers is countable. *Hint:* For every positive integer N there are only finitely many equations with
$$n + |a_0| + |a_1| + \cdots + |a_n| = N.$$
3. Prove that there exist real numbers which are not algebraic.
4. Is the set of all irrational real numbers countable?
5. Construct a bounded set of real numbers with exactly three limit points.
6. Let E' be the set of all limit points of a set E. Prove that E' is closed. Prove that E and $\bar{E}$ have the same limit points. (Recall that $\bar{E} = E \cup E'$.) Do E and E' always have the same limit points?
7. Let $A_1, A_2, A_3, \ldots$ be subsets of a metric space.
   (a) If $B_n = \bigcup_{i=1}^{n} A_i$, prove that $\bar{B}_n = \bigcup_{i=1}^{n} \bar{A}_i$, for $n = 1, 2, 3, \ldots$.
   (b) If $B = \bigcup_{i=1}^{\infty} A_i$, prove that $\bar{B} \supset \bigcup_{i=1}^{\infty} \bar{A}_i$.
   Show, by an example, that this inclusion can be proper.
8. Is every point of every open set $E \subset R^2$ a limit point of E? Answer the same question for closed sets in R^2.
9. Let $E°$ denote the set of all interior points of a set E. [See Definition 2.18(e); $E°$ is called the *interior* of E.]
   (a) Prove that $E°$ is always open.
   (b) Prove that E is open if and only if $E° = E$.
   (c) If $G \subset E$ and G is open, prove that $G \subset E°$.
   (d) Prove that the complement of $E°$ is the closure of the complement of E.
   (e) Do E and $\bar{E}$ always have the same interiors?
   (f) Do E and $E°$ always have the same closures?

10. Let X be an infinite set. For $p \in X$ and $q \in X$, define

$$d(p, q) = \begin{cases} 1 & (\text{if } p \neq q) \\ 0 & (\text{if } p = q). \end{cases}$$

Prove that this is a metric. Which subsets of the resulting metric space are open? Which are closed? Which are compact?

11. For $x \in R^1$ and $y \in R^1$, define

$$d_1(x, y) = (x - y)^2,$$
$$d_2(x, y) = \sqrt{|x - y|},$$
$$d_3(x, y) = |x^2 - y^2|,$$
$$d_4(x, y) = |x - 2y|,$$
$$d_5(x, y) = \frac{|x - y|}{1 + |x - y|}.$$

Determine, for each of these, whether it is a metric or not.

12. Let $K \subset R^1$ consist of 0 and the numbers $1/n$, for $n = 1, 2, 3, \ldots$. Prove that K is compact directly from the definition (without using the Heine-Borel theorem).

13. Construct a compact set of real numbers whose limit points form a countable set.

14. Give an example of an open cover of the segment $(0, 1)$ which has no finite subcover.

15. Show that Theorem 2.36 and its Corollary become false (in R^1, for example) if the word "compact" is replaced by "closed" or by "bounded."

16. Regard Q, the set of all rational numbers, as a metric space, with $d(p, q) = |p - q|$. Let E be the set of all $p \in Q$ such that $2 < p^2 < 3$. Show that E is closed and bounded in Q, but that E is not compact. Is E open in Q?

17. Let E be the set of all $x \in [0, 1]$ whose decimal expansion contains only the digits 4 and 7. Is E countable? Is E dense in $[0, 1]$? Is E compact? Is E perfect?

18. Is there a nonempty perfect set in R^1 which contains no rational number?

19. (a) If A and B are disjoint closed sets in some metric space X, prove that they are separated.

    (b) Prove the same for disjoint open sets.

    (c) Fix $p \in X$, $\delta > 0$, define A to be the set of all $q \in X$ for which $d(p, q) < \delta$, define B similarly, with $>$ in place of $<$. Prove that A and B are separated.

    (d) Prove that every connected metric space with at least two points is uncountable. *Hint*: Use (c).

20. Are closures and interiors of connected sets always connected? (Look at subsets of R^2.)

21. Let A and B be separated subsets of some R^k, suppose $\mathbf{a} \in A$, $\mathbf{b} \in B$, and define

$$\mathbf{p}(t) = (1 - t)\mathbf{a} + t\mathbf{b}$$

for $t \in R^1$. Put $A_0 = \mathbf{p}^{-1}(A)$, $B_0 = \mathbf{p}^{-1}(B)$. [Thus $t \in A_0$ if and only if $\mathbf{p}(t) \in A$.]

(a) Prove that A_0 and B_0 are separated subsets of R^1.
(b) Prove that there exists $t_0 \in (0, 1)$ such that $\mathbf{p}(t_0) \notin A \cup B$.
(c) Prove that every convex subset of R^k is connected.

22. A metric space is called *separable* if it contains a countable dense subset. Show that R^k is separable. *Hint:* Consider the set of points which have only rational coordinates.

23. A collection $\{V_\alpha\}$ of open subsets of X is said to be a *base* for X if the following is true: For every $x \in X$ and every open set $G \subset X$ such that $x \in G$, we have $x \in V_\alpha \subset G$ for some α. In other words, every open set in X is the union of a subcollection of $\{V_\alpha\}$.

    Prove that every separable metric space has a *countable* base. *Hint:* Take all neighborhoods with rational radius and center in some countable dense subset of X.

24. Let X be a metric space in which every infinite subset has a limit point. Prove that X is separable. *Hint:* Fix $\delta > 0$, and pick $x_1 \in X$. Having chosen $x_1, \ldots, x_j \in X$, choose $x_{j+1} \in X$, if possible, so that $d(x_i, x_{j+1}) \geq \delta$ for $i = 1, \ldots, j$. Show that this process must stop after a finite number of steps, and that X can therefore be covered by finitely many neighborhoods of radius δ. Take $\delta = 1/n$ ($n = 1, 2, 3, \ldots$), and consider the centers of the corresponding neighborhoods.

25. Prove that every compact metric space K has a countable base, and that K is therefore separable. *Hint:* For every positive integer n, there are finitely many neighborhoods of radius $1/n$ whose union covers K.

26. Let X be a metric space in which every infinite subset has a limit point. Prove that X is compact. *Hint:* By Exercises 23 and 24, X has a countable base. It follows that every open cover of X has a *countable* subcover $\{G_n\}$, $n = 1, 2, 3, \ldots$. If no finite subcollection of $\{G_n\}$ covers X, then the complement F_n of $G_1 \cup \cdots \cup G_n$ is nonempty for each n, but $\bigcap F_n$ is empty. If E is a set which contains a point from each F_n, consider a limit point of E, and obtain a contradiction.

27. Define a point p in a metric space X to be a *condensation point* of a set $E \subset X$ if every neighborhood of p contains uncountably many points of E.

    Suppose $E \subset R^k$, E is uncountable, and let P be the set of all condensation points of E. Prove that P is perfect and that at most countably many points of E are not in P. In other words, show that $P^c \cap E$ is at most countable. *Hint:* Let $\{V_n\}$ be a countable base of R^k, let W be the union of those V_n for which $E \cap V_n$ is at most countable, and show that $P = W^c$.

28. Prove that every closed set in a separable metric space is the union of a (possibly empty) perfect set and a set which is at most countable. (*Corollary:* Every countable closed set in R^k has isolated points.) *Hint:* Use Exercise 27.

29. Prove that every open set in R^1 is the union of an at most countable collection of disjoint segments. *Hint:* Use Exercise 22.

30. Imitate the proof of Theorem 2.43 to obtain the following result:

If $R^k = \bigcup_1^\infty F_n$, where each F_n is a closed subset of R^k, then at least one F_n has a nonempty interior.

Equivalent statement: If G_n is a dense open subset of R^k, for $n = 1, 2, 3, \ldots$, then $\bigcap_1^\infty G_n$ is not empty (in fact, it is dense in R^k).

(This is a special case of Baire's theorem; see Exercise 22, Chap. 3, for the general case.)

3
NUMERICAL SEQUENCES AND SERIES

As the title indicates, this chapter will deal primarily with sequences and series of complex numbers. The basic facts about convergence, however, are just as easily explained in a more general setting. The first three sections will therefore be concerned with sequences in euclidean spaces, or even in metric spaces.

CONVERGENT SEQUENCES

3.1 Definition A sequence $\{p_n\}$ in a metric space X is said to *converge* if there is a point $p \in X$ with the following property: For every $\varepsilon > 0$ there is an integer N such that $n \geq N$ implies that $d(p_n, p) < \varepsilon$. (Here d denotes the distance in X.)

In this case we also say that $\{p_n\}$ converges to p, or that p is the limit of $\{p_n\}$ [see Theorem 3.2(b)], and we write $p_n \to p$, or

$$\lim_{n \to \infty} p_n = p.$$

If $\{p_n\}$ does not converge, it is said to *diverge*.

It might be well to point out that our definition of "convergent sequence" depends not only on $\{p_n\}$ but also on X; for instance, the sequence $\{1/n\}$ converges in R^1 (to 0), but fails to converge in the set of all positive real numbers [with $d(x, y) = |x - y|$]. In cases of possible ambiguity, we can be more precise and specify "convergent in X" rather than "convergent."

We recall that the set of all points p_n ($n = 1, 2, 3, \ldots$) is the *range* of $\{p_n\}$. The range of a sequence may be a finite set, or it may be infinite. The sequence $\{p_n\}$ is said to be *bounded* if its range is bounded.

As examples, consider the following sequences of complex numbers (that is, $X = R^2$):

(a) If $s_n = 1/n$, then $\lim_{n \to \infty} s_n = 0$; the range is infinite, and the sequence is bounded.
(b) If $s_n = n^2$, the sequence $\{s_n\}$ is unbounded, is divergent, and has infinite range.
(c) If $s_n = 1 + [(-1)^n/n]$, the sequence $\{s_n\}$ converges to 1, is bounded, and has infinite range.
(d) If $s_n = i^n$, the sequence $\{s_n\}$ is divergent, is bounded, and has finite range.
(e) If $s_n = 1$ ($n = 1, 2, 3, \ldots$), then $\{s_n\}$ converges to 1, is bounded, and has finite range.

We now summarize some important properties of convergent sequences in metric spaces.

3.2 Theorem *Let $\{p_n\}$ be a sequence in a metric space X.*

(a) *$\{p_n\}$ converges to $p \in X$ if and only if every neighborhood of p contains p_n for all but finitely many n.*
(b) *If $p \in X$, $p' \in X$, and if $\{p_n\}$ converges to p and to p', then $p' = p$.*
(c) *If $\{p_n\}$ converges, then $\{p_n\}$ is bounded.*
(d) *If $E \subset X$ and if p is a limit point of E, then there is a sequence $\{p_n\}$ in E such that $p = \lim_{n \to \infty} p_n$.*

Proof (a) Suppose $p_n \to p$ and let V be a neighborhood of p. For some $\varepsilon > 0$, the conditions $d(q, p) < \varepsilon$, $q \in X$ imply $q \in V$. Corresponding to this ε, there exists N such that $n \geq N$ implies $d(p_n, p) < \varepsilon$. Thus $n \geq N$ implies $p_n \in V$.

Conversely, suppose every neighborhood of p contains all but finitely many of the p_n. Fix $\varepsilon > 0$, and let V be the set of all $q \in X$ such that $d(p, q) < \varepsilon$. By assumption, there exists N (corresponding to this V) such that $p_n \in V$ if $n \geq N$. Thus $d(p_n, p) < \varepsilon$ if $n \geq N$; hence $p_n \to p$.

(b) Let $\varepsilon > 0$ be given. There exist integers N, N' such that

$$n \geq N \quad \text{implies} \quad d(p_n, p) < \frac{\varepsilon}{2},$$

$$n \geq N' \quad \text{implies} \quad d(p_n, p') < \frac{\varepsilon}{2}.$$

Hence if $n \geq \max(N, N')$, we have

$$d(p, p') \leq d(p, p_n) + d(p_n, p') < \varepsilon.$$

Since ε was arbitrary, we conclude that $d(p, p') = 0$.

(c) Suppose $p_n \to p$. There is an integer N such that $n > N$ implies $d(p_n, p) < 1$. Put

$$r = \max\{1, d(p_1, p), \ldots, d(p_N, p)\}.$$

Then $d(p_n, p) \leq r$ for $n = 1, 2, 3, \ldots$.

(d) For each positive integer n, there is a point $p_n \in E$ such that $d(p_n, p) < 1/n$. Given $\varepsilon > 0$, choose N so that $N\varepsilon > 1$. If $n > N$, it follows that $d(p_n, p) < \varepsilon$. Hence $p_n \to p$.

This completes the proof.

For sequences in R^k we can study the relation between convergence, on the one hand, and the algebraic operations on the other. We first consider sequences of complex numbers.

3.3 Theorem *Suppose* $\{s_n\}$, $\{t_n\}$ *are complex sequences, and* $\lim_{n \to \infty} s_n = s$, $\lim_{n \to \infty} t_n = t$. *Then*

(a) $\lim_{n \to \infty} (s_n + t_n) = s + t$;

(b) $\lim_{n \to \infty} cs_n = cs$, $\lim_{n \to \infty} (c + s_n) = c + s$, *for any number* c;

(c) $\lim_{n \to \infty} s_n t_n = st$;

(d) $\lim_{n \to \infty} \frac{1}{s_n} = \frac{1}{s}$, *provided* $s_n \neq 0$ $(n = 1, 2, 3, \ldots)$, *and* $s \neq 0$.

Proof

(a) Given $\varepsilon > 0$, there exist integers N_1, N_2 such that

$$n \geq N_1 \quad \text{implies} \quad |s_n - s| < \frac{\varepsilon}{2},$$

$$n \geq N_2 \quad \text{implies} \quad |t_n - t| < \frac{\varepsilon}{2}.$$

If $N = \max(N_1, N_2)$, then $n \geq N$ implies
$$|(s_n + t_n) - (s + t)| \leq |s_n - s| + |t_n - t| < \varepsilon.$$
This proves (a). The proof of (b) is trivial.

(c) We use the identity
$$(1) \qquad s_n t_n - st = (s_n - s)(t_n - t) + s(t_n - t) + t(s_n - s).$$
Given $\varepsilon > 0$, there are integers N_1, N_2 such that
$$n \geq N_1 \text{ implies } |s_n - s| < \sqrt{\varepsilon},$$
$$n \geq N_2 \text{ implies } |t_n - t| < \sqrt{\varepsilon}.$$
If we take $N = \max(N_1, N_2)$, $n \geq N$ implies
$$|(s_n - s)(t_n - t)| < \varepsilon,$$
so that
$$\lim_{n \to \infty} (s_n - s)(t_n - t) = 0.$$
We now apply (a) and (b) to (1), and conclude that
$$\lim_{n \to \infty} (s_n t_n - st) = 0.$$

(d) Choosing m such that $|s_n - s| < \tfrac{1}{2}|s|$ if $n \geq m$, we see that
$$|s_n| > \tfrac{1}{2}|s| \qquad (n \geq m).$$
Given $\varepsilon > 0$, there is an integer $N > m$ such that $n \geq N$ implies
$$|s_n - s| < \tfrac{1}{2}|s|^2 \varepsilon.$$
Hence, for $n \geq N$,
$$\left|\frac{1}{s_n} - \frac{1}{s}\right| = \left|\frac{s_n - s}{s_n s}\right| < \frac{2}{|s|^2}|s_n - s| < \varepsilon.$$

3.4 Theorem

(a) Suppose $\mathbf{x}_n \in R^k$ ($n = 1, 2, 3, \ldots$) and
$$\mathbf{x}_n = (\alpha_{1,n}, \ldots, \alpha_{k,n}).$$
Then $\{\mathbf{x}_n\}$ converges to $\mathbf{x} = (\alpha_1, \ldots, \alpha_k)$ if and only if
$$(2) \qquad \lim_{n \to \infty} \alpha_{j,n} = \alpha_j \qquad (1 \leq j \leq k).$$

(b) Suppose $\{\mathbf{x}_n\}$, $\{\mathbf{y}_n\}$ are sequences in R^k, $\{\beta_n\}$ is a sequence of real numbers, and $\mathbf{x}_n \to \mathbf{x}$, $\mathbf{y}_n \to \mathbf{y}$, $\beta_n \to \beta$. Then

$$\lim_{n\to\infty} (\mathbf{x}_n + \mathbf{y}_n) = \mathbf{x} + \mathbf{y}, \qquad \lim_{n\to\infty} \mathbf{x}_n \cdot \mathbf{y}_n = \mathbf{x} \cdot \mathbf{y}, \qquad \lim_{n\to\infty} \beta_n \mathbf{x}_n = \beta \mathbf{x}.$$

Proof

(a) If $\mathbf{x}_n \to \mathbf{x}$, the inequalities

$$|\alpha_{j,n} - \alpha_j| \le |\mathbf{x}_n - \mathbf{x}|,$$

which follow immediately from the definition of the norm in R^k, show that (2) holds.

Conversely, if (2) holds, then to each $\varepsilon > 0$ there corresponds an integer N such that $n \ge N$ implies

$$|\alpha_{j,n} - \alpha_j| < \frac{\varepsilon}{\sqrt{k}} \qquad (1 \le j \le k).$$

Hence $n \ge N$ implies

$$|\mathbf{x}_n - \mathbf{x}| = \left\{ \sum_{j=1}^{k} |\alpha_{j,n} - \alpha_j|^2 \right\}^{1/2} < \varepsilon,$$

so that $\mathbf{x}_n \to \mathbf{x}$. This proves (a).

Part (b) follows from (a) and Theorem 3.3.

SUBSEQUENCES

3.5 Definition Given a sequence $\{p_n\}$, consider a sequence $\{n_k\}$ of positive integers, such that $n_1 < n_2 < n_3 < \cdots$. Then the sequence $\{p_{n_i}\}$ is called a *subsequence* of $\{p_n\}$. If $\{p_{n_i}\}$ converges, its limit is called a *subsequential limit* of $\{p_n\}$.

It is clear that $\{p_n\}$ converges to p if and only if every subsequence of $\{p_n\}$ converges to p. We leave the details of the proof to the reader.

3.6 Theorem

(a) *If $\{p_n\}$ is a sequence in a compact metric space X, then some subsequence of $\{p_n\}$ converges to a point of X.*

(b) *Every bounded sequence in R^k contains a convergent subsequence.*

Proof

(a) Let E be the range of $\{p_n\}$. If E is finite then there is a $p \in E$ and a sequence $\{n_i\}$ with $n_1 < n_2 < n_3 < \cdots$, such that
$$p_{n_1} = p_{n_2} = \cdots = p.$$
The subsequence $\{p_{n_i}\}$ so obtained converges evidently to p.

If E is infinite, Theorem 2.37 shows that E has a limit point $p \in X$. Choose n_1 so that $d(p, p_{n_1}) < 1$. Having chosen $n_1, \ldots, n_{i-1}$, we see from Theorem 2.20 that there is an integer $n_i > n_{i-1}$ such that $d(p, p_{n_i}) < 1/i$. Then $\{p_{n_i}\}$ converges to p.

(b) This follows from (a), since Theorem 2.41 implies that every bounded subset of R^k lies in a compact subset of R^k.

3.7 Theorem *The subsequential limits of a sequence $\{p_n\}$ in a metric space X form a closed subset of X.*

Proof Let E^* be the set of all subsequential limits of $\{p_n\}$ and let q be a limit point of E^*. We have to show that $q \in E^*$.

Choose n_1 so that $p_{n_1} \neq q$. (If no such n_1 exists, then E^* has only one point, and there is nothing to prove.) Put $\delta = d(q, p_{n_1})$. Suppose $n_1, \ldots, n_{i-1}$ are chosen. Since q is a limit point of E^*, there is an $x \in E^*$ with $d(x, q) < 2^{-i}\delta$. Since $x \in E^*$, there is an $n_i > n_{i-1}$ such that $d(x, p_{n_i}) < 2^{-i}\delta$. Thus
$$d(q, p_{n_i}) \leq 2^{1-i}\delta$$
for $i = 1, 2, 3, \ldots$. This says that $\{p_{n_i}\}$ converges to q. Hence $q \in E^*$.

CAUCHY SEQUENCES

3.8 Definition A sequence $\{p_n\}$ in a metric space X is said to be a *Cauchy sequence* if for every $\varepsilon > 0$ there is an integer N such that $d(p_n, p_m) < \varepsilon$ if $n \geq N$ and $m \geq N$.

In our discussion of Cauchy sequences, as well as in other situations which will arise later, the following geometric concept will be useful.

3.9 Definition Let E be a nonempty subset of a metric space X, and let S be the set of all real numbers of the form $d(p, q)$, with $p \in E$ and $q \in E$. The sup of S is called the *diameter* of E.

If $\{p_n\}$ is a sequence in X and if E_N consists of the points $p_N, p_{N+1}, p_{N+2}, \ldots$, it is clear from the two preceding definitions that $\{p_n\}$ is a Cauchy sequence if and only if
$$\lim_{N \to \infty} \operatorname{diam} E_N = 0.$$

3.10 Theorem

(a) If $\bar{E}$ is the closure of a set E in a metric space X, then
$$\operatorname{diam} \bar{E} = \operatorname{diam} E.$$

(b) If K_n is a sequence of compact sets in X such that $K_n \supset K_{n+1}$ $(n = 1, 2, 3, \ldots)$ and if
$$\lim_{n \to \infty} \operatorname{diam} K_n = 0,$$
then $\bigcap_1^\infty K_n$ consists of exactly one point.

Proof

(a) Since $E \subset \bar{E}$, it is clear that
$$\operatorname{diam} E \leq \operatorname{diam} \bar{E}.$$

Fix $\varepsilon > 0$, and choose $p \in \bar{E}, q \in \bar{E}$. By the definition of $\bar{E}$, there are points p', q', in E such that $d(p, p') < \varepsilon, d(q, q') < \varepsilon$. Hence
$$d(p, q) \leq d(p, p') + d(p'\, q') + d(q', q)$$
$$< 2\varepsilon + d(p', q') \leq 2\varepsilon + \operatorname{diam} E.$$

It follows that
$$\operatorname{diam} \bar{E} \leq 2\varepsilon + \operatorname{diam} E,$$
and since ε was arbitrary, (a) is proved.

(b) Put $K = \bigcap_1^\infty K_n$. By Theorem 2.36, K is not empty. If K contains more than one point, then $\operatorname{diam} K > 0$. But for each n, $K_n \supset K$, so that $\operatorname{diam} K_n \geq \operatorname{diam} K$. This contradicts the assumption that $\operatorname{diam} K_n \to 0$.

3.11 Theorem

(a) In any metric space X, every convergent sequence is a Cauchy sequence.
(b) If X is a compact metric space and if $\{p_n\}$ is a Cauchy sequence in X, then $\{p_n\}$ converges to some point of X.
(c) In R^k, every Cauchy sequence converges.

Note: The difference between the definition of convergence and the definition of a Cauchy sequence is that the limit is explicitly involved in the former, but not in the latter. Thus Theorem 3.11(b) may enable us to decide whether or not a given sequence converges without knowledge of the limit to which it may converge.

The fact (contained in Theorem 3.11) that a sequence converges in R^k if and only if it is a Cauchy sequence is usually called the *Cauchy criterion* for convergence.

Proof

(a) If $p_n \to p$ and if $\varepsilon > 0$, there is an integer N such that $d(p, p_n) < \varepsilon$ for all $n \geq N$. Hence

$$d(p_n, p_m) \leq d(p_n, p) + d(p, p_m) < 2\varepsilon$$

as soon as $n \geq N$ and $m \geq N$. Thus $\{p_n\}$ is a Cauchy sequence.

(b) Let $\{p_n\}$ be a Cauchy sequence in the compact space X. For $N = 1, 2, 3, \ldots$, let E_N be the set consisting of $p_N, p_{N+1}, p_{N+2}, \ldots$. Then

(3) $$\lim_{N \to \infty} \operatorname{diam} \bar{E}_N = 0,$$

by Definition 3.9 and Theorem 3.10(a). Being a closed subset of the compact space X, each $\bar{E}_N$ is compact (Theorem 2.35). Also $E_N \supset E_{N+1}$, so that $\bar{E}_N \supset \bar{E}_{N+1}$.

Theorem 3.10(b) shows now that there is a unique $p \in X$ which lies in every $\bar{E}_N$.

Let $\varepsilon > 0$ be given. By (3) there is an integer N_0 such that $\operatorname{diam} \bar{E}_N < \varepsilon$ if $N \geq N_0$. Since $p \in \bar{E}_N$, it follows that $d(p, q) < \varepsilon$ for every $q \in \bar{E}_N$, hence for every $q \in E_N$. In other words, $d(p, p_n) < \varepsilon$ if $n \geq N_0$. This says precisely that $p_n \to p$.

(c) Let $\{\mathbf{x}_n\}$ be a Cauchy sequence in R^k. Define E_N as in (b), with $\mathbf{x}_i$ in place of p_i. For some N, $\operatorname{diam} E_N < 1$. The range of $\{\mathbf{x}_n\}$ is the union of E_N and the finite set $\{\mathbf{x}_1, \ldots, \mathbf{x}_{N-1}\}$. Hence $\{\mathbf{x}_n\}$ is bounded. Since every bounded subset of R^k has compact closure in R^k (Theorem 2.41), (c) follows from (b).

3.12 Definition A metric space in which every Cauchy sequence converges is said to be *complete*.

Thus Theorem 3.11 says that *all compact metric spaces and all Euclidean spaces are complete*. Theorem 3.11 implies also that *every closed subset E of a complete metric space X is complete*. (Every Cauchy sequence in E is a Cauchy sequence in X, hence it converges to some $p \in X$, and actually $p \in E$ since E is closed.) An example of a metric space which is not complete is the space of all rational numbers, with $d(x, y) = |x - y|$.

Theorem 3.2(c) and example (d) of Definition 3.1 show that convergent sequences are bounded, but that bounded sequences in R^k need not converge. However, there is one important case in which convergence *is* equivalent to boundedness; this happens for monotonic sequences in R^1.

3.13 Definition A sequence $\{s_n\}$ of real numbers is said to be

(a) *monotonically increasing* if $s_n \leq s_{n+1}$ ($n = 1, 2, 3, \ldots$);
(b) *monotonically decreasing* if $s_n \geq s_{n+1}$ ($n = 1, 2, 3, \ldots$).

The class of monotonic sequences consists of the increasing and the decreasing sequences.

3.14 Theorem *Suppose $\{s_n\}$ is monotonic. Then $\{s_n\}$ converges if and only if it is bounded.*

Proof Suppose $s_n \leq s_{n+1}$ (the proof is analogous in the other case). Let E be the range of $\{s_n\}$. If $\{s_n\}$ is bounded, let s be the least upper bound of E. Then

$$s_n \leq s \quad (n = 1, 2, 3, \ldots).$$

For every $\varepsilon > 0$, there is an integer N such that

$$s - \varepsilon < s_N \leq s,$$

for otherwise $s - \varepsilon$ would be an upper bound of E. Since $\{s_n\}$ increases, $n \geq N$ therefore implies

$$s - \varepsilon < s_n \leq s,$$

which shows that $\{s_n\}$ converges (to s).

The converse follows from Theorem 3.2(c).

UPPER AND LOWER LIMITS

3.15 Definition Let $\{s_n\}$ be a sequence of real numbers with the following property: For every real M there is an integer N such that $n \geq N$ implies $s_n \geq M$. We then write

$$s_n \to +\infty.$$

Similarly, if for every real M there is an integer N such that $n \geq N$ implies $s_n \leq M$, we write

$$s_n \to -\infty.$$

It should be noted that we now use the symbol → (introduced in Definition 3.1) for certain types of divergent sequences, as well as for convergent sequences, but that the definitions of convergence and of limit, given in Definition 3.1, are in no way changed.

3.16 Definition Let $\{s_n\}$ be a sequence of real numbers. Let E be the set of numbers x (in the extended real number system) such that $s_{n_k} \to x$ for some subsequence $\{s_{n_k}\}$. This set E contains all subsequential limits as defined in Definition 3.5, plus possibly the numbers $+\infty$, $-\infty$.

We now recall Definitions 1.8 and 1.23 and put

$$s^* = \sup E,$$
$$s_* = \inf E.$$

The numbers s^*, s_* are called the *upper* and *lower limits* of $\{s_n\}$; we use the notation

$$\limsup_{n\to\infty} s_n = s^*, \qquad \liminf_{n\to\infty} s_n = s_*.$$

3.17 Theorem *Let $\{s_n\}$ be a sequence of real numbers. Let E and s^* have the same meaning as in Definition 3.16. Then s^* has the following two properties:*

(a) $s^* \in E$.
(b) *If $x > s^*$, there is an integer N such that $n \geq N$ implies $s_n < x$.*

Moreover, s^* is the only number with the properties (a) and (b).

Of course, an analogous result is true for s_*.

Proof

(a) If $s^* = +\infty$, then E is not bounded above; hence $\{s_n\}$ is not bounded above, and there is a subsequence $\{s_{n_k}\}$ such that $s_{n_k} \to +\infty$.

If s^* is real, then E is bounded above, and at least one subsequential limit exists, so that (a) follows from Theorems 3.7 and 2.28.

If $s^* = -\infty$, then E contains only one element, namely $-\infty$, and there is no subsequential limit. Hence, for any real M, $s_n > M$ for at most a finite number of values of n, so that $s_n \to -\infty$.

This establishes (a) in all cases.

(b) Suppose there is a number $x > s^*$ such that $s_n \geq x$ for infinitely many values of n. In that case, there is a number $y \in E$ such that $y \geq x > s^*$, contradicting the definition of s^*.

Thus s^* satisfies (a) and (b).

To show the uniqueness, suppose there are two numbers, p and q, which satisfy (a) and (b), and suppose $p < q$. Choose x such that $p < x < q$. Since p satisfies (b), we have $s_n < x$ for $n \geq N$. But then q cannot satisfy (a).

3.18 Examples

(a) Let $\{s_n\}$ be a sequence containing all rationals. Then every real number is a subsequential limit, and

$$\limsup_{n \to \infty} s_n = +\infty, \quad \liminf_{n \to \infty} s_n = -\infty.$$

(b) Let $s_n = (-1^n)/[1 + (1/n)]$. Then

$$\limsup_{n \to \infty} s_n = 1, \quad \liminf_{n \to \infty} s_n = -1.$$

(c) For a real-valued sequence $\{s_n\}$, $\lim_{n \to \infty} s_n = s$ if and only if

$$\limsup_{n \to \infty} s_n = \liminf_{n \to \infty} s_n = s.$$

We close this section with a theorem which is useful, and whose proof is quite trivial:

3.19 Theorem *If $s_n \leq t_n$ for $n \geq N$, where N is fixed, then*

$$\liminf_{n \to \infty} s_n \leq \liminf_{n \to \infty} t_n,$$

$$\limsup_{n \to \infty} s_n \leq \limsup_{n \to \infty} t_n.$$

SOME SPECIAL SEQUENCES

We shall now compute the limits of some sequences which occur frequently. The proofs will all be based on the following remark: If $0 \leq x_n \leq s_n$ for $n \geq N$, where N is some fixed number, and if $s_n \to 0$, then $x_n \to 0$.

3.20 Theorem

(a) If $p > 0$, then $\displaystyle\lim_{n \to \infty} \frac{1}{n^p} = 0$.

(b) If $p > 0$, then $\displaystyle\lim_{n \to \infty} \sqrt[n]{p} = 1$.

(c) $\displaystyle\lim_{n \to \infty} \sqrt[n]{n} = 1$.

(d) If $p > 0$ and α is real, then $\displaystyle\lim_{n \to \infty} \frac{n^\alpha}{(1 + p)^n} = 0$.

(e) If $|x| < 1$, then $\displaystyle\lim_{n \to \infty} x^n = 0$.

Proof

(a) Take $n > (1/\varepsilon)^{1/p}$. (Note that the archimedean property of the real number system is used here.)

(b) If $p > 1$, put $x_n = \sqrt[n]{p} - 1$. Then $x_n > 0$, and, by the binomial theorem,
$$1 + nx_n \leq (1 + x_n)^n = p,$$
so that
$$0 < x_n \leq \frac{p-1}{n}.$$
Hence $x_n \to 0$. If $p = 1$, (b) is trivial, and if $0 < p < 1$, the result is obtained by taking reciprocals.

(c) Put $x_n = \sqrt[n]{n} - 1$. Then $x_n \geq 0$, and, by the binomial theorem,
$$n = (1 + x_n)^n \geq \frac{n(n-1)}{2} x_n^2.$$
Hence
$$0 \leq x_n \leq \sqrt{\frac{2}{n-1}} \quad (n \geq 2).$$

(d) Let k be an integer such that $k > \alpha$, $k > 0$. For $n > 2k$,
$$(1 + p)^n > \binom{n}{k} p^k = \frac{n(n-1) \cdots (n-k+1)}{k!} p^k > \frac{n^k p^k}{2^k k!}.$$
Hence
$$0 < \frac{n^\alpha}{(1+p)^n} < \frac{2^k k!}{p^k} n^{\alpha-k} \quad (n > 2k).$$
Since $\alpha - k < 0$, $n^{\alpha-k} \to 0$, by (a).

(e) Take $\alpha = 0$ in (d).

SERIES

In the remainder of this chapter, all sequences and series under consideration will be complex-valued, unless the contrary is explicitly stated. Extensions of some of the theorems which follow, to series with terms in R^k, are mentioned in Exercise 15.

3.21 Definition Given a sequence $\{a_n\}$, we use the notation

$$\sum_{n=p}^{q} a_n \quad (p \leq q)$$

to denote the sum $a_p + a_{p+1} + \cdots + a_q$. With $\{a_n\}$ we associate a sequence $\{s_n\}$, where

$$s_n = \sum_{k=1}^{n} a_k.$$

For $\{s_n\}$ we also use the symbolic expression

$$a_1 + a_2 + a_3 + \cdots$$

or, more concisely,

(4) $$\sum_{n=1}^{\infty} a_n.$$

The symbol (4) we call an *infinite series*, or just a *series*. The numbers s_n are called the *partial sums* of the series. If $\{s_n\}$ converges to s, we say that the series *converges*, and write

$$\sum_{n=1}^{\infty} a_n = s.$$

The number s is called the sum of the series; but it should be clearly understood that *s is the limit of a sequence of sums*, and is not obtained simply by addition.

If $\{s_n\}$ diverges, the series is said to diverge.

Sometimes, for convenience of notation, we shall consider series of the form

(5) $$\sum_{n=0}^{\infty} a_n.$$

And frequently, when there is no possible ambiguity, or when the distinction is immaterial, we shall simply write Σa_n in place of (4) or (5).

It is clear that every theorem about sequences can be stated in terms of series (putting $a_1 = s_1$, and $a_n = s_n - s_{n-1}$ for $n > 1$), and vice versa. But it is nevertheless useful to consider both concepts.

The Cauchy criterion (Theorem 3.11) can be restated in the following form:

3.22 Theorem Σa_n *converges if and only if for every* $\varepsilon > 0$ *there is an integer N such that*

(6) $$\left| \sum_{k=n}^{m} a_k \right| \leq \varepsilon$$

if $m \geq n \geq N$.

In particular, by taking $m = n$, (6) becomes
$$|a_n| \leq \varepsilon \quad (n \geq N).$$
In other words:

3.23 Theorem *If Σa_n converges, then $\lim_{n \to \infty} a_n = 0$.*

The condition $a_n \to 0$ is not, however, sufficient to ensure convergence of Σa_n. For instance, the series
$$\sum_{n=1}^{\infty} \frac{1}{n}$$
diverges; for the proof we refer to Theorem 3.28.

Theorem 3.14, concerning monotonic sequences, also has an immediate counterpart for series.

3.24 Theorem *A series of nonnegative[1] terms converges if and only if its partial sums form a bounded sequence.*

We now turn to a convergence test of a different nature, the so-called "comparison test."

3.25 Theorem

(a) *If $|a_n| \leq c_n$ for $n \geq N_0$, where N_0 is some fixed integer, and if Σc_n converges, then Σa_n converges.*
(b) *If $a_n \geq d_n \geq 0$ for $n \geq N_0$, and if Σd_n diverges, then Σa_n diverges.*

Note that (b) applies only to series of nonnegative terms a_n.

Proof Given $\varepsilon > 0$, there exists $N \geq N_0$ such that $m \geq n \geq N$ implies
$$\sum_{k=n}^{m} c_k \leq \varepsilon,$$
by the Cauchy criterion. Hence
$$\left| \sum_{k=n}^{m} a_k \right| \leq \sum_{k=n}^{m} |a_k| \leq \sum_{k=n}^{m} c_k \leq \varepsilon,$$
and (a) follows.

Next, (b) follows from (a), for if Σa_n converges, so must Σd_n [note that (b) also follows from Theorem 3.24].

[1] The expression "nonnegative" always refers to *real* numbers.

The comparison test is a very useful one; to use it efficiently, we have to become familiar with a number of series of nonnegative terms whose convergence or divergence is known.

SERIES OF NONNEGATIVE TERMS

The simplest of all is perhaps the geometric series.

3.26 Theorem *If $0 \leq x < 1$, then*
$$\sum_{n=0}^{\infty} x^n = \frac{1}{1-x}.$$
If $x \geq 1$, the series diverges.

Proof If $x \neq 1$,
$$s_n = \sum_{k=0}^{n} x^k = \frac{1 - x^{n+1}}{1 - x}.$$
The result follows if we let $n \to \infty$. For $x = 1$, we get
$$1 + 1 + 1 + \cdots,$$
which evidently diverges.

In many cases which occur in applications, the terms of the series decrease monotonically. The following theorem of Cauchy is therefore of particular interest. The striking feature of the theorem is that a rather "thin" subsequence of $\{a_n\}$ determines the convergence or divergence of Σa_n.

3.27 Theorem *Suppose $a_1 \geq a_2 \geq a_3 \geq \cdots \geq 0$. Then the series $\sum_{n=1}^{\infty} a_n$ converges if and only if the series*

(7)
$$\sum_{k=0}^{\infty} 2^k a_{2^k} = a_1 + 2a_2 + 4a_4 + 8a_8 + \cdots$$

converges.

Proof By Theorem 3.24, it suffices to consider boundedness of the partial sums. Let
$$s_n = a_1 + a_2 + \cdots + a_n,$$
$$t_k = a_1 + 2a_2 + \cdots + 2^k a_{2^k}.$$

For $n < 2^k$,

$$s_n \leq a_1 + (a_2 + a_3) + \cdots + (a_{2^k} + \cdots + a_{2^{k+1}-1})$$
$$\leq a_1 + 2a_2 + \cdots + 2^k a_{2^k}$$
$$= t_k,$$

so that

(8) $$s_n \leq t_k.$$

On the other hand, if $n > 2^k$,

$$s_n \geq a_1 + a_2 + (a_3 + a_4) + \cdots + (a_{2^{k-1}+1} + \cdots + a_{2^k})$$
$$\geq \tfrac{1}{2}a_1 + a_2 + 2a_4 + \cdots + 2^{k-1}a_{2^k}$$
$$= \tfrac{1}{2}t_k,$$

so that

(9) $$2s_n \geq t_k.$$

By (8) and (9), the sequences $\{s_n\}$ and $\{t_k\}$ are either both bounded or both unbounded. This completes the proof.

3.28 Theorem $\sum \dfrac{1}{n^p}$ *converges if* $p > 1$ *and diverges if* $p \leq 1$.

Proof If $p \leq 0$, divergence follows from Theorem 3.23. If $p > 0$, Theorem 3.27 is applicable, and we are led to the series

$$\sum_{k=0}^{\infty} 2^k \cdot \frac{1}{2^{kp}} = \sum_{k=0}^{\infty} 2^{(1-p)k}.$$

Now, $2^{1-p} < 1$ if and only if $1 - p < 0$, and the result follows by comparison with the geometric series (take $x = 2^{1-p}$ in Theorem 3.26).

As a further application of Theorem 3.27, we prove:

3.29 Theorem *If* $p > 1$,

(10) $$\sum_{n=2}^{\infty} \frac{1}{n(\log n)^p}$$

converges; if $p \leq 1$, *the series diverges.*

Remark "$\log n$" denotes the logarithm of n to the base e (compare Exercise 7, Chap. 1); the number e will be defined in a moment (see Definition 3.30). We let the series start with $n = 2$, since $\log 1 = 0$.

Proof The monotonicity of the logarithmic function (which will be discussed in more detail in Chap. 8) implies that $\{\log n\}$ increases. Hence $\{1/n \log n\}$ decreases, and we can apply Theorem 3.27 to (10); this leads us to the series

$$\text{(11)} \quad \sum_{k=1}^{\infty} 2^k \cdot \frac{1}{2^k (\log 2^k)^p} = \sum_{k=1}^{\infty} \frac{1}{(k \log 2)^p} = \frac{1}{(\log 2)^p} \sum_{k=1}^{\infty} \frac{1}{k^p},$$

and Theorem 3.29 follows from Theorem 3.28.

This procedure may evidently be continued. For instance,

$$\text{(12)} \quad \sum_{n=3}^{\infty} \frac{1}{n \log n \log \log n}$$

diverges, whereas

$$\text{(13)} \quad \sum_{n=3}^{\infty} \frac{1}{n \log n (\log \log n)^2}$$

converges.

We may now observe that the terms of the series (12) differ very little from those of (13). Still, one diverges, the other converges. If we continue the process which led us from Theorem 3.28 to Theorem 3.29, and then to (12) and (13), we get pairs of convergent and divergent series whose terms differ even less than those of (12) and (13). One might thus be led to the conjecture that there is a limiting situation of some sort, a "boundary" with all convergent series on one side, all divergent series on the other side—at least as far as series with monotonic coefficients are concerned. This notion of "boundary" is of course quite vague. The point we wish to make is this: No matter how we make this notion precise, the conjecture is false. Exercises 11(*b*) and 12(*b*) may serve as illustrations.

We do not wish to go any deeper into this aspect of convergence theory, and refer the reader to Knopp's "Theory and Application of Infinite Series," Chap. IX, particularly Sec. 41.

THE NUMBER *e*

3.30 Definition $e = \sum_{n=0}^{\infty} \frac{1}{n!}.$

Here $n! = 1 \cdot 2 \cdot 3 \cdots n$ if $n \geq 1$, and $0! = 1$.

Since
$$s_n = 1 + 1 + \frac{1}{1\cdot 2} + \frac{1}{1\cdot 2 \cdot 3} + \cdots + \frac{1}{1\cdot 2 \cdots n}$$
$$< 1 + 1 + \frac{1}{2} + \frac{1}{2^2} + \cdots + \frac{1}{2^{n-1}} < 3,$$

the series converges, and the definition makes sense. In fact, the series converges very rapidly and allows us to compute e with great accuracy.

It is of interest to note that e can also be defined by means of another limit process; the proof provides a good illustration of operations with limits:

3.31 Theorem $\lim_{n\to\infty} \left(1 + \frac{1}{n}\right)^n = e.$

Proof Let
$$s_n = \sum_{k=0}^{n} \frac{1}{k!}, \qquad t_n = \left(1 + \frac{1}{n}\right)^n.$$

By the binomial theorem,
$$t_n = 1 + 1 + \frac{1}{2!}\left(1 - \frac{1}{n}\right) + \frac{1}{3!}\left(1 - \frac{1}{n}\right)\left(1 - \frac{2}{n}\right) + \cdots$$
$$+ \frac{1}{n!}\left(1 - \frac{1}{n}\right)\left(1 - \frac{2}{n}\right)\cdots\left(1 - \frac{n-1}{n}\right).$$

Hence $t_n \le s_n$, so that

(14) $$\limsup_{n\to\infty} t_n \le e,$$

by Theorem 3.19. Next, if $n \ge m$,
$$t_n \ge 1 + 1 + \frac{1}{2!}\left(1 - \frac{1}{n}\right) + \cdots + \frac{1}{m!}\left(1 - \frac{1}{n}\right)\cdots\left(1 - \frac{m-1}{n}\right).$$

Let $n \to \infty$, keeping m fixed. We get
$$\liminf_{n\to\infty} t_n \ge 1 + 1 + \frac{1}{2!} + \cdots + \frac{1}{m!},$$

so that
$$s_m \le \liminf_{n\to\infty} t_n.$$

Letting $m \to \infty$, we finally get

(15) $$e \le \liminf_{n\to\infty} t_n.$$

The theorem follows from (14) and (15).

The rapidity with which the series $\sum \frac{1}{n!}$ converges can be estimated as follows: If s_n has the same meaning as above, we have

$$e - s_n = \frac{1}{(n+1)!} + \frac{1}{(n+2)!} + \frac{1}{(n+3)!} + \cdots$$

$$< \frac{1}{(n+1)!}\left\{1 + \frac{1}{n+1} + \frac{1}{(n+1)^2} + \cdots\right\} = \frac{1}{n!n}$$

so that

(16) $$0 < e - s_n < \frac{1}{n!n}.$$

Thus s_{10}, for instance, approximates e with an error less than 10^{-7}. The inequality (16) is of theoretical interest as well, since it enables us to prove the irrationality of e very easily.

3.32 Theorem *e is irrational.*

Proof Suppose e is rational. Then $e = p/q$, where p and q are positive integers. By (16),

(17) $$0 < q!(e - s_q) < \frac{1}{q}.$$

By our assumption, $q!e$ is an integer. Since

$$q!s_q = q!\left(1 + 1 + \frac{1}{2!} + \cdots + \frac{1}{q!}\right)$$

is an integer, we see that $q!(e - s_q)$ is an integer.

Since $q \geq 1$, (17) implies the existence of an integer between 0 and 1. We have thus reached a contradiction.

Actually, e is not even an algebraic number. For a simple proof of this, see page 25 of Niven's book, or page 176 of Herstein's, cited in the Bibliography.

THE ROOT AND RATIO TESTS

3.33 Theorem (Root Test) *Given Σa_n, put $\alpha = \limsup\limits_{n \to \infty} \sqrt[n]{|a_n|}$.*

Then

(a) *if $\alpha < 1$, Σa_n converges;*
(b) *if $\alpha > 1$, Σa_n diverges;*
(c) *if $\alpha = 1$, the test gives no information.*

Proof If $\alpha < 1$, we can choose β so that $\alpha < \beta < 1$, and an integer N such that

$$\sqrt[n]{|a_n|} < \beta$$

for $n \geq N$ [by Theorem 3.17(b)]. That is, $n \geq N$ implies

$$|a_n| < \beta^n.$$

Since $0 < \beta < 1$, $\Sigma \beta^n$ converges. Convergence of Σa_n follows now from the comparison test.

If $\alpha > 1$, then, again by Theorem 3.17, there is a sequence $\{n_k\}$ such that

$$\sqrt[n_k]{|a_{n_k}|} \to \alpha.$$

Hence $|a_n| > 1$ for infinitely many values of n, so that the condition $a_n \to 0$, necessary for convergence of Σa_n, does not hold (Theorem 3.23).

To prove (c), we consider the series

$$\sum \frac{1}{n}, \quad \sum \frac{1}{n^2}.$$

For each of these series $\alpha = 1$, but the first diverges, the second converges.

3.34 Theorem (Ratio Test) *The series Σa_n*

(a) *converges if* $\displaystyle\limsup_{n \to \infty} \left|\frac{a_{n+1}}{a_n}\right| < 1,$

(b) *diverges if* $\left|\dfrac{a_{n+1}}{a_n}\right| \geq 1$ *for all $n \geq n_0$, where n_0 is some fixed integer.*

Proof If condition (a) holds, we can find $\beta < 1$, and an integer N, such that

$$\left|\frac{a_{n+1}}{a_n}\right| < \beta$$

for $n \geq N$. In particular,

$$|a_{N+1}| < \beta |a_N|,$$
$$|a_{N+2}| < \beta |a_{N+1}| < \beta^2 |a_N|,$$
$$\dotsb\dotsb\dotsb\dotsb\dotsb\dotsb$$
$$|a_{N+p}| < \beta^p |a_N|.$$

That is,
$$|a_n| < |a_N|\beta^{-N} \cdot \beta^n$$
for $n \geq N$, and (a) follows from the comparison test, since $\Sigma \beta^n$ converges.

If $|a_{n+1}| \geq |a_n|$ for $n \geq n_0$, it is easily seen that the condition $a_n \to 0$ does not hold, and (b) follows.

Note: The knowledge that $\lim a_{n+1}/a_n = 1$ implies nothing about the convergence of Σa_n. The series $\Sigma 1/n$ and $\Sigma 1/n^2$ demonstrate this.

3.35 Examples

(a) Consider the series
$$\frac{1}{2} + \frac{1}{3} + \frac{1}{2^2} + \frac{1}{3^2} + \frac{1}{2^3} + \frac{1}{3^3} + \frac{1}{2^4} + \frac{1}{3^4} + \cdots,$$
for which
$$\liminf_{n \to \infty} \frac{a_{n+1}}{a_n} = \lim_{n \to \infty} \left(\frac{2}{3}\right)^n = 0,$$
$$\liminf_{n \to \infty} \sqrt[n]{a_n} = \lim_{n \to \infty} \sqrt[2n]{\frac{1}{3^n}} = \frac{1}{\sqrt{3}},$$
$$\limsup_{n \to \infty} \sqrt[n]{a_n} = \lim_{n \to \infty} \sqrt[2n]{\frac{1}{2^n}} = \frac{1}{\sqrt{2}},$$
$$\limsup_{n \to \infty} \frac{a_{n+1}}{a_n} = \lim_{n \to \infty} \frac{1}{2}\left(\frac{3}{2}\right)^n = +\infty.$$

The root test indicates convergence; the ratio test does not apply.

(b) The same is true for the series
$$\frac{1}{2} + 1 + \frac{1}{8} + \frac{1}{4} + \frac{1}{32} + \frac{1}{16} + \frac{1}{128} + \frac{1}{64} + \cdots,$$
where
$$\liminf_{n \to \infty} \frac{a_{n+1}}{a_n} = \frac{1}{8},$$
$$\limsup_{n \to \infty} \frac{a_{n+1}}{a_n} = 2,$$
but
$$\lim \sqrt[n]{a_n} = \tfrac{1}{2}.$$

3.36 Remarks The ratio test is frequently easier to apply than the root test, since it is usually easier to compute ratios than nth roots. However, the root test has wider scope. More precisely: Whenever the ratio test shows convergence, the root test does too; whenever the root test is inconclusive, the ratio test is too. This is a consequence of Theorem 3.37, and is illustrated by the above examples.

Neither of the two tests is subtle with regard to divergence. Both deduce divergence from the fact that a_n does not tend to zero as $n \to \infty$.

3.37 Theorem *For any sequence $\{c_n\}$ of positive numbers,*

$$\liminf_{n \to \infty} \frac{c_{n+1}}{c_n} \leq \liminf_{n \to \infty} \sqrt[n]{c_n},$$

$$\limsup_{n \to \infty} \sqrt[n]{c_n} \leq \limsup_{n \to \infty} \frac{c_{n+1}}{c_n}.$$

Proof We shall prove the second inequality; the proof of the first is quite similar. Put

$$\alpha = \limsup_{n \to \infty} \frac{c_{n+1}}{c_n}.$$

If $\alpha = +\infty$, there is nothing to prove. If α is finite, choose $\beta > \alpha$. There is an integer N such that

$$\frac{c_{n+1}}{c_n} \leq \beta$$

for $n \geq N$. In particular, for any $p > 0$,

$$c_{N+k+1} \leq \beta c_{N+k} \qquad (k = 0, 1, \ldots, p-1).$$

Multiplying these inequalities, we obtain

$$c_{N+p} \leq \beta^p c_N,$$

or

$$c_n \leq c_N \beta^{-N} \cdot \beta^n \qquad (n \geq N).$$

Hence

$$\sqrt[n]{c_n} \leq \sqrt[n]{c_N \beta^{-N}} \cdot \beta,$$

so that

(18) $$\limsup_{n \to \infty} \sqrt[n]{c_n} \leq \beta,$$

by Theorem 3.20(b). Since (18) is true for every $\beta > \alpha$, we have
$$\limsup_{n\to\infty} \sqrt[n]{c_n} \leq \alpha.$$

POWER SERIES

3.38 Definition Given a sequence $\{c_n\}$ of complex numbers, the series

(19) $$\sum_{n=0}^{\infty} c_n z^n$$

is called a *power series*. The numbers c_n are called the *coefficients* of the series; z is a complex number.

In general, the series will converge or diverge, depending on the choice of z. More specifically, with every power series there is associated a circle, the circle of convergence, such that (19) converges if z is in the interior of the circle and diverges if z is in the exterior (to cover all cases, we have to consider the plane as the interior of a circle of infinite radius, and a point as a circle of radius zero). The behavior on the circle of convergence is much more varied and cannot be described so simply.

3.39 Theorem *Given the power series $\Sigma c_n z^n$, put*
$$\alpha = \limsup_{n\to\infty} \sqrt[n]{|c_n|}, \qquad R = \frac{1}{\alpha}.$$

(If $\alpha = 0$, $R = +\infty$; if $\alpha = +\infty$, $R = 0$.) Then $\Sigma c_n z^n$ converges if $|z| < R$, and diverges if $|z| > R$.

Proof Put $a_n = c_n z^n$, and apply the root test:
$$\limsup_{n\to\infty} \sqrt[n]{|a_n|} = |z| \limsup_{n\to\infty} \sqrt[n]{|c_n|} = \frac{|z|}{R}.$$

Note: R is called the *radius of convergence* of $\Sigma c_n z^n$.

3.40 Examples

(a) The series $\Sigma n^n z^n$ has $R = 0$.

(b) The series $\sum \dfrac{z^n}{n!}$ has $R = +\infty$. (In this case the ratio test is easier to apply than the root test.)

(c) The series Σz^n has $R = 1$. If $|z| = 1$, the series diverges, since $\{z^n\}$ does not tend to 0 as $n \to \infty$.

(d) The series $\sum \dfrac{z^n}{n}$ has $R = 1$. It diverges if $z = 1$. It converges for all other z with $|z| = 1$. (The last assertion will be proved in Theorem 3.44.)

(e) The series $\sum \dfrac{z^n}{n^2}$ has $R = 1$. It converges for all z with $|z| = 1$, by the comparison test, since $|z^n/n^2| = 1/n^2$.

SUMMATION BY PARTS

3.41 Theorem *Given two sequences $\{a_n\}$, $\{b_n\}$, put*

$$A_n = \sum_{k=0}^{n} a_k$$

if $n \geq 0$; put $A_{-1} = 0$. Then, if $0 \leq p \leq q$, we have

(20) $$\sum_{n=p}^{q} a_n b_n = \sum_{n=p}^{q-1} A_n(b_n - b_{n+1}) + A_q b_q - A_{p-1} b_p.$$

Proof

$$\sum_{n=p}^{q} a_n b_n = \sum_{n=p}^{q} (A_n - A_{n-1}) b_n = \sum_{n=p}^{q} A_n b_n - \sum_{n=p-1}^{q-1} A_n b_{n+1},$$

and the last expression on the right is clearly equal to the right side of (20).

Formula (20), the so-called "partial summation formula," is useful in the investigation of series of the form $\Sigma a_n b_n$, particularly when $\{b_n\}$ is monotonic. We shall now give applications.

3.42 Theorem *Suppose*

(a) *the partial sums A_n of Σa_n form a bounded sequence;*
(b) $b_0 \geq b_1 \geq b_2 \geq \cdots$;
(c) $\lim_{n \to \infty} b_n = 0$.

Then $\Sigma a_n b_n$ converges.

Proof Choose M such that $|A_n| \leq M$ for all n. Given $\varepsilon > 0$, there is an integer N such that $b_N \leq (\varepsilon/2M)$. For $N \leq p \leq q$, we have

$$\left| \sum_{n=p}^{q} a_n b_n \right| = \left| \sum_{n=p}^{q-1} A_n(b_n - b_{n+1}) + A_q b_q - A_{p-1} b_p \right|$$

$$\leq M \left| \sum_{n=p}^{q-1} (b_n - b_{n+1}) + b_q + b_p \right|$$

$$= 2Mb_p \leq 2Mb_N \leq \varepsilon.$$

Convergence now follows from the Cauchy criterion. We note that the first inequality in the above chain depends of course on the fact that $b_n - b_{n+1} \geq 0$.

3.43 Theorem *Suppose*

(a) $|c_1| \geq |c_2| \geq |c_3| \geq \cdots$;
(b) $c_{2m-1} \geq 0, c_{2m} \leq 0 \quad (m = 1, 2, 3, \ldots)$;
(c) $\lim_{n \to \infty} c_n = 0$.

Then Σc_n converges.

Series for which (b) holds are called "alternating series"; the theorem was known to Leibnitz.

Proof Apply Theorem 3.42, with $a_n = (-1)^{n+1}$, $b_n = |c_n|$.

3.44 Theorem *Suppose the radius of convergence of $\Sigma c_n z^n$ is 1, and suppose $c_0 \geq c_1 \geq c_2 \geq \cdots$, $\lim_{n \to \infty} c_n = 0$. Then $\Sigma c_n z^n$ converges at every point on the circle $|z| = 1$, except possibly at $z = 1$.*

Proof Put $a_n = z^n$, $b_n = c_n$. The hypotheses of Theorem 3.42 are then satisfied, since

$$|A_n| = \left| \sum_{m=0}^{n} z^m \right| = \left| \frac{1 - z^{n+1}}{1 - z} \right| \leq \frac{2}{|1 - z|},$$

if $|z| = 1$, $z \neq 1$.

ABSOLUTE CONVERGENCE

The series Σa_n is said to *converge absolutely* if the series $\Sigma |a_n|$ converges.

3.45 Theorem *If Σa_n converges absolutely, then Σa_n converges.*

Proof The assertion follows from the inequality

$$\left|\sum_{k=n}^{m} a_k\right| \leq \sum_{k=n}^{m} |a_k|,$$

plus the Cauchy criterion.

3.46 Remarks For series of positive terms, absolute convergence is the same as convergence.

If Σa_n converges, but $\Sigma |a_n|$ diverges, we say that Σa_n converges *nonabsolutely*. For instance, the series

$$\sum \frac{(-1)^n}{n}$$

converges nonabsolutely (Theorem 3.43).

The comparison test, as well as the root and ratio tests, is really a test for absolute convergence, and therefore cannot give any information about nonabsolutely convergent series. Summation by parts can sometimes be used to handle the latter. In particular, power series converge absolutely in the interior of the circle of convergence.

We shall see that we may operate with absolutely convergent series very much as with finite sums. We may multiply them term by term and we may change the order in which the additions are carried out, without affecting the sum of the series. But for nonabsolutely convergent series this is no longer true, and more care has to be taken when dealing with them.

ADDITION AND MULTIPLICATION OF SERIES

3.47 Theorem *If $\Sigma a_n = A$, and $\Sigma b_n = B$, then $\Sigma (a_n + b_n) = A + B$, and $\Sigma c a_n = cA$, for any fixed c.*

Proof Let

$$A_n = \sum_{k=0}^{n} a_k, \qquad B_n = \sum_{k=0}^{n} b_k.$$

Then

$$A_n + B_n = \sum_{k=0}^{n} (a_k + b_k).$$

Since $\lim_{n \to \infty} A_n = A$ and $\lim_{n \to \infty} B_n = B$, we see that

$$\lim_{n \to \infty} (A_n + B_n) = A + B.$$

The proof of the second assertion is even simpler.

Thus two convergent series may be added term by term, and the resulting series converges to the sum of the two series. The situation becomes more complicated when we consider multiplication of two series. To begin with, we have to define the product. This can be done in several ways; we shall consider the so-called "Cauchy product."

3.48 Definition Given Σa_n and Σb_n, we put

$$c_n = \sum_{k=0}^{n} a_k b_{n-k} \quad (n = 0, 1, 2, \ldots)$$

and call Σc_n the *product* of the two given series.

This definition may be motivated as follows. If we take two power series $\Sigma a_n z^n$ and $\Sigma b_n z^n$, multiply them term by term, and collect terms containing the same power of z, we get

$$\sum_{n=0}^{\infty} a_n z^n \cdot \sum_{n=0}^{\infty} b_n z^n = (a_0 + a_1 z + a_2 z^2 + \cdots)(b_0 + b_1 z + b_2 z^2 + \cdots)$$
$$= a_0 b_0 + (a_0 b_1 + a_1 b_0)z + (a_0 b_2 + a_1 b_1 + a_2 b_0)z^2 + \cdots$$
$$= c_0 + c_1 z + c_2 z^2 + \cdots.$$

Setting $z = 1$, we arrive at the above definition.

3.49 Example If

$$A_n = \sum_{k=0}^{n} a_k, \quad B_n = \sum_{k=0}^{n} b_k, \quad C_n = \sum_{k=0}^{n} c_k,$$

and $A_n \to A$, $B_n \to B$, then it is not at all clear that $\{C_n\}$ will converge to AB, since we do not have $C_n = A_n B_n$. The dependence of $\{C_n\}$ on $\{A_n\}$ and $\{B_n\}$ is quite a complicated one (see the proof of Theorem 3.50). We shall now show that the product of two convergent series may actually diverge.

The series

$$\sum_{n=0}^{\infty} \frac{(-1)^n}{\sqrt{n+1}} = 1 - \frac{1}{\sqrt{2}} + \frac{1}{\sqrt{3}} - \frac{1}{\sqrt{4}} + \cdots$$

converges (Theorem 3.43). We form the product of this series with itself and obtain

$$\sum_{n=0}^{\infty} c_n = 1 - \left(\frac{1}{\sqrt{2}} + \frac{1}{\sqrt{2}}\right) + \left(\frac{1}{\sqrt{3}} + \frac{1}{\sqrt{2}\sqrt{2}} + \frac{1}{\sqrt{3}}\right)$$
$$- \left(\frac{1}{\sqrt{4}} + \frac{1}{\sqrt{3}\sqrt{2}} + \frac{1}{\sqrt{2}\sqrt{3}} + \frac{1}{\sqrt{4}}\right) + \cdots,$$

so that
$$c_n = (-1)^n \sum_{k=0}^{n} \frac{1}{\sqrt{(n-k+1)(k+1)}}.$$
Since
$$(n-k+1)(k+1) = \left(\frac{n}{2}+1\right)^2 - \left(\frac{n}{2}-k\right)^2 \leq \left(\frac{n}{2}+1\right)^2.$$
we have
$$|c_n| \geq \sum_{k=0}^{n} \frac{2}{n+2} = \frac{2(n+1)}{n+2},$$
so that the condition $c_n \to 0$, which is necessary for the convergence of Σc_n, is not satisfied.

In view of the next theorem, due to Mertens, we note that we have here considered the product of two nonabsolutely convergent series.

3.50 Theorem *Suppose*

(a) $\sum_{n=0}^{\infty} a_n$ *converges absolutely,*

(b) $\sum_{n=0}^{\infty} a_n = A,$

(c) $\sum_{n=0}^{\infty} b_n = B,$

(d) $c_n = \sum_{k=0}^{n} a_k b_{n-k}$ $(n = 0, 1, 2, \ldots).$

Then
$$\sum_{n=0}^{\infty} c_n = AB.$$

That is, the product of two convergent series converges, and to the right value, if at least one of the two series converges absolutely.

Proof Put
$$A_n = \sum_{k=0}^{n} a_k, \qquad B_n = \sum_{k=0}^{n} b_k, \qquad C_n = \sum_{k=0}^{n} c_k, \qquad \beta_n = B_n - B.$$
Then
$$\begin{aligned}
C_n &= a_0 b_0 + (a_0 b_1 + a_1 b_0) + \cdots + (a_0 b_n + a_1 b_{n-1} + \cdots + a_n b_0) \\
&= a_0 B_n + a_1 B_{n-1} + \cdots + a_n B_0 \\
&= a_0 (B + \beta_n) + a_1 (B + \beta_{n-1}) + \cdots + a_n (B + \beta_0) \\
&= A_n B + a_0 \beta_n + a_1 \beta_{n-1} + \cdots + a_n \beta_0
\end{aligned}$$

Put
$$\gamma_n = a_0 \beta_n + a_1 \beta_{n-1} + \cdots + a_n \beta_0.$$

We wish to show that $C_n \to AB$. Since $A_n B \to AB$, it suffices to show that

(21) $$\lim_{n \to \infty} \gamma_n = 0.$$

Put
$$\alpha = \sum_{n=0}^{\infty} |a_n|.$$

[It is here that we use (a).] Let $\varepsilon > 0$ be given. By (c), $\beta_n \to 0$. Hence we can choose N such that $|\beta_n| \leq \varepsilon$ for $n \geq N$, in which case

$$|\gamma_n| \leq |\beta_0 a_n + \cdots + \beta_N a_{n-N}| + |\beta_{N+1} a_{n-N-1} + \cdots + \beta_n a_0|$$
$$\leq |\beta_0 a_n + \cdots + \beta_N a_{n-N}| + \varepsilon \alpha.$$

Keeping N fixed, and letting $n \to \infty$, we get

$$\limsup_{n \to \infty} |\gamma_n| \leq \varepsilon \alpha,$$

since $a_k \to 0$ as $k \to \infty$. Since ε is arbitrary, (21) follows.

Another question which may be asked is whether the series Σc_n, if convergent, must have the sum AB. Abel showed that the answer is in the affirmative.

3.51 Theorem *If the series Σa_n, Σb_n, Σc_n converge to A, B, C, and $c_n = a_0 b_n + \cdots + a_n b_0$, then $C = AB$.*

Here no assumption is made concerning absolute convergence. We shall give a simple proof (which depends on the continuity of power series) after Theorem 8.2.

REARRANGEMENTS

3.52 Definition Let $\{k_n\}$, $n = 1, 2, 3, \ldots$, be a sequence in which every positive integer appears once and only once (that is, $\{k_n\}$ is a 1-1 function from J onto J, in the notation of Definition 2.2). Putting

$$a'_n = a_{k_n} \quad (n = 1, 2, 3, \ldots),$$

we say that $\Sigma a'_n$ is a *rearrangement* of Σa_n.

If $\{s_n\}$, $\{s_n'\}$ are the sequences of partial sums of Σa_n, $\Sigma a_n'$, it is easily seen that, in general, these two sequences consist of entirely different numbers. We are thus led to the problem of determining under what conditions all rearrangements of a convergent series will converge and whether the sums are necessarily the same.

3.53 Example Consider the convergent series

(22) $$1 - \tfrac{1}{2} + \tfrac{1}{3} - \tfrac{1}{4} + \tfrac{1}{5} - \tfrac{1}{6} + \cdots$$

and one of its rearrangements

(23) $$1 + \tfrac{1}{3} - \tfrac{1}{2} + \tfrac{1}{5} + \tfrac{1}{7} - \tfrac{1}{4} + \tfrac{1}{9} + \tfrac{1}{11} - \tfrac{1}{6} + \cdots$$

in which two positive terms are always followed by one negative. If s is the sum of (22), then

$$s < 1 - \tfrac{1}{2} + \tfrac{1}{3} = \tfrac{5}{6}.$$

Since

$$\frac{1}{4k-3} + \frac{1}{4k-1} - \frac{1}{2k} > 0$$

for $k \geq 1$, we see that $s_3' < s_6' < s_9' < \cdots$, where s_n' is nth partial sum of (23). Hence

$$\limsup_{n \to \infty} s_n' > s_3' = \tfrac{5}{6},$$

so that (23) certainly does not converge to s [we leave it to the reader to verify that (23) does, however, converge].

This example illustrates the following theorem, due to Riemann.

3.54 Theorem *Let Σa_n be a series of real numbers which converges, but not absolutely. Suppose*

$$-\infty \leq \alpha \leq \beta \leq \infty.$$

Then there exists a rearrangement $\Sigma a_n'$ with partial sums s_n' such that

(24) $$\liminf_{n \to \infty} s_n' = \alpha, \qquad \limsup_{n \to \infty} s_n' = \beta.$$

Proof Let

$$p_n = \frac{|a_n| + a_n}{2}, \qquad q_n = \frac{|a_n| - a_n}{2} \qquad (n = 1, 2, 3, \ldots).$$

Then $p_n - q_n = a_n$, $p_n + q_n = |a_n|$, $p_n \geq 0$, $q_n \geq 0$. The series Σp_n, Σq_n must both diverge.

For if both were convergent, then

$$\Sigma(p_n + q_n) = \Sigma|a_n|$$

would converge, contrary to hypothesis. Since

$$\sum_{n=1}^{N} a_n = \sum_{n=1}^{N} (p_n - q_n) = \sum_{n=1}^{N} p_n - \sum_{n=1}^{N} q_n,$$

divergence of Σp_n and convergence of Σq_n (or vice versa) implies divergence of Σa_n, again contrary to hypothesis.

Now let $P_1, P_2, P_3, \ldots$ denote the nonnegative terms of Σa_n, in the order in which they occur, and let $Q_1, Q_2, Q_3, \ldots$ be the absolute values of the negative terms of Σa_n, also in their original order.

The series ΣP_n, ΣQ_n differ from Σp_n, Σq_n only by zero terms, and are therefore divergent.

We shall construct sequences $\{m_n\}$, $\{k_n\}$, such that the series

(25) $P_1 + \cdots + P_{m_1} - Q_1 - \cdots - Q_{k_1} + P_{m_1+1} + \cdots$
$$+ P_{m_2} - Q_{k_1+1} - \cdots - Q_{k_2} + \cdots,$$

which clearly is a rearrangement of Σa_n, satisfies (24).

Choose real-valued sequences $\{\alpha_n\}$, $\{\beta_n\}$ such that $\alpha_n \to \alpha$, $\beta_n \to \beta$, $\alpha_n < \beta_n$, $\beta_1 > 0$.

Let m_1, k_1 be the smallest integers such that

$$P_1 + \cdots + P_{m_1} > \beta_1,$$
$$P_1 + \cdots + P_{m_1} - Q_1 - \cdots - Q_{k_1} < \alpha_1;$$

let m_2, k_2 be the smallest integers such that

$$P_1 + \cdots + P_{m_1} - Q_1 - \cdots - Q_{k_1} + P_{m_1+1} + \cdots + P_{m_2} > \beta_2,$$
$$P_1 + \cdots + P_{m_1} - Q_1 - \cdots - Q_{k_1} + P_{m_1+1} + \cdots + P_{m_2} - Q_{k_1+1}$$
$$- \cdots - Q_{k_2} < \alpha_2;$$

and continue in this way. This is possible since ΣP_n and ΣQ_n diverge.

If x_n, y_n denote the partial sums of (25) whose last terms are P_{m_n}, $-Q_{k_n}$, then

$$|x_n - \beta_n| \leq P_{m_n}, \qquad |y_n - \alpha_n| \leq Q_{k_n}.$$

Since $P_n \to 0$ and $Q_n \to 0$ as $n \to \infty$, we see that $x_n \to \beta$, $y_n \to \alpha$.

Finally, it is clear that no number less than α or greater than β can be a subsequential limit of the partial sums of (25).

3.55 Theorem *If Σa_n is a series of complex numbers which converges absolutely, then every rearrangement of Σa_n converges, and they all converge to the same sum.*

Proof Let $\Sigma a_n'$ be a rearrangement, with partial sums s_n'. Given $\varepsilon > 0$, there exists an integer N such that $m \geq n \geq N$ implies

$$\sum_{i=n}^{m} |a_i| \leq \varepsilon. \tag{26}$$

Now choose p such that the integers $1, 2, \ldots, N$ are all contained in the set $k_1, k_2, \ldots, k_p$ (we use the notation of Definition 3.52). Then if $n > p$, the numbers $a_1, \ldots, a_N$ will cancel in the difference $s_n - s_n'$, so that $|s_n - s_n'| \leq \varepsilon$, by (26). Hence $\{s_n'\}$ converges to the same sum as $\{s_n\}$.

EXERCISES

1. Prove that convergence of $\{s_n\}$ implies convergence of $\{|s_n|\}$. Is the converse true?
2. Calculate $\lim_{n \to \infty} (\sqrt{n^2 + n} - n)$.
3. If $s_1 = \sqrt{2}$, and

$$s_{n+1} = \sqrt{2 + \sqrt{s_n}} \qquad (n = 1, 2, 3, \ldots),$$

prove that $\{s_n\}$ converges, and that $s_n < 2$ for $n = 1, 2, 3, \ldots$.
4. Find the upper and lower limits of the sequence $\{s_n\}$ defined by

$$s_1 = 0; \qquad s_{2m} = \frac{s_{2m-1}}{2}; \qquad s_{2m+1} = \frac{1}{2} + s_{2m}.$$

5. For any two real sequences $\{a_n\}, \{b_n\}$, prove that

$$\limsup_{n \to \infty} (a_n + b_n) \leq \limsup_{n \to \infty} a_n + \limsup_{n \to \infty} b_n,$$

provided the sum on the right is not of the form $\infty - \infty$.
6. Investigate the behavior (convergence or divergence) of Σa_n if

   (a) $a_n = \sqrt{n+1} - \sqrt{n}$;
   (b) $a_n = \dfrac{\sqrt{n+1} - \sqrt{n}}{n}$;
   (c) $a_n = (\sqrt[n]{n} - 1)^n$;
   (d) $a_n = \dfrac{1}{1+z^n}$, for complex values of z.

7. Prove that the convergence of Σa_n implies the convergence of

$$\sum \frac{\sqrt{a_n}}{n},$$

if $a_n \geq 0$.

8. If Σa_n converges, and if $\{b_n\}$ is monotonic and bounded, prove that $\Sigma a_n b_n$ converges.

9. Find the radius of convergence of each of the following power series:

   (a) $\sum n^3 z^n$, (b) $\sum \dfrac{2^n}{n!} z^n$,

   (c) $\sum \dfrac{2^n}{n^2} z^n$, (d) $\sum \dfrac{n^3}{3^n} z^n$.

10. Suppose that the coefficients of the power series $\sum a_n z^n$ are integers, infinitely many of which are distinct from zero. Prove that the radius of convergence is at most 1.

11. Suppose $a_n > 0$, $s_n = a_1 + \cdots + a_n$, and Σa_n diverges.

    (a) Prove that $\sum \dfrac{a_n}{1 + a_n}$ diverges.

    (b) Prove that
    $$\frac{a_{N+1}}{s_{N+1}} + \cdots + \frac{a_{N+k}}{s_{N+k}} \geq 1 - \frac{s_N}{s_{N+k}}$$
    and deduce that $\sum \dfrac{a_n}{s_n}$ diverges.

    (c) Prove that
    $$\frac{a_n}{s_n^2} \leq \frac{1}{s_{n-1}} - \frac{1}{s_n}$$
    and deduce that $\sum \dfrac{a_n}{s_n^2}$ converges.

    (d) What can be said about
    $$\sum \frac{a_n}{1 + na_n} \quad \text{and} \quad \sum \frac{a_n}{1 + n^2 a_n}?$$

12. Suppose $a_n > 0$ and Σa_n converges. Put
    $$r_n = \sum_{m=n}^{\infty} a_m.$$

    (a) Prove that
    $$\frac{a_m}{r_m} + \cdots + \frac{a_n}{r_n} > 1 - \frac{r_n}{r_m}$$
    if $m < n$, and deduce that $\sum \dfrac{a_n}{r_n}$ diverges.

(b) Prove that

$$\frac{a_n}{\sqrt{r_n}} < 2(\sqrt{r_n} - \sqrt{r_{n+1}})$$

and deduce that $\sum \dfrac{a_n}{\sqrt{r_n}}$ converges.

13. Prove that the Cauchy product of two absolutely convergent series converges absolutely.

14. If $\{s_n\}$ is a complex sequence, define its arithmetic means σ_n by

$$\sigma_n = \frac{s_0 + s_1 + \cdots + s_n}{n+1} \qquad (n = 0, 1, 2, \ldots).$$

(a) If $\lim s_n = s$, prove that $\lim \sigma_n = s$.
(b) Construct a sequence $\{s_n\}$ which does not converge, although $\lim \sigma_n = 0$.
(c) Can it happen that $s_n > 0$ for all n and that $\limsup s_n = \infty$, although $\lim \sigma_n = 0$?
(d) Put $a_n = s_n - s_{n-1}$, for $n \geq 1$. Show that

$$s_n - \sigma_n = \frac{1}{n+1} \sum_{k=1}^{n} k a_k.$$

Assume that $\lim (na_n) = 0$ and that $\{\sigma_n\}$ converges. Prove that $\{s_n\}$ converges. [This gives a converse of (a), but under the additional assumption that $na_n \to 0$.]
(e) Derive the last conclusion from a weaker hypothesis: Assume $M < \infty$, $|na_n| \leq M$ for all n, and $\lim \sigma_n = \sigma$. Prove that $\lim s_n = \sigma$, by completing the following outline:

If $m < n$, then

$$s_n - \sigma_n = \frac{m+1}{n-m}(\sigma_n - \sigma_m) + \frac{1}{n-m}\sum_{i=m+1}^{n}(s_n - s_i).$$

For these i,

$$|s_n - s_i| \leq \frac{(n-i)M}{i+1} \leq \frac{(n-m-1)M}{m+2}.$$

Fix $\varepsilon > 0$ and associate with each n the integer m that satisfies

$$m \leq \frac{n-\varepsilon}{1+\varepsilon} < m+1.$$

Then $(m+1)/(n-m) \leq 1/\varepsilon$ and $|s_n - s_i| < M\varepsilon$. Hence

$$\limsup_{n \to \infty} |s_n - \sigma| \leq M\varepsilon.$$

Since ε was arbitrary, $\lim s_n = \sigma$.

15. Definition 3.21 can be extended to the case in which the a_n lie in some fixed R^k. Absolute convergence is defined as convergence of $\Sigma|a_n|$. Show that Theorems 3.22, 3.23, 3.25(a), 3.33, 3.34, 3.42, 3.45, 3.47, and 3.55 are true in this more general setting. (Only slight modifications are required in any of the proofs.)

16. Fix a positive number α. Choose $x_1 > \sqrt{\alpha}$, and define $x_2, x_3, x_4, \ldots$, by the recursion formula

$$x_{n+1} = \frac{1}{2}\left(x_n + \frac{\alpha}{x_n}\right).$$

(a) Prove that $\{x_n\}$ decreases monotonically and that $\lim x_n = \sqrt{\alpha}$.

(b) Put $\varepsilon_n = x_n - \sqrt{\alpha}$, and show that

$$\varepsilon_{n+1} = \frac{\varepsilon_n^2}{2x_n} < \frac{\varepsilon_n^2}{2\sqrt{\alpha}}$$

so that, setting $\beta = 2\sqrt{\alpha}$,

$$\varepsilon_{n+1} < \beta\left(\frac{\varepsilon_1}{\beta}\right)^{2^n} \quad (n = 1, 2, 3, \ldots).$$

(c) This is a good algorithm for computing square roots, since the recursion formula is simple and the convergence is extremely rapid. For example, if $\alpha = 3$ and $x_1 = 2$, show that $\varepsilon_1/\beta < \frac{1}{10}$ and that therefore

$$\varepsilon_5 < 4 \cdot 10^{-16}, \qquad \varepsilon_6 < 4 \cdot 10^{-32}.$$

17. Fix $\alpha > 1$. Take $x_1 > \sqrt{\alpha}$, and define

$$x_{n+1} = \frac{\alpha + x_n}{1 + x_n} = x_n + \frac{\alpha - x_n^2}{1 + x_n}.$$

(a) Prove that $x_1 > x_3 > x_5 > \cdots$.
(b) Prove that $x_2 < x_4 < x_6 < \cdots$.
(c) Prove that $\lim x_n = \sqrt{\alpha}$.
(d) Compare the rapidity of convergence of this process with the one described in Exercise 16.

18. Replace the recursion formula of Exercise 16 by

$$x_{n+1} = \frac{p-1}{p}x_n + \frac{\alpha}{p}x_n^{-p+1}$$

where p is a fixed positive integer, and describe the behavior of the resulting sequences $\{x_n\}$.

19. Associate to each sequence $a = \{\alpha_n\}$, in which α_n is 0 or 2, the real number

$$x(a) = \sum_{n=1}^{\infty} \frac{\alpha_n}{3^n}.$$

Prove that the set of all $x(a)$ is precisely the Cantor set described in Sec. 2.44.

20. Suppose $\{p_n\}$ is a Cauchy sequence in a metric space X, and some subsequence $\{p_{n_i}\}$ converges to a point $p \in X$. Prove that the full sequence $\{p_n\}$ converges to p.

21. Prove the following analogue of Theorem 3.10(b): If $\{E_n\}$ is a sequence of closed nonempty and bounded sets in a *complete* metric space X, if $E_n \supset E_{n+1}$, and if
$$\lim_{n \to \infty} \text{diam } E_n = 0,$$
then $\bigcap_1^\infty E_n$ consists of exactly one point.

22. Suppose X is a nonempty complete metric space, and $\{G_n\}$ is a sequence of dense open subsets of X. Prove Baire's theorem, namely, that $\bigcap_1^\infty G_n$ is not empty. (In fact, it is dense in X.) *Hint:* Find a shrinking sequence of neighborhoods E_n such that $\overline{E}_n \subset G_n$, and apply Exercise 21.

23. Suppose $\{p_n\}$ and $\{q_n\}$ are Cauchy sequences in a metric space X. Show that the sequence $\{d(p_n, q_n)\}$ converges. *Hint:* For any m, n,
$$d(p_n, q_n) \leq d(p_n, p_m) + d(p_m, q_m) + d(q_m, q_n);$$
it follows that
$$|d(p_n, q_n) - d(p_m, q_m)|$$
is small if m and n are large.

24. Let X be a metric space.

(a) Call two Cauchy sequences $\{p_n\}, \{q_n\}$ in X *equivalent* if
$$\lim_{n \to \infty} d(p_n, q_n) = 0.$$
Prove that this is an equivalence relation.

(b) Let X^* be the set of all equivalence classes so obtained. If $P \in X^*$, $Q \in X^*$, $\{p_n\} \in P$, $\{q_n\} \in Q$, define
$$\Delta(P, Q) = \lim_{n \to \infty} d(p_n, q_n);$$
by Exercise 23, this limit exists. Show that the number $\Delta(P, Q)$ is unchanged if $\{p_n\}$ and $\{q_n\}$ are replaced by equivalent sequences, and hence that Δ is a distance function in X^*.

(c) Prove that the resulting metric space X^* is complete.

(d) For each $p \in X$, there is a Cauchy sequence all of whose terms are p; let P_p be the element of X^* which contains this sequence. Prove that
$$\Delta(P_p, P_q) = d(p, q)$$
for all $p, q \in X$. In other words, the mapping φ defined by $\varphi(p) = P_p$ is an isometry (i.e., a distance-preserving mapping) of X into X^*.

(e) Prove that $\varphi(X)$ is dense in X^*, and that $\varphi(X) = X^*$ if X is complete. By (d), we may identify X and $\varphi(X)$ and thus regard X as embedded in the complete metric space X^*. We call X^* the *completion* of X.

25. Let X be the metric space whose points are the rational numbers, with the metric $d(x, y) = |x - y|$. What is the completion of this space? (Compare Exercise 24.)

4
CONTINUITY

The function concept and some of the related terminology were introduced in Definitions 2.1 and 2.2. Although we shall (in later chapters) be mainly interested in real and complex functions (i.e., in functions whose values are real or complex numbers) we shall also discuss vector-valued functions (i.e., functions with values in R^k) and functions with values in an arbitrary metric space. The theorems we shall discuss in this general setting would not become any easier if we restricted ourselves to real functions, for instance, and it actually simplifies and clarifies the picture to discard unnecessary hypotheses and to state and prove theorems in an appropriately general context.

The domains of definition of our functions will also be metric spaces, suitably specialized in various instances.

LIMITS OF FUNCTIONS

4.1 Definition Let X and Y be metric spaces; suppose $E \subset X$, f maps E into Y, and p is a limit point of E. We write $f(x) \to q$ as $x \to p$, or

(1) $$\lim_{x \to p} f(x) = q$$

if there is a point $q \in Y$ with the following property: For every $\varepsilon > 0$ there exists a $\delta > 0$ such that

$$\text{(2)} \qquad d_Y(f(x), q) < \varepsilon$$

for all points $x \in E$ for which

$$\text{(3)} \qquad 0 < d_X(x, p) < \delta.$$

The symbols d_X and d_Y refer to the distances in X and Y, respectively.

If X and/or Y are replaced by the real line, the complex plane, or by some euclidean space R^k, the distances d_X, d_Y are of course replaced by absolute values, or by norms of differences (see Sec. 2.16).

It should be noted that $p \in X$, but that p need not be a point of E in the above definition. Moreover, even if $p \in E$, we may very well have $f(p) \neq \lim_{x \to p} f(x)$.

We can recast this definition in terms of limits of sequences:

4.2 Theorem *Let X, Y, E, f, and p be as in Definition 4.1. Then*

$$\text{(4)} \qquad \lim_{x \to p} f(x) = q$$

if and only if

$$\text{(5)} \qquad \lim_{n \to \infty} f(p_n) = q$$

for every sequence $\{p_n\}$ in E such that

$$\text{(6)} \qquad p_n \neq p, \qquad \lim_{n \to \infty} p_n = p.$$

Proof Suppose (4) holds. Choose $\{p_n\}$ in E satisfying (6). Let $\varepsilon > 0$ be given. Then there exists $\delta > 0$ such that $d_Y(f(x), q) < \varepsilon$ if $x \in E$ and $0 < d_X(x, p) < \delta$. Also, there exists N such that $n > N$ implies $0 < d_X(p_n, p) < \delta$. Thus, for $n > N$, we have $d_Y(f(p_n), q) < \varepsilon$, which shows that (5) holds.

Conversely, suppose (4) is false. Then there exists some $\varepsilon > 0$ such that for every $\delta > 0$ there exists a point $x \in E$ (depending on δ), for which $d_Y(f(x), q) \geq \varepsilon$ but $0 < d_X(x, p) < \delta$. Taking $\delta_n = 1/n$ ($n = 1, 2, 3, \ldots$), we thus find a sequence in E satisfying (6) for which (5) is false.

Corollary *If f has a limit at p, this limit is unique.*

This follows from Theorems 3.2(b) and 4.2.

4.3 Definition Suppose we have two complex functions, f and g, both defined on E. By $f + g$ we mean the function which assigns to each point x of E the number $f(x) + g(x)$. Similarly we define the difference $f - g$, the product fg, and the quotient f/g of the two functions, with the understanding that the quotient is defined only at those points x of E at which $g(x) \neq 0$. If f assigns to each point x of E the same number c, then f is said to be a constant function, or simply a constant, and we write $f = c$. If f and g are real functions, and if $f(x) \geq g(x)$ for every $x \in E$, we shall sometimes write $f \geq g$, for brevity.

Similarly, if **f** and **g** map E into R^k, we define $\mathbf{f} + \mathbf{g}$ and $\mathbf{f} \cdot \mathbf{g}$ by

$$(\mathbf{f} + \mathbf{g})(x) = \mathbf{f}(x) + \mathbf{g}(x), \qquad (\mathbf{f} \cdot \mathbf{g})(x) = \mathbf{f}(x) \cdot \mathbf{g}(x);$$

and if λ is a real number, $(\lambda \mathbf{f})(x) = \lambda \mathbf{f}(x)$.

4.4 Theorem *Suppose $E \subset X$, a metric space, p is a limit point of E, f and g are complex functions on E, and*

$$\lim_{x \to p} f(x) = A, \qquad \lim_{x \to p} g(x) = B.$$

Then (a) $\lim_{x \to p} (f + g)(x) = A + B$;

(b) $\lim_{x \to p} (fg)(x) = AB$;

(c) $\lim_{x \to p} \left(\dfrac{f}{g}\right)(x) = \dfrac{A}{B}$, *if* $B \neq 0$.

Proof In view of Theorem 4.2, these assertions follow immediately from the analogous properties of sequences (Theorem 3.3).

Remark If **f** and **g** map E into R^k, then (a) remains true, and (b) becomes

(b') $\lim_{x \to p} (\mathbf{f} \cdot \mathbf{g})(x) = \mathbf{A} \cdot \mathbf{B}.$

(Compare Theorem 3.4.)

CONTINUOUS FUNCTIONS

4.5 Definition Suppose X and Y are metric spaces, $E \subset X$, $p \in E$, and f maps E into Y. Then f is said to be *continuous at* p if for every $\varepsilon > 0$ there exists a $\delta > 0$ such that
$$d_Y(f(x), f(p)) < \varepsilon$$
for all points $x \in E$ for which $d_X(x, p) < \delta$.

If f is continuous at every point of E, then f is said to be *continuous on* E.

It should be noted that f has to be defined at the point p in order to be continuous at p. (Compare this with the remark following Definition 4.1.)

If p is an isolated point of E, then our definition implies that every function f which has E as its domain of definition is continuous at p. For, no matter which $\varepsilon > 0$ we choose, we can pick $\delta > 0$ so that the only point $x \in E$ for which $d_X(x, p) < \delta$ is $x = p$; then
$$d_Y(f(x), f(p)) = 0 < \varepsilon.$$

4.6 Theorem *In the situation given in Definition 4.5, assume also that p is a limit point of E. Then f is continuous at p if and only if $\lim_{x \to p} f(x) = f(p)$.*

Proof This is clear if we compare Definitions 4.1 and 4.5.

We now turn to compositions of functions. A brief statement of the following theorem is that a continuous function of a continuous function is continuous.

4.7 Theorem *Suppose X, Y, Z are metric spaces, $E \subset X$, f maps E into Y, g maps the range of f, $f(E)$, into Z, and h is the mapping of E into Z defined by*
$$h(x) = g(f(x)) \qquad (x \in E).$$
If f is continuous at a point $p \in E$ and if g is continuous at the point $f(p)$, then h is continuous at p.

This function h is called the *composition* or the *composite* of f and g. The notation
$$h = g \circ f$$
is frequently used in this context.

Proof Let $\varepsilon > 0$ be given. Since g is continuous at $f(p)$, there exists $\eta > 0$ such that
$$d_Z(g(y), g(f(p))) < \varepsilon \text{ if } d_Y(y, f(p)) < \eta \text{ and } y \in f(E).$$
Since f is continuous at p, there exists $\delta > 0$ such that
$$d_Y(f(x), f(p)) < \eta \text{ if } d_X(x, p) < \delta \text{ and } x \in E.$$
It follows that
$$d_Z(h(x), h(p)) = d_Z(g(f(x)), g(f(p))) < \varepsilon$$
if $d_X(x, p) < \delta$ and $x \in E$. Thus h is continuous at p.

4.8 Theorem *A mapping f of a metric space X into a metric space Y is continuous on X if and only if $f^{-1}(V)$ is open in X for every open set V in Y.*

(Inverse images are defined in Definition 2.2.) This is a very useful characterization of continuity.

Proof Suppose f is continuous on X and V is an open set in Y. We have to show that every point of $f^{-1}(V)$ is an interior point of $f^{-1}(V)$. So, suppose $p \in X$ and $f(p) \in V$. Since V is open, there exists $\varepsilon > 0$ such that $y \in V$ if $d_Y(f(p), y) < \varepsilon$; and since f is continuous at p, there exists $\delta > 0$ such that $d_Y(f(x), f(p)) < \varepsilon$ if $d_X(x, p) < \delta$. Thus $x \in f^{-1}(V)$ as soon as $d_X(x, p) < \delta$.

Conversely, suppose $f^{-1}(V)$ is open in X for every open set V in Y. Fix $p \in X$ and $\varepsilon > 0$, let V be the set of all $y \in Y$ such that $d_Y(y, f(p)) < \varepsilon$. Then V is open; hence $f^{-1}(V)$ is open; hence there exists $\delta > 0$ such that $x \in f^{-1}(V)$ as soon as $d_X(p, x) < \delta$. But if $x \in f^{-1}(V)$, then $f(x) \in V$, so that $d_Y(f(x), f(p)) < \varepsilon$.

This completes the proof.

Corollary *A mapping f of a metric space X into a metric space Y is continuous if and only if $f^{-1}(C)$ is closed in X for every closed set C in Y.*

This follows from the theorem, since a set is closed if and only if its complement is open, and since $f^{-1}(E^c) = [f^{-1}(E)]^c$ for every $E \subset Y$.

We now turn to complex-valued and vector-valued functions, and to functions defined on subsets of R^k.

4.9 Theorem *Let f and g be complex continuous functions on a metric space X. Then $f + g$, fg, and f/g are continuous on X.*

In the last case, we must of course assume that $g(x) \neq 0$, for all $x \in X$.

Proof At isolated points of X there is nothing to prove. At limit points, the statement follows from Theorems 4.4 and 4.6.

4.10 Theorem

(a) *Let $f_1, \ldots, f_k$ be real functions on a metric space X, and let $\mathbf{f}$ be the mapping of X into R^k defined by*

(7) $$\mathbf{f}(x) = (f_1(x), \ldots, f_k(x)) \qquad (x \in X);$$

then $\mathbf{f}$ is continuous if and only if each of the functions $f_1, \ldots, f_k$ is continuous.
(b) *If $\mathbf{f}$ and $\mathbf{g}$ are continuous mappings of X into R^k, then $\mathbf{f} + \mathbf{g}$ and $\mathbf{f} \cdot \mathbf{g}$ are continuous on X.*

The functions $f_1, \ldots, f_k$ are called the *components* of $\mathbf{f}$. Note that $\mathbf{f} + \mathbf{g}$ is a mapping into R^k, whereas $\mathbf{f} \cdot \mathbf{g}$ is a *real* function on X.

Proof Part (*a*) follows from the inequalities

$$|f_j(x) - f_j(y)| \leq |\mathbf{f}(x) - \mathbf{f}(y)| = \left\{ \sum_{i=1}^{k} |f_i(x) - f_i(y)|^2 \right\}^{\frac{1}{2}},$$

for $j = 1, \ldots, k$. Part (*b*) follows from (*a*) and Theorem 4.9.

4.11 Examples If $x_1, \ldots, x_k$ are the coordinates of the point $\mathbf{x} \in R^k$, the functions ϕ_i defined by

$$\phi_i(\mathbf{x}) = x_i \qquad (\mathbf{x} \in R^k) \tag{8}$$

are continuous on R^k, since the inequality

$$|\phi_i(\mathbf{x}) - \phi_i(\mathbf{y})| \leq |\mathbf{x} - \mathbf{y}|$$

shows that we may take $\delta = \varepsilon$ in Definition 4.5. The functions ϕ_i are sometimes called the *coordinate functions*.

Repeated application of Theorem 4.9 then shows that every monomial

$$x_1^{n_1} x_2^{n_2} \cdots x_k^{n_k} \tag{9}$$

where $n_1, \ldots, n_k$ are nonnegative integers, is continuous on R^k. The same is true of constant multiples of (9), since constants are evidently continuous. It follows that every polynomial P, given by

$$P(\mathbf{x}) = \Sigma c_{n_1 \cdots n_k} x_1^{n_1} \cdots x_k^{n_k} \qquad (\mathbf{x} \in R^k), \tag{10}$$

is continuous on R^k. Here the coefficients $c_{n_1 \cdots n_k}$ are complex numbers, $n_1, \ldots, n_k$ are nonnegative integers, and the sum in (10) has finitely many terms.

Furthermore, every rational function in $x_1, \ldots, x_k$, that is, every quotient of two polynomials of the form (10), is continuous on R^k wherever the denominator is different from zero.

From the triangle inequality one sees easily that

$$||\mathbf{x}| - |\mathbf{y}|| \leq |\mathbf{x} - \mathbf{y}| \qquad (\mathbf{x}, \mathbf{y} \in R^k). \tag{11}$$

Hence the mapping $\mathbf{x} \to |\mathbf{x}|$ is a continuous real function on R^k.

If now $\mathbf{f}$ is a continuous mapping from a metric space X into R^k, and if ϕ is defined on X by setting $\phi(p) = |\mathbf{f}(p)|$, it follows, by Theorem 4.7, that ϕ is a continuous real function on X.

4.12 Remark We defined the notion of continuity for functions defined on a *subset* E of a metric space X. However, the complement of E in X plays no role whatever in this definition (note that the situation was somewhat different for limits of functions). Accordingly, we lose nothing of interest by discarding the complement of the domain of f. This means that we may just as well talk only about continuous mappings of one metric space into another, rather than of mappings of subsets. This simplifies statements and proofs of some theorems. We have already made use of this principle in Theorems 4.8 to 4.10, and will continue to do so in the following section on compactness.

CONTINUITY AND COMPACTNESS

4.13 Definition A mapping $\mathbf{f}$ of a set E into R^k is said to be *bounded* if there is a real number M such that $|\mathbf{f}(x)| \leq M$ for all $x \in E$.

4.14 Theorem *Suppose f is a continuous mapping of a compact metric space X into a metric space Y. Then $f(X)$ is compact.*

Proof Let $\{V_\alpha\}$ be an open cover of $f(X)$. Since f is continuous, Theorem 4.8 shows that each of the sets $f^{-1}(V_\alpha)$ is open. Since X is compact, there are finitely many indices, say $\alpha_1, \ldots, \alpha_n$, such that

$$(12) \qquad X \subset f^{-1}(V_{\alpha_1}) \cup \cdots \cup f^{-1}(V_{\alpha_n}).$$

Since $f(f^{-1}(E)) \subset E$ for every $E \subset Y$, (12) implies that

$$(13) \qquad f(X) \subset V_{\alpha_1} \cup \cdots \cup V_{\alpha_n}.$$

This completes the proof.

Note: We have used the relation $f(f^{-1}(E)) \subset E$, valid for $E \subset Y$. If $E \subset X$, then $f^{-1}(f(E)) \supset E$; equality need not hold in either case.

We shall now deduce some consequences of Theorem 4.14.

4.15 Theorem *If $\mathbf{f}$ is a continuous mapping of a compact metric space X into R^k, then $\mathbf{f}(X)$ is closed and bounded. Thus, $\mathbf{f}$ is bounded.*

This follows from Theorem 2.41. The result is particularly important when f is real:

4.16 Theorem *Suppose f is a continuous real function on a compact metric space X, and*

$$(14) \qquad M = \sup_{p \in X} f(p), \qquad m = \inf_{p \in X} f(p).$$

Then there exist points $p, q \in X$ such that $f(p) = M$ and $f(q) = m$.

The notation in (14) means that M is the least upper bound of the set of all numbers $f(p)$, where p ranges over X, and that m is the greatest lower bound of this set of numbers.

The conclusion may also be stated as follows: *There exist points p and q in X such that $f(q) \leq f(x) \leq f(p)$ for all $x \in X$*; that is, f attains its maximum (at p) and its minimum (at q).

Proof By Theorem 4.15, $f(X)$ is a closed and bounded set of real numbers; hence $f(X)$ contains

$$M = \sup f(X) \quad \text{and} \quad m = \inf f(X),$$

by Theorem 2.28.

4.17 Theorem *Suppose f is a continuous 1-1 mapping of a compact metric space X onto a metric space Y. Then the inverse mapping f^{-1} defined on Y by*

$$f^{-1}(f(x)) = x \qquad (x \in X)$$

is a continuous mapping of Y onto X.

Proof Applying Theorem 4.8 to f^{-1} in place of f, we see that it suffices to prove that $f(V)$ is an open set in Y for every open set V in X. Fix such a set V.

The complement V^c of V is closed in X, hence compact (Theorem 2.35); hence $f(V^c)$ is a compact subset of Y (Theorem 4.14) and so is closed in Y (Theorem 2.34). Since f is one-to-one and onto, $f(V)$ is the complement of $f(V^c)$. Hence $f(V)$ is open.

4.18 Definition Let f be a mapping of a metric space X into a metric space Y. We say that f is *uniformly continuous* on X if for every $\varepsilon > 0$ there exists $\delta > 0$ such that

$$d_Y(f(p), f(q)) < \varepsilon \tag{15}$$

for all p and q in X for which $d_X(p, q) < \delta$.

Let us consider the differences between the concepts of continuity and of uniform continuity. First, uniform continuity is a property of a function on a set, whereas continuity can be defined at a single point. To ask whether a given function is uniformly continuous at a certain point is meaningless. Second, if f is continuous on X, then it is possible to find, for each $\varepsilon > 0$ and for each point p of X, a number $\delta > 0$ having the property specified in Definition 4.5. This δ depends on ε *and* on p. If f is, however, uniformly continuous on X, then it is possible, for each $\varepsilon > 0$, to find *one* number $\delta > 0$ which will do for *all* points p of X.

Evidently, every uniformly continuous function is continuous. That the two concepts are equivalent on compact sets follows from the next theorem.

4.19 Theorem *Let f be a continuous mapping of a compact metric space X into a metric space Y. Then f is uniformly continuous on X.*

Proof Let $\varepsilon > 0$ be given. Since f is continuous, we can associate to each point $p \in X$ a positive number $\phi(p)$ such that

(16) $$q \in X, d_X(p, q) < \phi(p) \text{ implies } d_Y(f(p), f(q)) < \frac{\varepsilon}{2}.$$

Let $J(p)$ be the set of all $q \in X$ for which

(17) $$d_X(p, q) < \tfrac{1}{2}\phi(p).$$

Since $p \in J(p)$, the collection of all sets $J(p)$ is an open cover of X; and since X is compact, there is a finite set of points $p_1, \ldots, p_n$ in X, such that

(18) $$X \subset J(p_1) \cup \cdots \cup J(p_n).$$

We put

(19) $$\delta = \tfrac{1}{2} \min\,[\phi(p_1), \ldots, \phi(p_n)].$$

Then $\delta > 0$. (This is one point where the finiteness of the covering, inherent in the definition of compactness, is essential. The minimum of a finite set of positive numbers is positive, whereas the inf of an infinite set of positive numbers may very well be 0.)

Now let q and p be points of X, such that $d_X(p, q) < \delta$. By (18), there is an integer m, $1 \leq m \leq n$, such that $p \in J(p_m)$; hence

(20) $$d_X(p, p_m) < \tfrac{1}{2}\phi(p_m),$$

and we also have

$$d_X(q, p_m) \leq d_X(p, q) + d_X(p, p_m) < \delta + \tfrac{1}{2}\phi(p_m) \leq \phi(p_m).$$

Finally, (16) shows that therefore

$$d_Y(f(p), f(q)) \leq d_Y(f(p), f(p_m)) + d_Y(f(q), f(p_m)) < \varepsilon.$$

This completes the proof.

An alternative proof is sketched in Exercise 10.

We now proceed to show that compactness is essential in the hypotheses of Theorems 4.14, 4.15, 4.16, and 4.19.

4.20 Theorem *Let E be a noncompact set in R^1. Then*

(a) there exists a continuous function on E which is not bounded;
(b) there exists a continuous and bounded function on E which has no maximum.

If, in addition, E is bounded, then

(c) there exists a continuous function on E which is not uniformly continuous.

Proof Suppose first that E is bounded, so that there exists a limit point x_0 of E which is not a point of E. Consider

$$(21) \qquad f(x) = \frac{1}{x - x_0} \qquad (x \in E).$$

This is continuous on E (Theorem 4.9), but evidently unbounded. To see that (21) is not uniformly continuous, let $\varepsilon > 0$ and $\delta > 0$ be arbitrary, and choose a point $x \in E$ such that $|x - x_0| < \delta$. Taking t close enough to x_0, we can then make the difference $|f(t) - f(x)|$ greater than ε, although $|t - x| < \delta$. Since this is true for every $\delta > 0$, f is not uniformly continuous on E.

The function g given by

$$(22) \qquad g(x) = \frac{1}{1 + (x - x_0)^2} \qquad (x \in E)$$

is continuous on E, and is bounded, since $0 < g(x) < 1$. It is clear that

$$\sup_{x \in E} g(x) = 1,$$

whereas $g(x) < 1$ for all $x \in E$. Thus g has no maximum on E.

Having proved the theorem for bounded sets E, let us now suppose that E is unbounded. Then $f(x) = x$ establishes (a), whereas

$$(23) \qquad h(x) = \frac{x^2}{1 + x^2} \qquad (x \in E)$$

establishes (b), since

$$\sup_{x \in E} h(x) = 1$$

and $h(x) < 1$ for all $x \in E$.

Assertion (c) would be false if boundedness were omitted from the hypotheses. For, let E be the set of all integers. Then every function defined on E is uniformly continuous on E. To see this, we need merely take $\delta < 1$ in Definition 4.18.

We conclude this section by showing that compactness is also essential in Theorem 4.17.

4.21 Example Let X be the half-open interval $[0, 2\pi)$ on the real line, and let **f** be the mapping of X onto the circle Y consisting of all points whose distance from the origin is 1, given by

(24) $$\mathbf{f}(t) = (\cos t, \sin t) \qquad (0 \leq t < 2\pi).$$

The continuity of the trigonometric functions cosine and sine, as well as their periodicity properties, will be established in Chap. 8. These results show that **f** is a continuous 1-1 mapping of X onto Y.

However, the inverse mapping (which exists, since **f** is one-to-one and onto) fails to be continuous at the point $(1, 0) = \mathbf{f}(0)$. Of course, X is not compact in this example. (It may be of interest to observe that $\mathbf{f}^{-1}$ fails to be continuous in spite of the fact that Y *is* compact!)

CONTINUITY AND CONNECTEDNESS

4.22 Theorem *If f is a continuous mapping of a metric space X into a metric space Y, and if E is a connected subset of X, then $f(E)$ is connected.*

> **Proof** Assume, on the contrary, that $f(E) = A \cup B$, where A and B are nonempty separated subsets of Y. Put $G = E \cap f^{-1}(A)$, $H = E \cap f^{-1}(B)$.
> Then $E = G \cup H$, and neither G nor H is empty.
> Since $A \subset \bar{A}$ (the closure of A), we have $G \subset f^{-1}(\bar{A})$; the latter set is closed, since f is continuous; hence $\bar{G} \subset f^{-1}(\bar{A})$. It follows that $f(\bar{G}) \subset \bar{A}$. Since $f(H) = B$ and $\bar{A} \cap B$ is empty, we conclude that $\bar{G} \cap H$ is empty.
> The same argument shows that $G \cap \bar{H}$ is empty. Thus G and H are separated. This is impossible if E is connected.

4.23 Theorem *Let f be a continuous real function on the interval $[a, b]$. If $f(a) < f(b)$ and if c is a number such that $f(a) < c < f(b)$, then there exists a point $x \in (a, b)$ such that $f(x) = c$.*

A similar result holds, of course, if $f(a) > f(b)$. Roughly speaking, the theorem says that a continuous real function assumes all intermediate values on an interval.

> **Proof** By Theorem 2.47, $[a, b]$ is connected; hence Theorem 4.22 shows that $f([a, b])$ is a connected subset of R^1, and the assertion follows if we appeal once more to Theorem 2.47.

4.24 Remark At first glance, it might seem that Theorem 4.23 has a converse. That is, one might think that if for any two points $x_1 < x_2$ and for any number c between $f(x_1)$ and $f(x_2)$ there is a point x in (x_1, x_2) such that $f(x) = c$, then f must be continuous.

That this is not so may be concluded from Example 4.27(d).

DISCONTINUITIES

If x is a point in the domain of definition of the function f at which f is not continuous, we say that f is *discontinuous* at x, or that *f has a discontinuity* at x. If f is defined on an interval or on a segment, it is customary to divide discontinuities into two types. Before giving this classification, we have to define the *right-hand* and the *left-hand limits* of f at x, which we denote by $f(x+)$ and $f(x-)$, respectively.

4.25 Definition Let f be defined on (a, b). Consider any point x such that $a \leq x < b$. We write

$$f(x+) = q$$

if $f(t_n) \to q$ as $n \to \infty$, for all sequences $\{t_n\}$ in (x, b) such that $t_n \to x$. To obtain the definition of $f(x-)$, for $a < x \leq b$, we restrict ourselves to sequences $\{t_n\}$ in (a, x).

It is clear that any point x of (a, b), $\lim_{t \to x} f(t)$ exists if and only if

$$f(x+) = f(x-) = \lim_{t \to x} f(t).$$

4.26 Definition Let f be defined on (a, b). If f is discontinuous at a point x, and if $f(x+)$ and $f(x-)$ exist, then f is said to have a *discontinuity of the first kind*, or a *simple discontinuity*, at x. Otherwise the discontinuity is said to be of the *second kind*.

There are two ways in which a function can have a simple discontinuity: either $f(x+) \neq f(x-)$ [in which case the value $f(x)$ is immaterial], or $f(x+) = f(x-) \neq f(x)$.

4.27 Examples
(a) Define

$$f(x) = \begin{cases} 1 & (x \text{ rational}), \\ 0 & (x \text{ irrational}). \end{cases}$$

Then f has a discontinuity of the second kind at every point x, since neither $f(x+)$ nor $f(x-)$ exists.
(b) Define

$$f(x) = \begin{cases} x & (x \text{ rational}), \\ 0 & (x \text{ irrational}). \end{cases}$$

Then f is continuous at $x = 0$ and has a discontinuity of the second kind at every other point.

(c) Define
$$f(x) = \begin{cases} x + 2 & (-3 < x < -2), \\ -x - 2 & (-2 \le x < 0), \\ x + 2 & (0 \le x < 1). \end{cases}$$

Then f has a simple discontinuity at $x = 0$ and is continuous at every other point of $(-3, 1)$.

(d) Define
$$f(x) = \begin{cases} \sin \dfrac{1}{x} & (x \ne 0), \\ 0 & (x = 0). \end{cases}$$

Since neither $f(0+)$ nor $f(0-)$ exists, f has a discontinuity of the second kind at $x = 0$. We have not yet shown that $\sin x$ is a continuous function. If we assume this result for the moment, Theorem 4.7 implies that f is continuous at every point $x \ne 0$.

MONOTONIC FUNCTIONS

We shall now study those functions which never decrease (or never increase) on a given segment.

4.28 Definition Let f be real on (a, b). Then f is said to be *monotonically increasing* on (a, b) if $a < x < y < b$ implies $f(x) \le f(y)$. If the last inequality is reversed, we obtain the definition of a *monotonically decreasing* function. The class of monotonic functions consists of both the increasing and the decreasing functions.

4.29 Theorem *Let f be monotonically increasing on (a, b). Then $f(x+)$ and $f(x-)$ exist at every point of x of (a, b). More precisely,*

(25) $$\sup_{a<t<x} f(t) = f(x-) \le f(x) \le f(x+) = \inf_{x<t<b} f(t).$$

Furthermore, if $a < x < y < b$, then

(26) $$f(x+) \le f(y-).$$

Analogous results evidently hold for monotonically decreasing functions.

Proof By hypothesis, the set of numbers $f(t)$, where $a < t < x$, is bounded above by the number $f(x)$, and therefore has a least upper bound which we shall denote by A. Evidently $A \leq f(x)$. We have to show that $A = f(x-)$.

Let $\varepsilon > 0$ be given. It follows from the definition of A as a least upper bound that there exists $\delta > 0$ such that $a < x - \delta < x$ and

(27) $$A - \varepsilon < f(x - \delta) \leq A.$$

Since f is monotonic, we have

(28) $$f(x - \delta) \leq f(t) \leq A \quad (x - \delta < t < x).$$

Combining (27) and (28), we see that

$$|f(t) - A| < \varepsilon \quad (x - \delta < t < x).$$

Hence $f(x-) = A$.

The second half of (25) is proved in precisely the same way.

Next, if $a < x < y < b$, we see from (25) that

(29) $$f(x+) = \inf_{x < t < b} f(t) = \inf_{x < t < y} f(t).$$

The last equality is obtained by applying (25) to (a, y) in place of (a, b). Similarly,

(30) $$f(y-) = \sup_{a < t < y} f(t) = \sup_{x < t < y} f(t).$$

Comparison of (29) and (30) gives (26).

Corollary *Monotonic functions have no discontinuities of the second kind.*

This corollary implies that every monotonic function is discontinuous at a countable set of points at most. Instead of appealing to the general theorem whose proof is sketched in Exercise 17, we give here a simple proof which is applicable to monotonic functions.

4.30 Theorem *Let f be monotonic on (a, b). Then the set of points of (a, b) at which f is discontinuous is at most countable.*

Proof Suppose, for the sake of definiteness, that f is increasing, and let E be the set of points at which f is discontinuous.

With every point x of E we associate a rational number $r(x)$ such that

$$f(x-) < r(x) < f(x+).$$

Since $x_1 < x_2$ implies $f(x_1+) \le f(x_2-)$, we see that $r(x_1) \ne r(x_2)$ if $x_1 \ne x_2$.

We have thus established a 1-1 correspondence between the set E and a subset of the set of rational numbers. The latter, as we know, is countable.

4.31 Remark It should be noted that the discontinuities of a monotonic function need not be isolated. In fact, given any countable subset E of (a, b), which may even be dense, we can construct a function f, monotonic on (a, b), discontinuous at every point of E, and at no other point of (a, b).

To show this, let the points of E be arranged in a sequence $\{x_n\}$, $n = 1, 2, 3, \ldots$. Let $\{c_n\}$ be a sequence of positive numbers such that Σc_n converges. Define

(31) $$f(x) = \sum_{x_n < x} c_n \qquad (a < x < b).$$

The summation is to be understood as follows: Sum over those indices n for which $x_n < x$. If there are no points x_n to the left of x, the sum is empty; following the usual convention, we define it to be zero. Since (31) converges absolutely, the order in which the terms are arranged is immaterial.

We leave the verification of the following properties of f to the reader:

(a) f is monotonically increasing on (a, b);
(b) f is discontinuous at every point of E; in fact,

$$f(x_n+) - f(x_n-) = c_n.$$

(c) f is continuous at every other point of (a, b).

Moreover, it is not hard to see that $f(x-) = f(x)$ at all points of (a, b). If a function satisfies this condition, we say that f is *continuous from the left*. If the summation in (31) were taken over all indices n for which $x_n \le x$, we would have $f(x+) = f(x)$ at every point of (a, b); that is, f would be *continuous from the right*.

Functions of this sort can also be defined by another method; for an example we refer to Theorem 6.16.

INFINITE LIMITS AND LIMITS AT INFINITY

To enable us to operate in the extended real number system, we shall now enlarge the scope of Definition 4.1, by reformulating it in terms of neighborhoods.

For any real number x, we have already defined a neighborhood of x to be any segment $(x - \delta, x + \delta)$.

4.32 Definition For any real c, the set of real numbers x such that $x > c$ is called a neighborhood of $+\infty$ and is written $(c, +\infty)$. Similarly, the set $(-\infty, c)$ is a neighborhood of $-\infty$.

4.33 Definition Let f be a real function defined on $E \subset R$. We say that
$$f(t) \to A \text{ as } t \to x,$$
where A and x are in the extended real number system, if for every neighborhood U of A there is a neighborhood V of x such that $V \cap E$ is not empty, and such that $f(t) \in U$ for all $t \in V \cap E$, $t \neq x$.

A moment's consideration will show that this coincides with Definition 4.1 when A and x are real.

The analogue of Theorem 4.4 is still true, and the proof offers nothing new. We state it, for the sake of completeness.

4.34 Theorem *Let f and g be defined on $E \subset R$. Suppose*
$$f(t) \to A, \quad g(t) \to B \quad \text{as } t \to x.$$
Then

(a) $f(t) \to A'$ implies $A' = A$.
(b) $(f+g)(t) \to A+B$,
(c) $(fg)(t) \to AB$,
(d) $(f/g)(t) \to A/B$,

provided the right members of (b), (c), and (d) are defined.

Note that $\infty - \infty$, $0 \cdot \infty$, ∞/∞, $A/0$ are not defined (see Definition 1.23).

EXERCISES

1. Suppose f is a real function defined on R^1 which satisfies
$$\lim_{h \to 0} [f(x+h) - f(x-h)] = 0$$
for every $x \in R^1$. Does this imply that f is continuous?

2. If f is a continuous mapping of a metric space X into a metric space Y, prove that
$$f(\bar{E}) \subset \overline{f(E)}$$
for every set $E \subset X$. ($\bar{E}$ denotes the closure of E.) Show, by an example, that $f(\bar{E})$ can be a proper subset of $\overline{f(E)}$.

3. Let f be a continuous real function on a metric space X. Let $Z(f)$ (the *zero set* of f) be the set of all $p \in X$ at which $f(p) = 0$. Prove that $Z(f)$ is closed.

4. Let f and g be continuous mappings of a metric space X into a metric space Y,

and let E be a dense subset of X. Prove that $f(E)$ is dense in $f(X)$. If $g(p) = f(p)$ for all $p \in E$, prove that $g(p) = f(p)$ for all $p \in X$. (In other words, a continuous mapping is determined by its values on a dense subset of its domain.)

5. If f is a real continuous function defined on a closed set $E \subset R^1$, prove that there exist continuous real functions g on R^1 such that $g(x) = f(x)$ for all $x \in E$. (Such functions g are called *continuous extensions* of f from E to R^1.) Show that the result becomes false if the word "closed" is omitted. Extend the result to vector-valued functions. *Hint:* Let the graph of g be a straight line on each of the segments which constitute the complement of E (compare Exercise 29, Chap. 2). The result remains true if R^1 is replaced by any metric space, but the proof is not so simple.

6. If f is defined on E, the *graph* of f is the set of points $(x, f(x))$, for $x \in E$. In particular, if E is a set of real numbers, and f is real-valued, the graph of f is a subset of the plane.

    Suppose E is compact, and prove that f is continuous on E if and only if its graph is compact.

7. If $E \subset X$ and if f is a function defined on X, the *restriction* of f to E is the function g whose domain of definition is E, such that $g(p) = f(p)$ for $p \in E$. Define f and g on R^2 by: $f(0, 0) = g(0, 0) = 0$, $f(x, y) = xy^2/(x^2 + y^4)$, $g(x, y) = xy^2/(x^2 + y^6)$ if $(x, y) \neq (0, 0)$. Prove that f is bounded on R^2, that g is unbounded in every neighborhood of $(0, 0)$, and that f is not continuous at $(0, 0)$; nevertheless, the restrictions of both f and g to every straight line in R^2 are continuous!

8. Let f be a real uniformly continuous function on the bounded set E in R^1. Prove that f is bounded on E.

    Show that the conclusion is false if boundedness of E is omitted from the hypothesis.

9. Show that the requirement in the definition of uniform continuity can be rephrased as follows, in terms of diameters of sets: To every $\varepsilon > 0$ there exists a $\delta > 0$ such that diam $f(E) < \varepsilon$ for all $E \subset X$ with diam $E < \delta$.

10. Complete the details of the following alternative proof of Theorem 4.19: If f is not uniformly continuous, then for some $\varepsilon > 0$ there are sequences $\{p_n\}$, $\{q_n\}$ in X such that $d_X(p_n, q_n) \to 0$ but $d_Y(f(p_n), f(q_n)) > \varepsilon$. Use Theorem 2.37 to obtain a contradiction.

11. Suppose f is a uniformly continuous mapping of a metric space X into a metric space Y and prove that $\{f(x_n)\}$ is a Cauchy sequence in Y for every Cauchy sequence $\{x_n\}$ in X. Use this result to give an alternative proof of the theorem stated in Exercise 13.

12. A uniformly continuous function of a uniformly continuous function is uniformly continuous.

    State this more precisely and prove it.

13. Let E be a dense subset of a metric space X, and let f be a uniformly continuous *real* function defined on E. Prove that f has a continuous extension from E to X

(see Exercise 5 for terminology). (Uniqueness follows from Exercise 4.) *Hint:* For each $p \in X$ and each positive integer n, let $V_n(p)$ be the set of all $q \in E$ with $d(p, q) < 1/n$. Use Exercise 9 to show that the intersection of the closures of the sets $f(V_1(p)), f(V_2(p)), \ldots$, consists of a single point, say $g(p)$, of R^1. Prove that the function g so defined on X is the desired extension of f.

Could the range space R^1 be replaced by R^k? By any compact metric space? By any complete metric space? By any metric space?

14. Let $I = [0, 1]$ be the closed unit interval. Suppose f is a continuous mapping of I into I. Prove that $f(x) = x$ for at least one $x \in I$.

15. Call a mapping of X into Y *open* if $f(V)$ is an open set in Y whenever V is an open set in X.

    Prove that every continuous open mapping of R^1 into R^1 is monotonic.

16. Let $[x]$ denote the largest integer contained in x, that is, $[x]$ is the integer such that $x - 1 < [x] \leq x$; and let $(x) = x - [x]$ denote the fractional part of x. What discontinuities do the functions $[x]$ and (x) have?

17. Let f be a real function defined on (a, b). Prove that the set of points at which f has a simple discontinuity is at most countable. *Hint:* Let E be the set on which $f(x-) < f(x+)$. With each point x of E, associate a triple (p, q, r) of rational numbers such that
    (a) $f(x-) < p < f(x+)$,
    (b) $a < q < t < x$ implies $f(t) < p$,
    (c) $x < t < r < b$ implies $f(t) > p$.
    The set of all such triples is countable. Show that each triple is associated with at most one point of E. Deal similarly with the other possible types of simple discontinuities.

18. Every rational x can be written in the form $x = m/n$, where $n > 0$, and m and n are integers without any common divisors. When $x = 0$, we take $n = 1$. Consider the function f defined on R^1 by

$$f(x) = \begin{cases} 0 & (x \text{ irrational}), \\ \dfrac{1}{n} & \left(x = \dfrac{m}{n}\right). \end{cases}$$

Prove that f is continuous at every irrational point, and that f has a simple discontinuity at every rational point.

19. Suppose f is a real function with domain R^1 which has the intermediate value property: If $f(a) < c < f(b)$, then $f(x) = c$ for some x between a and b.

    Suppose also, for every rational r, that the set of all x with $f(x) = r$ is closed.
    Prove that f is continuous.

    Hint: If $x_n \to x_0$ but $f(x_n) > r > f(x_0)$ for some r and all n, then $f(t_n) = r$ for some t_n between x_0 and x_n; thus $t_n \to x_0$. Find a contradiction. (N. J. Fine, *Amer. Math. Monthly*, vol. 73, 1966, p. 782.)

20. If E is a nonempty subset of a metric space X, define the distance from $x \in X$ to E by
$$\rho_E(x) = \inf_{z \in E} d(x, z).$$
(a) Prove that $\rho_E(x) = 0$ if and only if $x \in \bar{E}$.
(b) Prove that ρ_E is a uniformly continuous function on X, by showing that
$$|\rho_E(x) - \rho_E(y)| \leq d(x, y)$$
for all $x \in X$, $y \in X$.
  Hint: $\rho_E(x) \leq d(x, z) \leq d(x, y) + d(y, z)$, so that
$$\rho_E(x) \leq d(x, y) + \rho_E(y).$$

21. Suppose K and F are disjoint sets in a metric space X, K is compact, F is closed. Prove that there exists $\delta > 0$ such that $d(p, q) > \delta$ if $p \in K$, $q \in F$. Hint: ρ_F is a continuous positive function on K.

Show that the conclusion may fail for two disjoint closed sets if neither is compact.

22. Let A and B be disjoint nonempty closed sets in a metric space X, and define
$$f(p) = \frac{\rho_A(p)}{\rho_A(p) + \rho_B(p)} \quad (p \in X).$$
Show that f is a continuous function on X whose range lies in $[0, 1]$, that $f(p) = 0$ precisely on A and $f(p) = 1$ precisely on B. This establishes a converse of Exercise 3: Every closed set $A \subset X$ is $Z(f)$ for some continuous real f on X. Setting
$$V = f^{-1}([0, \tfrac{1}{2})), \quad W = f^{-1}((\tfrac{1}{2}, 1]),$$
show that V and W are open and disjoint, and that $A \subset V$, $B \subset W$. (Thus pairs of disjoint closed sets in a metric space can be covered by pairs of disjoint open sets. This property of metric spaces is called *normality*.)

23. A real-valued function f defined in (a, b) is said to be *convex* if
$$f(\lambda x + (1 - \lambda)y) \leq \lambda f(x) + (1 - \lambda)f(y)$$
whenever $a < x < b$, $a < y < b$, $0 < \lambda < 1$. Prove that every convex function is continuous. Prove that every increasing convex function of a convex function is convex. (For example, if f is convex, so is e^f.)

If f is convex in (a, b) and if $a < s < t < u < b$, show that
$$\frac{f(t) - f(s)}{t - s} \leq \frac{f(u) - f(s)}{u - s} \leq \frac{f(u) - f(t)}{u - t}.$$

24. Assume that f is a continuous real function defined in (a, b) such that
$$f\left(\frac{x + y}{2}\right) \leq \frac{f(x) + f(y)}{2}$$
for all $x, y \in (a, b)$. Prove that f is convex.

25. If $A \subset R^k$ and $B \subset R^k$, define $A + B$ to be the set of all sums $\mathbf{x} + \mathbf{y}$ with $\mathbf{x} \in A$, $\mathbf{y} \in B$.

   (a) If K is compact and C is closed in R^k, prove that $K + C$ is closed.

   Hint: Take $\mathbf{z} \notin K + C$, put $F = \mathbf{z} - C$, the set of all $\mathbf{z} - \mathbf{y}$ with $\mathbf{y} \in C$. Then K and F are disjoint. Choose δ as in Exercise 21. Show that the open ball with center $\mathbf{z}$ and radius δ does not intersect $K + C$.

   (b) Let α be an irrational real number. Let C_1 be the set of all integers, let C_2 be the set of all $n\alpha$ with $n \in C_1$. Show that C_1 and C_2 are closed subsets of R^1 whose sum $C_1 + C_2$ is *not* closed, by showing that $C_1 + C_2$ is a countable dense subset of R^1.

26. Suppose X, Y, Z are metric spaces, and Y is compact. Let f map X into Y, let g be a continuous one-to-one mapping of Y into Z, and put $h(x) = g(f(x))$ for $x \in X$.

   Prove that f is uniformly continuous if h is uniformly continuous.

   Hint: g^{-1} has compact domain $g(Y)$, and $f(x) = g^{-1}(h(x))$.

   Prove also that f is continuous if h is continuous.

   Show (by modifying Example 4.21, or by finding a different example) that the compactness of Y cannot be omitted from the hypotheses, even when X and Z are compact.

5
DIFFERENTIATION

In this chapter we shall (except in the final section) confine our attention to *real* functions defined on intervals or segments. This is not just a matter of convenience, since genuine differences appear when we pass from real functions to vector-valued ones. Differentiation of functions defined on R^k will be discussed in Chap. 9.

THE DERIVATIVE OF A REAL FUNCTION

5.1 Definition Let f be defined (and real-valued) on $[a, b]$. For any $x \in [a, b]$ form the quotient

$$\phi(t) = \frac{f(t) - f(x)}{t - x} \quad (a < t < b, t \neq x), \tag{1}$$

and define

$$f'(x) = \lim_{t \to x} \phi(t), \tag{2}$$

provided this limit exists in accordance with Definition 4.1.

We thus associate with the function f a function f' whose domain is the set of points x at which the limit (2) exists; f' is called the *derivative* of f.

If f' is defined at a point x, we say that f is *differentiable* at x. If f' is defined at every point of a set $E \subset [a, b]$, we say that f is differentiable on E.

It is possible to consider right-hand and left-hand limits in (2); this leads to the definition of right-hand and left-hand derivatives. In particular, at the endpoints a and b, the derivative, if it exists, is a right-hand or left-hand derivative, respectively. We shall not, however, discuss one-sided derivatives in any detail.

If f is defined on a segment (a, b) and if $a < x < b$, then $f'(x)$ is defined by (1) and (2), as above. But $f'(a)$ and $f'(b)$ are not defined in this case.

5.2 Theorem *Let f be defined on $[a, b]$. If f is differentiable at a point $x \in [a, b]$, then f is continuous at x.*

Proof As $t \to x$, we have, by Theorem 4.4,

$$f(t) - f(x) = \frac{f(t) - f(x)}{t - x} \cdot (t - x) \to f'(x) \cdot 0 = 0.$$

The converse of this theorem is not true. It is easy to construct continuous functions which fail to be differentiable at isolated points. In Chap. 7 we shall even become acquainted with a function which is continuous on the whole line without being differentiable at any point!

5.3 Theorem *Suppose f and g are defined on $[a, b]$ and are differentiable at a point $x \in [a, b]$. Then $f + g$, fg, and f/g are differentiable at x, and*

(a) $(f + g)'(x) = f'(x) + g'(x);$

(b) $(fg)'(x) = f'(x)g(x) + f(x)g'(x);$

(c) $\left(\dfrac{f}{g}\right)'(x) = \dfrac{g(x)f'(x) - g'(x)f(x)}{g^2(x)}.$

In (c), we assume of course that $g(x) \neq 0$.

Proof (a) is clear, by Theorem 4.4. Let $h = fg$. Then

$$h(t) - h(x) = f(t)[g(t) - g(x)] + g(x)[f(t) - f(x)].$$

If we divide this by $t - x$ and note that $f(t) \to f(x)$ as $t \to x$ (Theorem 5.2), (b) follows. Next, let $h = f/g$. Then

$$\frac{h(t) - h(x)}{t - x} = \frac{1}{g(t)g(x)} \left[g(x) \frac{f(t) - f(x)}{t - x} - f(x) \frac{g(t) - g(x)}{t - x} \right].$$

Letting $t \to x$, and applying Theorems 4.4 and 5.2, we obtain (c).

5.4 Examples The derivative of any constant is clearly zero. If f is defined by $f(x) = x$, then $f'(x) = 1$. Repeated application of (b) and (c) then shows that x^n is differentiable, and that its derivative is nx^{n-1}, for any integer n (if $n < 0$, we have to restrict ourselves to $x \neq 0$). Thus every polynomial is differentiable, and so is every rational function, except at the points where the denominator is zero.

The following theorem is known as the "chain rule" for differentiation. It deals with differentiation of composite functions and is probably the most important theorem about derivatives. We shall meet more general versions of it in Chap. 9.

5.5 Theorem *Suppose f is continuous on $[a, b]$, $f'(x)$ exists at some point $x \in [a, b]$, g is defined on an interval I which contains the range of f, and g is differentiable at the point $f(x)$. If*

$$h(t) = g(f(t)) \qquad (a \leq t \leq b),$$

then h is differentiable at x, and

(3) $$h'(x) = g'(f(x))f'(x).$$

Proof Let $y = f(x)$. By the definition of the derivative, we have

(4) $$f(t) - f(x) = (t - x)[f'(x) + u(t)],$$

(5) $$g(s) - g(y) = (s - y)[g'(y) + v(s)],$$

where $t \in [a, b]$, $s \in I$, and $u(t) \to 0$ as $t \to x$, $v(s) \to 0$ as $s \to y$. Let $s = f(t)$. Using first (5) and then (4), we obtain

$$h(t) - h(x) = g(f(t)) - g(f(x))$$
$$= [f(t) - f(x)] \cdot [g'(y) + v(s)]$$
$$= (t - x) \cdot [f'(x) + u(t)] \cdot [g'(y) + v(s)],$$

or, if $t \neq x$,

(6) $$\frac{h(t) - h(x)}{t - x} = [g'(y) + v(s)] \cdot [f'(x) + u(t)].$$

Letting $t \to x$, we see that $s \to y$, by the continuity of f, so that the right side of (6) tends to $g'(y)f'(x)$, which gives (3).

5.6 Examples

(a) Let f be defined by

(7)
$$f(x) = \begin{cases} x \sin \dfrac{1}{x} & (x \neq 0), \\ 0 & (x = 0). \end{cases}$$

Taking for granted that the derivative of $\sin x$ is $\cos x$ (we shall discuss the trigonometric functions in Chap. 8), we can apply Theorems 5.3 and 5.5 whenever $x \neq 0$, and obtain

(8)
$$f'(x) = \sin \frac{1}{x} - \frac{1}{x} \cos \frac{1}{x} \qquad (x \neq 0).$$

At $x = 0$, these theorems do not apply any longer, since $1/x$ is not defined there, and we appeal directly to the definition: for $t \neq 0$,

$$\frac{f(t) - f(0)}{t - 0} = \sin \frac{1}{t}.$$

As $t \to 0$, this does not tend to any limit, so that $f'(0)$ does not exist.

(b) Let f be defined by

(9)
$$f(x) = \begin{cases} x^2 \sin \dfrac{1}{x} & (x \neq 0), \\ 0 & (x = 0), \end{cases}$$

As above, we obtain

(10)
$$f'(x) = 2x \sin \frac{1}{x} - \cos \frac{1}{x} \qquad (x \neq 0).$$

At $x = 0$, we appeal to the definition, and obtain

$$\left| \frac{f(t) - f(0)}{t - 0} \right| = \left| t \sin \frac{1}{t} \right| \leq |t| \qquad (t \neq 0);$$

letting $t \to 0$, we see that

(11)
$$f'(0) = 0.$$

Thus f is differentiable at all points x, but f' is not a continuous function, since $\cos (1/x)$ in (10) does not tend to a limit as $x \to 0$.

MEAN VALUE THEOREMS

5.7 Definition Let f be a real function defined on a metric space X. We say that f has a *local maximum* at a point $p \in X$ if there exists $\delta > 0$ such that $f(q) \leq f(p)$ for all $q \in X$ with $d(p, q) < \delta$.

Local minima are defined likewise.

Our next theorem is the basis of many applications of differentiation.

5.8 Theorem *Let f be defined on $[a, b]$; if f has a local maximum at a point $x \in (a, b)$, and if $f'(x)$ exists, then $f'(x) = 0$.*

The analogous statement for local minima is of course also true.

Proof Choose δ in accordance with Definition 5.7, so that

$$a < x - \delta < x < x + \delta < b.$$

If $x - \delta < t < x$, then

$$\frac{f(t) - f(x)}{t - x} \geq 0.$$

Letting $t \to x$, we see that $f'(x) \geq 0$.

If $x < t < x + \delta$, then

$$\frac{f(t) - f(x)}{t - x} \leq 0,$$

which shows that $f'(x) \leq 0$. Hence $f'(x) = 0$.

5.9 Theorem *If f and g are continuous real functions on $[a, b]$ which are differentiable in (a, b), then there is a point $x \in (a, b)$ at which*

$$[f(b) - f(a)]g'(x) = [g(b) - g(a)]f'(x).$$

Note that differentiability is not required at the endpoints.

Proof Put

$$h(t) = [f(b) - f(a)]g(t) - [g(b) - g(a)]f(t) \qquad (a \leq t \leq b).$$

Then h is continuous on $[a, b]$, h is differentiable in (a, b), and

(12) $$h(a) = f(b)g(a) - f(a)g(b) = h(b).$$

To prove the theorem, we have to show that $h'(x) = 0$ for some $x \in (a, b)$.

If h is constant, this holds for every $x \in (a, b)$. If $h(t) > h(a)$ for some $t \in (a, b)$, let x be a point on $[a, b]$ at which h attains its maximum

(Theorem 4.16). By (12), $x \in (a, b)$, and Theorem 5.8 shows that $h'(x) = 0$. If $h(t) < h(a)$ for some $t \in (a, b)$, the same argument applies if we choose for x a point on $[a, b]$ where h attains its minimum.

This theorem is often called a *generalized mean value theorem*; the following special case is usually referred to as "the" mean value theorem:

5.10 Theorem *If f is a real continuous function on $[a, b]$ which is differentiable in (a, b), then there is a point $x \in (a, b)$ at which*

$$f(b) - f(a) = (b - a)f'(x).$$

Proof Take $g(x) = x$ in Theorem 5.9.

5.11 Theorem *Suppose f is differentiable in (a, b).*

(a) *If $f'(x) \geq 0$ for all $x \in (a, b)$, then f is monotonically increasing.*

(b) *If $f'(x) = 0$ for all $x \in (a, b)$, then f is constant.*

(c) *If $f'(x) \leq 0$ for all $x \in (a, b)$, then f is monotonically decreasing.*

Proof All conclusions can be read off from the equation

$$f(x_2) - f(x_1) = (x_2 - x_1)f'(x),$$

which is valid, for each pair of numbers x_1, x_2 in (a, b), for *some* x between x_1 and x_2.

THE CONTINUITY OF DERIVATIVES

We have already seen [Example 5.6(*b*)] that a function f may have a derivative f' which exists at every point, but is discontinuous at some point. However, not every function is a derivative. In particular, derivatives which exist at every point of an interval have one important property in common with functions which are continuous on an interval: Intermediate values are assumed (compare Theorem 4.23). The precise statement follows.

5.12 Theorem *Suppose f is a real differentiable function on $[a, b]$ and suppose $f'(a) < \lambda < f'(b)$. Then there is a point $x \in (a, b)$ such that $f'(x) = \lambda$.*

A similar result holds of course if $f'(a) > f'(b)$.

Proof Put $g(t) = f(t) - \lambda t$. Then $g'(a) < 0$, so that $g(t_1) < g(a)$ for some $t_1 \in (a, b)$, and $g'(b) > 0$, so that $g(t_2) < g(b)$ for some $t_2 \in (a, b)$. Hence g attains its minimum on $[a, b]$ (Theorem 4.16) at some point x such that $a < x < b$. By Theorem 5.8, $g'(x) = 0$. Hence $f'(x) = \lambda$.

Corollary *If f is differentiable on $[a, b]$, then f' cannot have any simple discontinuities on $[a, b]$.*

But f' may very well have discontinuities of the second kind.

L'HOSPITAL'S RULE

The following theorem is frequently useful in the evaluation of limits.

5.13 Theorem *Suppose f and g are real and differentiable in (a, b), and $g'(x) \neq 0$ for all $x \in (a, b)$, where $-\infty \leq a < b \leq +\infty$. Suppose*

$$\frac{f'(x)}{g'(x)} \to A \text{ as } x \to a. \tag{13}$$

If

$$f(x) \to 0 \text{ and } g(x) \to 0 \text{ as } x \to a, \tag{14}$$

or if

$$g(x) \to +\infty \text{ as } x \to a, \tag{15}$$

then

$$\frac{f(x)}{g(x)} \to A \text{ as } x \to a. \tag{16}$$

The analogous statement is of course also true if $x \to b$, or if $g(x) \to -\infty$ in (15). Let us note that we now use the limit concept in the extended sense of Definition 4.33.

Proof We first consider the case in which $-\infty \leq A < +\infty$. Choose a real number q such that $A < q$, and then choose r such that $A < r < q$. By (13) there is a point $c \in (a, b)$ such that $a < x < c$ implies

$$\frac{f'(x)}{g'(x)} < r. \tag{17}$$

If $a < x < y < c$, then Theorem 5.9 shows that there is a point $t \in (x, y)$ such that

$$\frac{f(x) - f(y)}{g(x) - g(y)} = \frac{f'(t)}{g'(t)} < r. \tag{18}$$

Suppose (14) holds. Letting $x \to a$ in (18), we see that

$$\frac{f(y)}{g(y)} \leq r < q \qquad (a < y < c). \tag{19}$$

Next, suppose (15) holds. Keeping y fixed in (18), we can choose a point $c_1 \in (a, y)$ such that $g(x) > g(y)$ and $g(x) > 0$ if $a < x < c_1$. Multiplying (18) by $[g(x) - g(y)]/g(x)$, we obtain

(20) $$\frac{f(x)}{g(x)} < r - r\frac{g(y)}{g(x)} + \frac{f(y)}{g(x)} \qquad (a < x < c_1).$$

If we let $x \to a$ in (20), (15) shows that there is a point $c_2 \in (a, c_1)$ such that

(21) $$\frac{f(x)}{g(x)} < q \qquad (a < x < c_2).$$

Summing up, (19) and (21) show that for any q, subject only to the condition $A < q$, there is a point c_2 such that $f(x)/g(x) < q$ if $a < x < c_2$.

In the same manner, if $-\infty < A \leq +\infty$, and p is chosen so that $p < A$, we can find a point c_3 such that

(22) $$p < \frac{f(x)}{g(x)} \qquad (a < x < c_3),$$

and (16) follows from these two statements.

DERIVATIVES OF HIGHER ORDER

5.14 Definition If f has a derivative f' on an interval, and if f' is itself differentiable, we denote the derivative of f' by f'' and call f'' the second derivative of f. Continuing in this manner, we obtain functions

$$f, f', f'', f^{(3)}, \ldots, f^{(n)},$$

each of which is the derivative of the preceding one. $f^{(n)}$ is called the nth derivative, or the derivative of order n, of f.

In order for $f^{(n)}(x)$ to exist at a point x, $f^{(n-1)}(t)$ must exist in a neighborhood of x (or in a one-sided neighborhood, if x is an endpoint of the interval on which f is defined), and $f^{(n-1)}$ must be differentiable at x. Since $f^{(n-1)}$ must exist in a neighborhood of x, $f^{(n-2)}$ must be differentiable in that neighborhood.

TAYLOR'S THEOREM

5.15 Theorem *Suppose f is a real function on $[a, b]$, n is a positive integer, $f^{(n-1)}$ is continuous on $[a, b]$, $f^{(n)}(t)$ exists for every $t \in (a, b)$. Let α, β be distinct points of $[a, b]$, and define*

(23) $$P(t) = \sum_{k=0}^{n-1} \frac{f^{(k)}(\alpha)}{k!} (t - \alpha)^k.$$

Then there exists a point x between α and β such that

$$f(\beta) = P(\beta) + \frac{f^{(n)}(x)}{n!}(\beta - \alpha)^n. \tag{24}$$

For $n = 1$, this is just the mean value theorem. In general, the theorem shows that f can be approximated by a polynomial of degree $n - 1$, and that (24) allows us to estimate the error, if we know bounds on $|f^{(n)}(x)|$.

Proof Let M be the number defined by

$$f(\beta) = P(\beta) + M(\beta - \alpha)^n \tag{25}$$

and put

$$g(t) = f(t) - P(t) - M(t - \alpha)^n \quad (a \le t \le b). \tag{26}$$

We have to show that $n!M = f^{(n)}(x)$ for some x between α and β. By (23) and (26),

$$g^{(n)}(t) = f^{(n)}(t) - n!M \quad (a < t < b). \tag{27}$$

Hence the proof will be complete if we can show that $g^{(n)}(x) = 0$ for some x between α and β.

Since $P^{(k)}(\alpha) = f^{(k)}(\alpha)$ for $k = 0, \ldots, n - 1$, we have

$$g(\alpha) = g'(\alpha) = \cdots = g^{(n-1)}(\alpha) = 0. \tag{28}$$

Our choice of M shows that $g(\beta) = 0$, so that $g'(x_1) = 0$ for some x_1 between α and β, by the mean value theorem. Since $g'(\alpha) = 0$, we conclude similarly that $g''(x_2) = 0$ for some x_2 between α and x_1. After n steps we arrive at the conclusion that $g^{(n)}(x_n) = 0$ for some x_n between α and x_{n-1}, that is, between α and β.

DIFFERENTIATION OF VECTOR-VALUED FUNCTIONS

5.16 Remarks Definition 5.1 applies without any change to complex functions f defined on $[a, b]$, and Theorems 5.2 and 5.3, as well as their proofs, remain valid. If f_1 and f_2 are the real and imaginary parts of f, that is, if

$$f(t) = f_1(t) + if_2(t)$$

for $a \le t \le b$, where $f_1(t)$ and $f_2(t)$ are real, then we clearly have

$$f'(x) = f_1'(x) + if_2'(x); \tag{29}$$

also, f is differentiable at x if and only if both f_1 and f_2 are differentiable at x.

Passing to vector-valued functions in general, i.e., to functions **f** which map $[c, b]$ into some R^k, we may still apply Definition 5.1 to define $\mathbf{f}'(x)$. The term $\phi(t)$ in (1) is now, for each t, a point in R^k, and the limit in (2) is taken with respect to the norm of R^k. In other words, $\mathbf{f}'(x)$ is that point of R^k (if there is one) for which

$$(30) \qquad \lim_{t \to x} \left| \frac{\mathbf{f}(t) - \mathbf{f}(x)}{t - x} - \mathbf{f}'(x) \right| = 0,$$

and $\mathbf{f}'$ is again a function with values in R^k.

If $f_1, \ldots, f_k$ are the components of **f**, as defined in Theorem 4.10, then

$$(31) \qquad \mathbf{f}' = (f_1', \ldots, f_k'),$$

and **f** is differentiable at a point x if and only if each of the functions $f_1, \ldots, f_k$ is differentiable at x.

Theorem 5.2 is true in this context as well, and so is Theorem 5.3(a) and (b), if fg is replaced by the inner product $\mathbf{f} \cdot \mathbf{g}$ (see Definition 4.3).

When we turn to the mean value theorem, however, and to one of its consequences, namely, L'Hospital's rule, the situation changes. The next two examples will show that each of these results fails to be true for complex-valued functions.

5.17 Example Define, for real x,

$$(32) \qquad f(x) = e^{ix} = \cos x + i \sin x.$$

(The last expression may be taken as the definition of the complex exponential e^{ix}; see Chap. 8 for a full discussion of these functions.) Then

$$(33) \qquad f(2\pi) - f(0) = 1 - 1 = 0,$$

but

$$(34) \qquad f'(x) = ie^{ix},$$

so that $|f'(x)| = 1$ for all real x.

Thus Theorem 5.10 fails to hold in this case.

5.18 Example On the segment $(0, 1)$, define $f(x) = x$ and

$$(35) \qquad g(x) = x + x^2 e^{i/x^2}.$$

Since $|e^{it}| = 1$ for all real t, we see that

$$(36) \qquad \lim_{x \to 0} \frac{f(x)}{g(x)} = 1.$$

Next,

(37) $$g'(x) = 1 + \left(2x - \frac{2i}{x}\right)e^{i/x^2} \quad (0 < x < 1),$$

so that

(38) $$|g'(x)| \geq \left|2x - \frac{2i}{x}\right| - 1 \geq \frac{2}{x} - 1.$$

Hence

(39) $$\left|\frac{f'(x)}{g'(x)}\right| = \frac{1}{|g'(x)|} \leq \frac{x}{2 - x}$$

and so

(40) $$\lim_{x \to 0} \frac{f'(x)}{g'(x)} = 0.$$

By (36) and (40), L'Hospital's rule fails in this case. Note also that $g'(x) \neq 0$ on $(0, 1)$, by (38).

However, there *is* a consequence of the mean value theorem which, for purposes of applications, is almost as useful as Theorem 5.10, and which remains true for vector-valued functions: From Theorem 5.10 it follows that

(41) $$|f(b) - f(a)| \leq (b - a) \sup_{a < x < b} |f'(x)|.$$

5.19 Theorem *Suppose* **f** *is a continuous mapping of* $[a, b]$ *into* R^k *and* **f** *is differentiable in* (a, b). *Then there exists* $x \in (a, b)$ *such that*

$$|\mathbf{f}(b) - \mathbf{f}(a)| \leq (b - a)|\mathbf{f}'(x)|.$$

Proof[1] Put $\mathbf{z} = \mathbf{f}(b) - \mathbf{f}(a)$, and define

$$\varphi(t) = \mathbf{z} \cdot \mathbf{f}(t) \quad (a \leq t \leq b).$$

Then φ is a real-valued continuous function on $[a, b]$ which is differentiable in (a, b). The mean value theorem shows therefore that

$$\varphi(b) - \varphi(a) = (b - a)\varphi'(x) = (b - a)\mathbf{z} \cdot \mathbf{f}'(x)$$

for some $x \in (a, b)$. On the other hand,

$$\varphi(b) - \varphi(a) = \mathbf{z} \cdot \mathbf{f}(b) - \mathbf{z} \cdot \mathbf{f}(a) = \mathbf{z} \cdot \mathbf{z} = |\mathbf{z}|^2.$$

The Schwarz inequality now gives

$$|\mathbf{z}|^2 = (b - a)|\mathbf{z} \cdot \mathbf{f}'(x)| \leq (b - a)|\mathbf{z}||\mathbf{f}'(x)|.$$

Hence $|\mathbf{z}| \leq (b - a)|\mathbf{f}'(x)|$, which is the desired conclusion.

---

[1] V. P. Havin translated the second edition of this book into Russian and added this proof to the original one.

EXERCISES

1. Let f be defined for all real x, and suppose that
$$|f(x) - f(y)| \leq (x-y)^2$$
for all real x and y. Prove that f is constant.

2. Suppose $f'(x) > 0$ in (a, b). Prove that f is strictly increasing in (a, b), and let g be its inverse function. Prove that g is differentiable, and that
$$g'(f(x)) = \frac{1}{f'(x)} \qquad (a < x < b).$$

3. Suppose g is a real function on R^1, with bounded derivative (say $|g'| \leq M$). Fix $\varepsilon > 0$, and define $f(x) = x + \varepsilon g(x)$. Prove that f is one-to-one if ε is small enough. (A set of admissible values of ε can be determined which depends only on M.)

4. If
$$C_0 + \frac{C_1}{2} + \cdots + \frac{C_{n-1}}{n} + \frac{C_n}{n+1} = 0,$$
where $C_0, \ldots, C_n$ are real constants, prove that the equation
$$C_0 + C_1 x + \cdots + C_{n-1} x^{n-1} + C_n x^n = 0$$
has at least one real root between 0 and 1.

5. Suppose f is defined and differentiable for every $x > 0$, and $f'(x) \to 0$ as $x \to +\infty$. Put $g(x) = f(x+1) - f(x)$. Prove that $g(x) \to 0$ as $x \to +\infty$.

6. Suppose
    (a) f is continuous for $x \geq 0$,
    (b) $f'(x)$ exists for $x > 0$,
    (c) $f(0) = 0$,
    (d) f' is monotonically increasing.
    Put
    $$g(x) = \frac{f(x)}{x} \qquad (x > 0)$$
    and prove that g is monotonically increasing.

7. Suppose $f'(x)$, $g'(x)$ exist, $g'(x) \neq 0$, and $f(x) = g(x) = 0$. Prove that
$$\lim_{t \to x} \frac{f(t)}{g(t)} = \frac{f'(x)}{g'(x)}.$$
(This holds also for complex functions.)

8. Suppose f' is continuous on $[a, b]$ and $\varepsilon > 0$. Prove that there exists $\delta > 0$ such that
$$\left| \frac{f(t) - f(x)}{t - x} - f'(x) \right| < \varepsilon$$

whenever $0 < |t - x| < \delta$, $a \leq x \leq b$, $a \leq t \leq b$. (This could be expressed by saying that f is *uniformly differentiable* on $[a, b]$ if f' is continuous on $[a, b]$.) Does this hold for vector-valued functions too?

9. Let f be a continuous real function on R^1, of which it is known that $f'(x)$ exists for all $x \neq 0$ and that $f'(x) \to 3$ as $x \to 0$. Does it follow that $f'(0)$ exists?

10. Suppose f and g are complex differentiable functions on $(0, 1)$, $f(x) \to 0$, $g(x) \to 0$, $f'(x) \to A$, $g'(x) \to B$ as $x \to 0$, where A and B are complex numbers, $B \neq 0$. Prove that

$$\lim_{x \to 0} \frac{f(x)}{g(x)} = \frac{A}{B}.$$

Compare with Example 5.18. *Hint*:

$$\frac{f(x)}{g(x)} = \left\{\frac{f(x)}{x} - A\right\} \cdot \frac{x}{g(x)} + A \cdot \frac{x}{g(x)}.$$

Apply Theorem 5.13 to the real and imaginary parts of $f(x)/x$ and $g(x)/x$.

11. Suppose f is defined in a neighborhood of x, and suppose $f''(x)$ exists. Show that

$$\lim_{h \to 0} \frac{f(x+h) + f(x-h) - 2f(x)}{h^2} = f''(x).$$

Show by an example that the limit may exist even if $f''(x)$ does not.

Hint: Use Theorem 5.13.

12. If $f(x) = |x|^3$, compute $f'(x)$, $f''(x)$ for all real x, and show that $f^{(3)}(0)$ does not exist.

13. Suppose a and c are real numbers, $c > 0$, and f is defined on $[-1, 1]$ by

$$f(x) = \begin{cases} x^a \sin(|x|^{-c}) & \text{(if } x \neq 0\text{),} \\ 0 & \text{(if } x = 0\text{).} \end{cases}$$

Prove the following statements:

(a) f is continuous if and only if $a > 0$.
(b) $f'(0)$ exists if and only if $a > 1$.
(c) f' is bounded if and only if $a \geq 1 + c$.
(d) f' is continuous if and only if $a > 1 + c$.
(e) $f''(0)$ exists if and only if $a > 2 + c$.
(f) f'' is bounded if and only if $a \geq 2 + 2c$.
(g) f'' is continuous if and only if $a > 2 + 2c$.

14. Let f be a differentiable real function defined in (a, b). Prove that f is convex if and only if f' is monotonically increasing. Assume next that $f''(x)$ exists for every $x \in (a, b)$, and prove that f is convex if and only if $f''(x) \geq 0$ for all $x \in (a, b)$.

15. Suppose $a \in R^1$, f is a twice-differentiable real function on (a, ∞), and M_0, M_1, M_2 are the least upper bounds of $|f(x)|$, $|f'(x)|$, $|f''(x)|$, respectively, on (a, ∞). Prove that

$$M_1^2 \leq 4 M_0 M_2.$$

Hint: If $h > 0$, Taylor's theorem shows that
$$f'(x) = \frac{1}{2h}[f(x+2h) - f(x)] - hf''(\xi)$$
for some $\xi \in (x, x+2h)$. Hence
$$|f'(x)| \leq hM_2 + \frac{M_0}{h}.$$

To show that $M_1^2 = 4M_0 M_2$ can actually happen, take $a = -1$, define
$$f(x) = \begin{cases} 2x^2 - 1 & (-1 < x < 0), \\ \dfrac{x^2 - 1}{x^2 + 1} & (0 \leq x < \infty), \end{cases}$$
and show that $M_0 = 1$, $M_1 = 4$, $M_2 = 4$.

Does $M_1^2 \leq 4M_0 M_2$ hold for vector-valued functions too?

16. Suppose f is twice-differentiable on $(0, \infty)$, f'' is bounded on $(0, \infty)$, and $f(x) \to 0$ as $x \to \infty$. Prove that $f'(x) \to 0$ as $x \to \infty$.

Hint: Let $a \to \infty$ in Exercise 15.

17. Suppose f is a real, three times differentiable function on $[-1, 1]$, such that
$$f(-1) = 0, \quad f(0) = 0, \quad f(1) = 1, \quad f'(0) = 0.$$
Prove that $f^{(3)}(x) \geq 3$ for some $x \in (-1, 1)$.

Note that equality holds for $\frac{1}{2}(x^3 + x^2)$.

Hint: Use Theorem 5.15, with $\alpha = 0$ and $\beta = \pm 1$, to show that there exist $s \in (0, 1)$ and $t \in (-1, 0)$ such that
$$f^{(3)}(s) + f^{(3)}(t) = 6.$$

18. Suppose f is a real function on $[a, b]$, n is a positive integer, and $f^{(n-1)}$ exists for every $t \in [a, b]$. Let α, β, and P be as in Taylor's theorem (5.15). Define
$$Q(t) = \frac{f(t) - f(\beta)}{t - \beta}$$
for $t \in [a, b]$, $t \neq \beta$, differentiate
$$f(t) - f(\beta) = (t - \beta)Q(t)$$
$n - 1$ times at $t = \alpha$, and derive the following version of Taylor's theorem:
$$f(\beta) = P(\beta) + \frac{Q^{(n-1)}(\alpha)}{(n-1)!}(\beta - \alpha)^n.$$

19. Suppose f is defined in $(-1, 1)$ and $f'(0)$ exists. Suppose $-1 < \alpha_n < \beta_n < 1$, $\alpha_n \to 0$, and $\beta_n \to 0$ as $n \to \infty$. Define the difference quotients
$$D_n = \frac{f(\beta_n) - f(\alpha_n)}{\beta_n - \alpha_n}.$$

Prove the following statements:

(a) If $\alpha_n < 0 < \beta_n$, then $\lim D_n = f'(0)$.

(b) If $0 < \alpha_n < \beta_n$ and $\{\beta_n/(\beta_n - \alpha_n)\}$ is bounded, then $\lim D_n = f'(0)$.

(c) If f' is continuous in $(-1, 1)$, then $\lim D_n = f'(0)$.

Give an example in which f is differentiable in $(-1, 1)$ (but f' is not continuous at 0) and in which α_n, β_n tend to 0 in such a way that $\lim D_n$ exists but is different from $f'(0)$.

20. Formulate and prove an inequality which follows from Taylor's theorem and which remains valid for vector-valued functions.

21. Let E be a closed subset of R^1. We saw in Exercise 22, Chap. 4, that there is a real continuous function f on R^1 whose zero set is E. Is it possible, for each closed set E, to find such an f which is differentiable on R^1, or one which is n times differentiable, or even one which has derivatives of all orders on R^1?

22. Suppose f is a real function on $(-\infty, \infty)$. Call x a *fixed point* of f if $f(x) = x$.

(a) If f is differentiable and $f'(t) \neq 1$ for every real t, prove that f has at most one fixed point.

(b) Show that the function f defined by

$$f(t) = t + (1 + e^t)^{-1}$$

has no fixed point, although $0 < f'(t) < 1$ for all real t.

(c) However, if there is a constant $A < 1$ such that $|f'(t)| \leq A$ for all real t, prove that a fixed point x of f exists, and that $x = \lim x_n$, where x_1 is an arbitrary real number and

$$x_{n+1} = f(x_n)$$

for $n = 1, 2, 3, \ldots$.

(d) Show that the process described in (c) can be visualized by the zig-zag path

$$(x_1, x_2) \to (x_2, x_2) \to (x_2, x_3) \to (x_3, x_3) \to (x_3, x_4) \to \cdots.$$

23. The function f defined by

$$f(x) = \frac{x^3 + 1}{3}$$

has three fixed points, say α, β, γ, where

$$-2 < \alpha < -1, \quad 0 < \beta < 1, \quad 1 < \gamma < 2.$$

For arbitrarily chosen x_1, define $\{x_n\}$ by setting $x_{n+1} = f(x_n)$.

(a) If $x_1 < \alpha$, prove that $x_n \to -\infty$ as $n \to \infty$.

(b) If $\alpha < x_1 < \gamma$, prove that $x_n \to \beta$ as $n \to \infty$.

(c) If $\gamma < x_1$, prove that $x_n \to +\infty$ as $n \to \infty$.

Thus β can be located by this method, but α and γ cannot.

24. The process described in part (c) of Exercise 22 can of course also be applied to functions that map $(0, \infty)$ to $(0, \infty)$.

Fix some $\alpha > 1$, and put

$$f(x) = \frac{1}{2}\left(x + \frac{\alpha}{x}\right), \qquad g(x) = \frac{\alpha + x}{1 + x}.$$

Both f and g have $\sqrt{\alpha}$ as their only fixed point in $(0, \infty)$. Try to explain, on the basis of properties of f and g, why the convergence in Exercise 16, Chap. 3, is so much more rapid than it is in Exercise 17. (Compare f' and g', draw the zig-zags suggested in Exercise 22.)

Do the same when $0 < \alpha < 1$.

25. Suppose f is twice differentiable on $[a, b]$, $f(a) < 0$, $f(b) > 0$, $f'(x) \geq \delta > 0$, and $0 \leq f''(x) \leq M$ for all $x \in [a, b]$. Let ξ be the unique point in (a, b) at which $f(\xi) = 0$.

Complete the details in the following outline of *Newton's method* for computing ξ.

(a) Choose $x_1 \in (\xi, b)$, and define $\{x_n\}$ by

$$x_{n+1} = x_n - \frac{f(x_n)}{f'(x_n)}.$$

Interpret this geometrically, in terms of a tangent to the graph of f.

(b) Prove that $x_{n+1} < x_n$ and that

$$\lim_{n \to \infty} x_n = \xi.$$

(c) Use Taylor's theorem to show that

$$x_{n+1} - \xi = \frac{f''(t_n)}{2f'(x_n)}(x_n - \xi)^2$$

for some $t_n \in (\xi, x_n)$.

(d) If $A = M/2\delta$, deduce that

$$0 \leq x_{n+1} - \xi \leq \frac{1}{A}[A(x_1 - \xi)]^{2^n}.$$

(Compare with Exercises 16 and 18, Chap. 3.)

(e) Show that Newton's method amounts to finding a fixed point of the function g defined by

$$g(x) = x - \frac{f(x)}{f'(x)}.$$

How does $g'(x)$ behave for x near ξ?

(f) Put $f(x) = x^{1/3}$ on $(-\infty, \infty)$ and try Newton's method. What happens?

26. Suppose f is differentiable on $[a, b]$, $f(a) = 0$, and there is a real number A such that $|f'(x)| \leq A|f(x)|$ on $[a, b]$. Prove that $f(x) = 0$ for all $x \in [a, b]$. *Hint:* Fix $x_0 \in [a, b]$, let
$$M_0 = \sup |f(x)|, \qquad M_1 = \sup |f'(x)|$$
for $a \leq x \leq x_0$. For any such x,
$$|f(x)| \leq M_1(x_0 - a) \leq A(x_0 - a)M_0.$$
Hence $M_0 = 0$ if $A(x_0 - a) < 1$. That is, $f = 0$ on $[a, x_0]$. Proceed.

27. Let ϕ be a real function defined on a rectangle R in the plane, given by $a \leq x \leq b$, $\alpha \leq y \leq \beta$. A *solution* of the initial-value problem
$$y' = \phi(x, y), \qquad y(a) = c \qquad (\alpha \leq c \leq \beta)$$
is, by definition, a differentiable function f on $[a, b]$ such that $f(a) = c$, $\alpha \leq f(x) \leq \beta$, and
$$f'(x) = \phi(x, f(x)) \qquad (a \leq x \leq b).$$
Prove that such a problem has at most one solution if there is a constant A such that
$$|\phi(x, y_2) - \phi(x, y_1)| \leq A|y_2 - y_1|$$
whenever $(x, y_1) \in R$ and $(x, y_2) \in R$.

Hint: Apply Exercise 26 to the difference of two solutions. Note that this uniqueness theorem does not hold for the initial-value problem
$$y' = y^{1/2}, \qquad y(0) = 0,$$
which has two solutions: $f(x) = 0$ and $f(x) = x^2/4$. Find all other solutions.

28. Formulate and prove an analogous uniqueness theorem for systems of differential equations of the form
$$y'_j = \phi_j(x, y_1, \ldots, y_k), \qquad y_j(a) = c_j \qquad (j = 1, \ldots, k).$$
Note that this can be rewritten in the form
$$\mathbf{y}' = \boldsymbol{\phi}(x, \mathbf{y}), \qquad \mathbf{y}(a) = \mathbf{c}$$
where $\mathbf{y} = (y_1, \ldots, y_k)$ ranges over a k-cell, $\boldsymbol{\phi}$ is the mapping of a $(k+1)$-cell into the Euclidean k-space whose components are the functions $\phi_1, \ldots, \phi_k$, and $\mathbf{c}$ is the vector $(c_1, \ldots, c_k)$. Use Exercise 26, for vector-valued functions.

29. Specialize Exercise 28 by considering the system
$$y'_j = y_{j+1} \qquad (j = 1, \ldots, k-1),$$
$$y'_k = f(x) - \sum_{j=1}^{k} g_j(x) y_j,$$
where $f, g_1, \ldots, g_k$ are continuous real functions on $[a, b]$, and derive a uniqueness theorem for solutions of the equation
$$y^{(k)} + g_k(x) y^{(k-1)} + \cdots + g_2(x) y' + g_1(x) y = f(x),$$
subject to initial conditions
$$y(a) = c_1, \qquad y'(a) = c_2, \qquad \ldots, \qquad y^{(k-1)}(a) = c_k.$$

6

THE RIEMANN-STIELTJES INTEGRAL

The present chapter is based on a definition of the Riemann integral which depends very explicitly on the order structure of the real line. Accordingly, we begin by discussing integration of real-valued functions on intervals. Extensions to complex- and vector-valued functions on intervals follow in later sections. Integration over sets other than intervals is discussed in Chaps. 10 and 11.

DEFINITION AND EXISTENCE OF THE INTEGRAL

6.1 Definition Let $[a, b]$ be a given interval. By a *partition* P of $[a, b]$ we mean a finite set of points $x_0, x_1, \ldots, x_n$, where

$$a = x_0 \leq x_1 \leq \cdots \leq x_{n-1} \leq x_n = b.$$

We write

$$\Delta x_i = x_i - x_{i-1} \qquad (i = 1, \ldots, n).$$

Now suppose f is a bounded real function defined on $[a, b]$. Corresponding to each partition P of $[a, b]$ we put

$$M_i = \sup f(x) \quad (x_{i-1} \leq x \leq x_i),$$
$$m_i = \inf f(x) \quad (x_{i-1} \leq x \leq x_i),$$
$$U(P, f) = \sum_{i=1}^{n} M_i \Delta x_i,$$
$$L(P, f) = \sum_{i=1}^{n} m_i \Delta x_i,$$

and finally

(1) $$\overline{\int_a^b} f \, dx = \inf U(P, f),$$

(2) $$\underline{\int_a^b} f \, dx = \sup L(P, f),$$

where the inf and the sup are taken over all partitions P of $[a, b]$. The left members of (1) and (2) are called the *upper* and *lower Riemann integrals* of f over $[a, b]$, respectively.

If the upper and lower integrals are equal, we say that f is *Riemann-integrable* on $[a, b]$, we write $f \in \mathscr{R}$ (that is, $\mathscr{R}$ denotes the set of Riemann-integrable functions), and we denote the common value of (1) and (2) by

(3) $$\int_a^b f \, dx,$$

or by

(4) $$\int_a^b f(x) \, dx.$$

This is the *Riemann integral* of f over $[a, b]$. Since f is bounded, there exist two numbers, m and M, such that

$$m \leq f(x) \leq M \quad (a \leq x \leq b).$$

Hence, for every P,

$$m(b - a) \leq L(P, f) \leq U(P, f) \leq M(b - a),$$

so that the numbers $L(P, f)$ and $U(P, f)$ form a bounded set. This shows that *the upper and lower integrals are defined* for *every* bounded function f. The question of their equality, and hence the question of the integrability of f, is a more delicate one. Instead of investigating it separately for the Riemann integral, we shall immediately consider a more general situation.

6.2 Definition Let α be a monotonically increasing function on $[a, b]$ (since $\alpha(a)$ and $\alpha(b)$ are finite, it follows that α is bounded on $[a, b]$). Corresponding to each partition P of $[a, b]$, we write

$$\Delta\alpha_i = \alpha(x_i) - \alpha(x_{i-1}).$$

It is clear that $\Delta\alpha_i \geq 0$. For any real function f which is bounded on $[a, b]$ we put

$$U(P, f, \alpha) = \sum_{i=1}^{n} M_i \, \Delta\alpha_i,$$

$$L(P, f, \alpha) = \sum_{i=1}^{n} m_i \, \Delta\alpha_i,$$

where M_i, m_i have the same meaning as in Definition 6.1, and we define

(5) $$\overline{\int_a^b} f \, d\alpha = \inf U(P, f, \alpha),$$

(6) $$\underline{\int_a^b} f \, d\alpha = \sup L(P, f, \alpha),$$

the inf and sup again being taken over all partitions.

If the left members of (5) and (6) are equal, we denote their common value by

(7) $$\int_a^b f \, d\alpha$$

or sometimes by

(8) $$\int_a^b f(x) \, d\alpha(x).$$

This is the *Riemann-Stieltjes integral* (or simply the *Stieltjes integral*) of f with respect to α, over $[a, b]$.

If (7) exists, i.e., if (5) and (6) are equal, we say that f is integrable with respect to α, in the Riemann sense, and write $f \in \mathscr{R}(\alpha)$.

By taking $\alpha(x) = x$, the Riemann integral is seen to be a special case of the Riemann-Stieltjes integral. Let us mention explicitly, however, that in the general case α need not even be continuous.

A few words should be said about the notation. We prefer (7) to (8), since the letter x which appears in (8) adds nothing to the content of (7). It is immaterial which letter we use to represent the so-called "variable of integration." For instance, (8) is the same as

$$\int_a^b f(y) \, d\alpha(y).$$

The integral depends on f, α, a and b, but not on the variable of integration, which may as well be omitted.

The role played by the variable of integration is quite analogous to that of the index of summation: The two symbols

$$\sum_{i=1}^{n} c_i, \qquad \sum_{k=1}^{n} c_k$$

mean the same thing, since each means $c_1 + c_2 + \cdots + c_n$.

Of course, no harm is done by inserting the variable of integration, and in many cases it is actually convenient to do so.

We shall now investigate the existence of the integral (7). Without saying so every time, f will be assumed real and bounded, and α monotonically increasing on $[a, b]$; and, when there can be no misunderstanding, we shall write $\int$ in place of $\int_a^b$.

6.3 Definition We say that the partition P^* is a *refinement* of P if $P^* \supset P$ (that is, if every point of P is a point of P^*). Given two partitions, P_1 and P_2, we say that P^* is their *common refinement* if $P^* = P_1 \cup P_2$.

6.4 Theorem *If P^* is a refinement of P, then*

(9) $$L(P, f, \alpha) \leq L(P^*, f, \alpha)$$

and

(10) $$U(P^*, f, \alpha) \leq U(P, f, \alpha).$$

Proof To prove (9), suppose first that P^* contains just one point more than P. Let this extra point be x^*, and suppose $x_{i-1} < x^* < x_i$, where x_{i-1} and x_i are two consecutive points of P. Put

$$w_1 = \inf f(x) \qquad (x_{i-1} \leq x \leq x^*),$$
$$w_2 = \inf f(x) \qquad (x^* \leq x \leq x_i).$$

Clearly $w_1 \geq m_i$ and $w_2 \geq m_i$, where, as before,

$$m_i = \inf f(x) \qquad (x_{i-1} \leq x \leq x_i).$$

Hence

$$L(P^*, f, \alpha) - L(P, f, \alpha)$$
$$= w_1[\alpha(x^*) - \alpha(x_{i-1})] + w_2[\alpha(x_i) - \alpha(x^*)] - m_i[\alpha(x_i) - \alpha(x_{i-1})]$$
$$= (w_1 - m_i)[\alpha(x^*) - \alpha(x_{i-1})] + (w_2 - m_i)[\alpha(x_i) - \alpha(x^*)] \geq 0.$$

If P^* contains k points more than P, we repeat this reasoning k times, and arrive at (9). The proof of (10) is analogous.

6.5 Theorem $\underline{\int_a^b} f\,d\alpha \le \overline{\int_a^b} f\,d\alpha.$

Proof Let P^* be the common refinement of two partitions P_1 and P_2. By Theorem 6.4,

$$L(P_1, f, \alpha) \le L(P^*, f, \alpha) \le U(P^*, f, \alpha) \le U(P_2, f, \alpha).$$

Hence

(11) $$L(P_1, f, \alpha) \le U(P_2, f, \alpha).$$

If P_2 is fixed and the sup is taken over all P_1, (11) gives

(12) $$\underline{\int} f\,d\alpha \le U(P_2, f, \alpha).$$

The theorem follows by taking the inf over all P_2 in (12).

6.6 Theorem $f \in \mathscr{R}(\alpha)$ on $[a, b]$ if and only if for every $\varepsilon > 0$ there exists a partition P such that

(13) $$U(P, f, \alpha) - L(P, f, \alpha) < \varepsilon.$$

Proof For every P we have

$$L(P, f, \alpha) \le \underline{\int} f\,d\alpha \le \overline{\int} f\,d\alpha \le U(P, f, \alpha).$$

Thus (13) implies

$$0 \le \overline{\int} f\,d\alpha - \underline{\int} f\,d\alpha < \varepsilon.$$

Hence, if (13) can be satisfied for every $\varepsilon > 0$, we have

$$\overline{\int} f\,d\alpha = \underline{\int} f\,d\alpha,$$

that is, $f \in \mathscr{R}(\alpha)$.

Conversely, suppose $f \in \mathscr{R}(\alpha)$, and let $\varepsilon > 0$ be given. Then there exist partitions P_1 and P_2 such that

(14) $$U(P_2, f, \alpha) - \int f\,d\alpha < \frac{\varepsilon}{2},$$

(15) $$\int f\,d\alpha - L(P_1, f, \alpha) < \frac{\varepsilon}{2}.$$

We choose P to be the common refinement of P_1 and P_2. Then Theorem 6.4, together with (14) and (15), shows that

$$U(P, f, \alpha) \leq U(P_2, f, \alpha) < \int f\, d\alpha + \frac{\varepsilon}{2} < L(P_1, f, \alpha) + \varepsilon \leq L(P, f, \alpha) + \varepsilon,$$

so that (13) holds for this partition P.

Theorem 6.6 furnishes a convenient criterion for integrability. Before we apply it, we state some closely related facts.

6.7 Theorem
(a) *If (13) holds for some P and some ε, then (13) holds (with the same ε) for every refinement of P.*
(b) *If (13) holds for $P = \{x_0, \ldots, x_n\}$ and if s_i, t_i are arbitrary points in $[x_{i-1}, x_i]$, then*

$$\sum_{i=1}^{n} |f(s_i) - f(t_i)|\, \Delta\alpha_i < \varepsilon.$$

(c) *If $f \in \mathscr{R}(\alpha)$ and the hypotheses of (b) hold, then*

$$\left| \sum_{i=1}^{n} f(t_i)\, \Delta\alpha_i - \int_a^b f\, d\alpha \right| < \varepsilon.$$

Proof Theorem 6.4 implies (a). Under the assumptions made in (b), both $f(s_i)$ and $f(t_i)$ lie in $[m_i, M_i]$, so that $|f(s_i) - f(t_i)| \leq M_i - m_i$. Thus

$$\sum_{i=1}^{n} |f(s_i) - f(t_i)|\, \Delta\alpha_i \leq U(P, f, \alpha) - L(P, f, \alpha),$$

which proves (b). The obvious inequalities

$$L(P, f, \alpha) \leq \sum f(t_i)\, \Delta\alpha_i \leq U(P, f, \alpha)$$

and

$$L(P, f, \alpha) \leq \int f\, d\alpha \leq U(P, f, \alpha)$$

prove (c).

6.8 Theorem *If f is continuous on $[a, b]$ then $f \in \mathscr{R}(\alpha)$ on $[a, b]$.*

Proof Let $\varepsilon > 0$ be given. Choose $\eta > 0$ so that

$$[\alpha(b) - \alpha(a)]\eta < \varepsilon.$$

Since f is uniformly continuous on $[a, b]$ (Theorem 4.19), there exists a $\delta > 0$ such that

(16) $$|f(x) - f(t)| < \eta$$

if $x \in [a, b]$, $t \in [a, b]$, and $|x - t| < \delta$.

If P is any partition of $[a, b]$ such that $\Delta x_i < \delta$ for all i, then (16) implies that

(17) $$M_i - m_i \leq \eta \quad (i = 1, \ldots, n)$$

and therefore

$$U(P, f, \alpha) - L(P, f, \alpha) = \sum_{i=1}^{n} (M_i - m_i) \Delta \alpha_i$$

$$\leq \eta \sum_{i=1}^{n} \Delta \alpha_i = \eta[\alpha(b) - \alpha(a)] < \varepsilon.$$

By Theorem 6.6, $f \in \mathscr{R}(\alpha)$.

6.9 Theorem *If f is monotonic on $[a, b]$, and if α is continuous on $[a, b]$, then $f \in \mathscr{R}(\alpha)$. (We still assume, of course, that α is monotonic.)*

Proof Let $\varepsilon > 0$ be given. For any positive integer n, choose a partition such that

$$\Delta \alpha_i = \frac{\alpha(b) - \alpha(a)}{n} \quad (i = 1, \ldots, n).$$

This is possible since α is continuous (Theorem 4.23).

We suppose that f is monotonically increasing (the proof is analogous in the other case). Then

$$M_i = f(x_i), \quad m_i = f(x_{i-1}) \quad (i = 1, \ldots, n),$$

so that

$$U(P, f, \alpha) - L(P, f, \alpha) = \frac{\alpha(b) - \alpha(a)}{n} \sum_{i=1}^{n} [f(x_i) - f(x_{i-1})]$$

$$= \frac{\alpha(b) - \alpha(a)}{n} \cdot [f(b) - f(a)] < \varepsilon$$

if n is taken large enough. By Theorem 6.6, $f \in \mathscr{R}(\alpha)$.

6.10 Theorem *Suppose f is bounded on $[a, b]$, f has only finitely many points of discontinuity on $[a, b]$, and α is continuous at every point at which f is discontinuous. Then $f \in \mathscr{R}(\alpha)$.*

Proof Let $\varepsilon > 0$ be given. Put $M = \sup |f(x)|$, let E be the set of points at which f is discontinuous. Since E is finite and α is continuous at every point of E, we can cover E by finitely many disjoint intervals $[u_j, v_j] \subset [a, b]$ such that the sum of the corresponding differences $\alpha(v_j) - \alpha(u_j)$ is less than ε. Furthermore, we can place these intervals in such a way that every point of $E \cap (a, b)$ lies in the interior of some $[u_j, v_j]$.

Remove the segments (u_j, v_j) from $[a, b]$. The remaining set K is compact. Hence f is uniformly continuous on K, and there exists $\delta > 0$ such that $|f(s) - f(t)| < \varepsilon$ if $s \in K$, $t \in K$, $|s - t| < \delta$.

Now form a partition $P = \{x_0, x_1, \ldots, x_n\}$ of $[a, b]$, as follows: Each u_j occurs in P. Each v_j occurs in P. No point of any segment (u_j, v_j) occurs in P. If x_{i-1} is not one of the u_j, then $\Delta x_i < \delta$.

Note that $M_i - m_i \leq 2M$ for every i, and that $M_i - m_i \leq \varepsilon$ unless x_{i-1} is one of the u_j. Hence, as in the proof of Theorem 6.8,

$$U(P, f, \alpha) - L(P, f, \alpha) \leq [\alpha(b) - \alpha(a)]\varepsilon + 2M\varepsilon.$$

Since ε is arbitrary, Theorem 6.6 shows that $f \in \mathcal{R}(\alpha)$.

Note: If f and α have a common point of discontinuity, then f need not be in $\mathcal{R}(\alpha)$. Exercise 3 shows this.

6.11 Theorem *Suppose $f \in \mathcal{R}(\alpha)$ on $[a, b]$, $m \leq f \leq M$, ϕ is continuous on $[m, M]$, and $h(x) = \phi(f(x))$ on $[a, b]$. Then $h \in \mathcal{R}(\alpha)$ on $[a, b]$.*

Proof Choose $\varepsilon > 0$. Since ϕ is uniformly continuous on $[m, M]$, there exists $\delta > 0$ such that $\delta < \varepsilon$ and $|\phi(s) - \phi(t)| < \varepsilon$ if $|s - t| \leq \delta$ and $s, t \in [m, M]$.

Since $f \in \mathcal{R}(\alpha)$, there is a partition $P = \{x_0, x_1, \ldots, x_n\}$ of $[a, b]$ such that

(18) $$U(P, f, \alpha) - L(P, f, \alpha) < \delta^2.$$

Let M_i, m_i have the same meaning as in Definition 6.1, and let M_i^*, m_i^* be the analogous numbers for h. Divide the numbers $1, \ldots, n$ into two classes: $i \in A$ if $M_i - m_i < \delta$, $i \in B$ if $M_i - m_i \geq \delta$.

For $i \in A$, our choice of δ shows that $M_i^* - m_i^* \leq \varepsilon$.

For $i \in B$, $M_i^* - m_i^* \leq 2K$, where $K = \sup|\phi(t)|$, $m \leq t \leq M$. By (18), we have

(19) $$\delta \sum_{i \in B} \Delta\alpha_i \leq \sum_{i \in B} (M_i - m_i) \Delta\alpha_i < \delta^2$$

so that $\sum_{i \in B} \Delta\alpha_i < \delta$. It follows that

$$U(P, h, \alpha) - L(P, h, \alpha) = \sum_{i \in A} (M_i^* - m_i^*) \Delta\alpha_i + \sum_{i \in B} (M_i^* - m_i^*) \Delta\alpha_i$$

$$\leq \varepsilon[\alpha(b) - \alpha(a)] + 2K\delta < \varepsilon[\alpha(b) - \alpha(a) + 2K].$$

Since ε was arbitrary, Theorem 6.6 implies that $h \in \mathcal{R}(\alpha)$.

Remark: This theorem suggests the question: Just what functions are Riemann-integrable? The answer is given by Theorem 11.33(b).

PROPERTIES OF THE INTEGRAL

6.12 Theorem

(a) If $f_1 \in \mathscr{R}(\alpha)$ and $f_2 \in \mathscr{R}(\alpha)$ on $[a, b]$, then
$$f_1 + f_2 \in \mathscr{R}(\alpha),$$
$cf \in \mathscr{R}(\alpha)$ for every constant c, and
$$\int_a^b (f_1 + f_2) \, d\alpha = \int_a^b f_1 \, d\alpha + \int_a^b f_2 \, d\alpha,$$
$$\int_a^b cf \, d\alpha = c \int_a^b f \, d\alpha.$$

(b) If $f_1(x) \leq f_2(x)$ on $[a, b]$, then
$$\int_a^b f_1 \, d\alpha \leq \int_a^b f_2 \, d\alpha.$$

(c) If $f \in \mathscr{R}(\alpha)$ on $[a, b]$ and if $a < c < b$, then $f \in \mathscr{R}(\alpha)$ on $[a, c]$ and on $[c, b]$, and
$$\int_a^c f \, d\alpha + \int_c^b f \, d\alpha = \int_a^b f \, d\alpha.$$

(d) If $f \in \mathscr{R}(\alpha)$ on $[a, b]$ and if $|f(x)| \leq M$ on $[a, b]$, then
$$\left| \int_a^b f \, d\alpha \right| \leq M[\alpha(b) - \alpha(a)].$$

(e) If $f \in \mathscr{R}(\alpha_1)$ and $f \in \mathscr{R}(\alpha_2)$, then $f \in \mathscr{R}(\alpha_1 + \alpha_2)$ and
$$\int_a^b f \, d(\alpha_1 + \alpha_2) = \int_a^b f \, d\alpha_1 + \int_a^b f \, d\alpha_2;$$
if $f \in \mathscr{R}(\alpha)$ and c is a positive constant, then $f \in \mathscr{R}(c\alpha)$ and
$$\int_a^b f \, d(c\alpha) = c \int_a^b f \, d\alpha.$$

Proof If $f = f_1 + f_2$ and P is any partition of $[a, b]$, we have

(20) $\quad L(P, f_1, \alpha) + L(P, f_2, \alpha) \leq L(P, f, \alpha)$
$$\leq U(P, f, \alpha) \leq U(P, f_1, \alpha) + U(P, f_2, \alpha).$$

If $f_1 \in \mathscr{R}(\alpha)$ and $f_2 \in \mathscr{R}(\alpha)$, let $\varepsilon > 0$ be given. There are partitions P_j ($j = 1, 2$) such that
$$U(P_j, f_j, \alpha) - L(P_j, f_j, \alpha) < \varepsilon.$$

These inequalities persist if P_1 and P_2 are replaced by their common refinement P. Then (20) implies
$$U(P, f, \alpha) - L(P, f, \alpha) < 2\varepsilon,$$
which proves that $f \in \mathscr{R}(\alpha)$.

With this same P we have
$$U(P, f_j, \alpha) < \int f_j \, d\alpha + \varepsilon \quad (j = 1, 2);$$
hence (20) implies
$$\int f \, d\alpha \leq U(P, f, \alpha) < \int f_1 \, d\alpha + \int f_2 \, d\alpha + 2\varepsilon.$$

Since ε was arbitrary, we conclude that

(21)
$$\int f \, d\alpha \leq \int f_1 \, d\alpha + \int f_2 \, d\alpha.$$

If we replace f_1 and f_2 in (21) by $-f_1$ and $-f_2$, the inequality is reversed, and the equality is proved.

The proofs of the other assertions of Theorem 6.12 are so similar that we omit the details. In part (c) the point is that (by passing to refinements) we may restrict ourselves to partitions which contain the point c, in approximating $\int f \, d\alpha$.

6.13 Theorem *If $f \in \mathscr{R}(\alpha)$ and $g \in \mathscr{R}(\alpha)$ on $[a, b]$, then*

(a) $fg \in \mathscr{R}(\alpha)$;

(b) $|f| \in \mathscr{R}(\alpha)$ and $\left| \int_a^b f \, d\alpha \right| \leq \int_a^b |f| \, d\alpha.$

Proof If we take $\phi(t) = t^2$, Theorem 6.11 shows that $f^2 \in \mathscr{R}(\alpha)$ if $f \in \mathscr{R}(\alpha)$. The identity
$$4fg = (f + g)^2 - (f - g)^2$$
completes the proof of (a).

If we take $\phi(t) = |t|$, Theorem 6.11 shows similarly that $|f| \in \mathscr{R}(\alpha)$. Choose $c = \pm 1$, so that
$$c \int f \, d\alpha \geq 0.$$
Then
$$\left| \int f \, d\alpha \right| = c \int f \, d\alpha = \int cf \, d\alpha \leq \int |f| \, d\alpha,$$
since $cf \leq |f|$.

6.14 Definition The *unit step function* I is defined by
$$I(x) = \begin{cases} 0 & (x \leq 0), \\ 1 & (x > 0). \end{cases}$$

6.15 Theorem *If $a < s < b$, f is bounded on $[a, b]$, f is continuous at s, and $\alpha(x) = I(x - s)$, then*

$$\int_a^b f\,d\alpha = f(s).$$

Proof Consider partitions $P = \{x_0, x_1, x_2, x_3\}$, where $x_0 = a$, and $x_1 = s < x_2 < x_3 = b$. Then

$$U(P, f, \alpha) = M_2, \qquad L(P, f, \alpha) = m_2.$$

Since f is continuous at s, we see that M_2 and m_2 converge to $f(s)$ as $x_2 \to s$.

6.16 Theorem *Suppose $c_n \geq 0$ for $1, 2, 3, \ldots$, Σc_n converges, $\{s_n\}$ is a sequence of distinct points in (a, b), and*

(22) $$\alpha(x) = \sum_{n=1}^{\infty} c_n I(x - s_n).$$

Let f be continuous on $[a, b]$. Then

(23) $$\int_a^b f\,d\alpha = \sum_{n=1}^{\infty} c_n f(s_n).$$

Proof The comparison test shows that the series (22) converges for every x. Its sum $\alpha(x)$ is evidently monotonic, and $\alpha(a) = 0$, $\alpha(b) = \Sigma c_n$. (This is the type of function that occurred in Remark 4.31.)

Let $\varepsilon > 0$ be given, and choose N so that

$$\sum_{N+1}^{\infty} c_n < \varepsilon.$$

Put

$$\alpha_1(x) = \sum_{n=1}^{N} c_n I(x - s_n), \qquad \alpha_2(x) = \sum_{N+1}^{\infty} c_n I(x - s_n).$$

By Theorems 6.12 and 6.15,

(24) $$\int_a^b f\,d\alpha_1 = \sum_{i=1}^{N} c_n f(s_n).$$

Since $\alpha_2(b) - \alpha_2(a) < \varepsilon$,

(25) $$\left| \int_a^b f\,d\alpha_2 \right| \leq M\varepsilon,$$

where $M = \sup|f(x)|$. Since $\alpha = \alpha_1 + \alpha_2$, it follows from (24) and (25) that

(26) $$\left|\int_a^b f\, d\alpha - \sum_{i=1}^N c_n f(s_n)\right| \le M\varepsilon.$$

If we let $N \to \infty$, we obtain (23).

6.17 Theorem *Assume α increases monotonically and $\alpha' \in \mathscr{R}$ on $[a, b]$. Let f be a bounded real function on $[a, b]$.*

Then $f \in \mathscr{R}(\alpha)$ if and only if $f\alpha' \in \mathscr{R}$. In that case

(27) $$\int_a^b f\, d\alpha = \int_a^b f(x)\alpha'(x)\, dx.$$

Proof Let $\varepsilon > 0$ be given and apply Theorem 6.6 to α': There is a partition $P = \{x_0, \ldots, x_n\}$ of $[a, b]$ such that

(28) $$U(P, \alpha') - L(P, \alpha') < \varepsilon.$$

The mean value theorem furnishes points $t_i \in [x_{i-1}, x_i]$ such that

$$\Delta\alpha_i = \alpha'(t_i)\, \Delta x_i$$

for $i = 1, \ldots, n$. If $s_i \in [x_{i-1}, x_i]$, then

(29) $$\sum_{i=1}^n |\alpha'(s_i) - \alpha'(t_i)|\, \Delta x_i < \varepsilon,$$

by (28) and Theorem 6.7(b). Put $M = \sup|f(x)|$. Since

$$\sum_{i=1}^n f(s_i)\, \Delta\alpha_i = \sum_{i=1}^n f(s_i)\alpha'(t_i)\, \Delta x_i$$

it follows from (29) that

(30) $$\left|\sum_{i=1}^n f(s_i)\, \Delta\alpha_i - \sum_{i=1}^n f(s_i)\alpha'(s_i)\, \Delta x_i\right| \le M\varepsilon.$$

In particular,

$$\sum_{i=1}^n f(s_i)\, \Delta\alpha_i \le U(P, f\alpha') + M\varepsilon,$$

for all choices of $s_i \in [x_{i-1}, x_i]$, so that

$$U(P, f, \alpha) \le U(P, f\alpha') + M\varepsilon.$$

The same argument leads from (30) to

$$U(P, f\alpha') \le U(P, f, \alpha) + M\varepsilon.$$

Thus

(31) $$|U(P, f, \alpha) - U(P, f\alpha')| \le M\varepsilon.$$

Now note that (28) remains true if P is replaced by any refinement. Hence (31) also remains true. We conclude that

$$\left| \int_a^{\overline{b}} f\, d\alpha - \int_a^{\overline{b}} f(x)\alpha'(x)\, dx \right| \le M\varepsilon.$$

But ε is arbitrary. Hence

(32) $$\int_a^{\overline{b}} f\, d\alpha = \int_a^{\overline{b}} f(x)\alpha'(x)\, dx,$$

for *any* bounded f. The equality of the lower integrals follows from (30) in exactly the same way. The theorem follows.

6.18 Remark The two preceding theorems illustrate the generality and flexibility which are inherent in the Stieltjes process of integration. If α is a pure step function [this is the name often given to functions of the form (22)], the integral reduces to a finite or infinite series. If α has an integrable derivative, the integral reduces to an ordinary Riemann integral. This makes it possible in many cases to study series and integrals simultaneously, rather than separately.

To illustrate this point, consider a physical example. The moment of inertia of a straight wire of unit length, about an axis through an endpoint, at right angles to the wire, is

(33) $$\int_0^1 x^2\, dm$$

where $m(x)$ is the mass contained in the interval $[0, x]$. If the wire is regarded as having a continuous density ρ, that is, if $m'(x) = \rho(x)$, then (33) turns into

(34) $$\int_0^1 x^2\, \rho(x)\, dx.$$

On the other hand, if the wire is composed of masses m_i concentrated at points x_i, (33) becomes

(35) $$\sum_i x_i^2\, m_i.$$

Thus (33) contains (34) and (35) as special cases, but it contains much more; for instance, the case in which m is continuous but not everywhere differentiable.

6.19 Theorem (change of variable) *Suppose φ is a strictly increasing continuous function that maps an interval $[A, B]$ onto $[a, b]$. Suppose α is monotonically increasing on $[a, b]$ and $f \in \mathscr{R}(\alpha)$ on $[a, b]$. Define β and g on $[A, B]$ by*

(36) $$\beta(y) = \alpha(\varphi(y)), \qquad g(y) = f(\varphi(y)).$$

Then $g \in \mathscr{R}(\beta)$ and

(37)
$$\int_A^B g \, d\beta = \int_a^b f \, d\alpha.$$

Proof To each partition $P = \{x_0, \ldots, x_n\}$ of $[a, b]$ corresponds a partition $Q = \{y_0, \ldots, y_n\}$ of $[A, B]$, so that $x_i = \varphi(y_i)$. All partitions of $[A, B]$ are obtained in this way. Since the values taken by f on $[x_{i-1}, x_i]$ are exactly the same as those taken by g on $[y_{i-1}, y_i]$, we see that

(38)
$$U(Q, g, \beta) = U(P, f, \alpha), \quad L(Q, g, \beta) = L(P, f, \alpha).$$

Since $f \in \mathscr{R}(\alpha)$, P can be chosen so that both $U(P, f, \alpha)$ and $L(P, f, \alpha)$ are close to $\int f \, d\alpha$. Hence (38), combined with Theorem 6.6, shows that $g \in \mathscr{R}(\beta)$ and that (37) holds. This completes the proof.

Let us note the following special case:
Take $\alpha(x) = x$. Then $\beta = \varphi$. Assume $\varphi' \in \mathscr{R}$ on $[A, B]$. If Theorem 6.17 is applied to the left side of (37), we obtain

(39)
$$\int_a^b f(x) \, dx = \int_A^B f(\varphi(y))\varphi'(y) \, dy.$$

INTEGRATION AND DIFFERENTIATION

We still confine ourselves to real functions in this section. We shall show that integration and differentiation are, in a certain sense, inverse operations.

6.20 Theorem *Let $f \in \mathscr{R}$ on $[a, b]$. For $a \leq x \leq b$, put*

$$F(x) = \int_a^x f(t) \, dt.$$

Then F is continuous on $[a, b]$; furthermore, if f is continuous at a point x_0 of $[a, b]$, then F is differentiable at x_0, and

$$F'(x_0) = f(x_0).$$

Proof Since $f \in \mathscr{R}$, f is bounded. Suppose $|f(t)| \leq M$ for $a \leq t \leq b$. If $a \leq x < y \leq b$, then

$$|F(y) - F(x)| = \left| \int_x^y f(t) \, dt \right| \leq M(y - x),$$

by Theorem 6.12(c) and (d). Given $\varepsilon > 0$, we see that

$$|F(y) - F(x)| < \varepsilon,$$

provided that $|y - x| < \varepsilon/M$. This proves continuity (and, in fact, uniform continuity) of F.

Now suppose f is continuous at x_0. Given $\varepsilon > 0$, choose $\delta > 0$ such that
$$|f(t) - f(x_0)| < \varepsilon$$
if $|t - x_0| < \delta$, and $a \le t \le b$. Hence, if
$$x_0 - \delta < s \le x_0 \le t < x_0 + \delta \quad \text{and} \quad a \le s < t \le b,$$
we have, by Theorem 6.12(d),
$$\left| \frac{F(t) - F(s)}{t - s} - f(x_0) \right| = \left| \frac{1}{t - s} \int_s^t [f(u) - f(x_0)] \, du \right| < \varepsilon.$$
It follows that $F'(x_0) = f(x_0)$.

6.21 The fundamental theorem of calculus *If $f \in \mathscr{R}$ on $[a, b]$ and if there is a differentiable function F on $[a, b]$ such that $F' = f$, then*
$$\int_a^b f(x) \, dx = F(b) - F(a).$$

Proof Let $\varepsilon > 0$ be given. Choose a partition $P = \{x_0, \ldots, x_n\}$ of $[a, b]$ so that $U(P, f) - L(P, f) < \varepsilon$. The mean value theorem furnishes points $t_i \in [x_{i-1}, x_i]$ such that
$$F(x_i) - F(x_{i-1}) = f(t_i) \Delta x_i$$
for $i = 1, \ldots, n$. Thus
$$\sum_{i=1}^n f(t_i) \Delta x_i = F(b) - F(a).$$
It now follows from Theorem 6.7(c) that
$$\left| F(b) - F(a) - \int_a^b f(x) \, dx \right| < \varepsilon.$$
Since this holds for every $\varepsilon > 0$, the proof is complete.

6.22 Theorem (integration by parts) *Suppose F and G are differentiable functions on $[a, b]$, $F' = f \in \mathscr{R}$, and $G' = g \in \mathscr{R}$. Then*
$$\int_a^b F(x)g(x) \, dx = F(b)G(b) - F(a)G(a) - \int_a^b f(x)G(x) \, dx.$$

Proof Put $H(x) = F(x)G(x)$ and apply Theorem 6.21 to H and its derivative. Note that $H' \in \mathscr{R}$, by Theorem 6.13.

INTEGRATION OF VECTOR-VALUED FUNCTIONS

6.23 Definition Let $f_1, \ldots, f_k$ be real functions on $[a, b]$, and let $\mathbf{f} = (f_1, \ldots, f_k)$ be the corresponding mapping of $[a, b]$ into R^k. If α increases monotonically on $[a, b]$, to say that $\mathbf{f} \in \mathscr{R}(\alpha)$ means that $f_j \in \mathscr{R}(\alpha)$ for $j = 1, \ldots, k$. If this is the case, we define

$$\int_a^b \mathbf{f}\, d\alpha = \left(\int_a^b f_1\, d\alpha, \ldots, \int_a^b f_k\, d\alpha \right).$$

In other words, $\int \mathbf{f}\, d\alpha$ is the point in R^k whose jth coordinate is $\int f_j\, d\alpha$.

It is clear that parts (*a*), (*c*), and (*e*) of Theorem 6.12 are valid for these vector-valued integrals; we simply apply the earlier results to each coordinate. The same is true of Theorems 6.17, 6.20, and 6.21. To illustrate, we state the analogue of Theorem 6.21.

6.24 Theorem *If $\mathbf{f}$ and $\mathbf{F}$ map $[a, b]$ into R^k, if $\mathbf{f} \in \mathscr{R}$ on $[a, b]$, and if $\mathbf{F}' = \mathbf{f}$, then*

$$\int_a^b \mathbf{f}(t)\, dt = \mathbf{F}(b) - \mathbf{F}(a).$$

The analogue of Theorem 6.13(*b*) offers some new features, however, at least in its proof.

6.25 Theorem *If $\mathbf{f}$ maps $[a, b]$ into R^k and if $\mathbf{f} \in \mathscr{R}(\alpha)$ for some monotonically increasing function α on $[a, b]$, then $|\mathbf{f}| \in \mathscr{R}(\alpha)$, and*

(40) $$\left| \int_a^b \mathbf{f}\, d\alpha \right| \leq \int_a^b |\mathbf{f}|\, d\alpha.$$

Proof If $f_1, \ldots, f_k$ are the components of $\mathbf{f}$, then

(41) $$|\mathbf{f}| = (f_1^2 + \cdots + f_k^2)^{1/2}.$$

By Theorem 6.11, each of the functions f_i^2 belongs to $\mathscr{R}(\alpha)$; hence so does their sum. Since x^2 is a continuous function of x, Theorem 4.17 shows that the square-root function is continuous on $[0, M]$, for every real M. If we apply Theorem 6.11 once more, (41) shows that $|\mathbf{f}| \in \mathscr{R}(\alpha)$.

To prove (40), put $\mathbf{y} = (y_1, \ldots, y_k)$, where $y_j = \int f_j\, d\alpha$. Then we have $\mathbf{y} = \int \mathbf{f}\, d\alpha$, and

$$|\mathbf{y}|^2 = \sum y_i^2 = \sum y_j \int f_j\, d\alpha = \int \left(\sum y_j f_j \right) d\alpha.$$

By the Schwarz inequality,

(42) $$\sum y_j f_j(t) \leq |\mathbf{y}|\, |\mathbf{f}(t)| \qquad (a \leq t \leq b);$$

hence Theorem 6.12(b) implies

(43) $$|\mathbf{y}|^2 \le |\mathbf{y}| \int |\mathbf{f}|\, d\alpha.$$

If $\mathbf{y} = \mathbf{0}$, (40) is trivial. If $\mathbf{y} \ne \mathbf{0}$, division of (43) by $|\mathbf{y}|$ gives (40).

RECTIFIABLE CURVES

We conclude this chapter with a topic of geometric interest which provides an application of some of the preceding theory. The case $k = 2$ (i.e., the case of plane curves) is of considerable importance in the study of analytic functions of a complex variable.

6.26 Definition A continuous mapping γ of an interval $[a, b]$ into R^k is called a *curve* in R^k. To emphasize the parameter interval $[a, b]$, we may also say that γ is a curve on $[a, b]$.

If γ is one-to-one, γ is called an *arc*.
If $\gamma(a) = \gamma(b)$, γ is said to be a *closed curve*.

It should be noted that we define a curve to be a *mapping*, not a point set. Of course, with each curve γ in R^k there is associated a subset of R^k, namely the range of γ, but different curves may have the same range.

We associate to each partition $P = \{x_0, \ldots, x_n\}$ of $[a, b]$ and to each curve γ on $[a, b]$ the number

$$\Lambda(P, \gamma) = \sum_{i=1}^{n} |\gamma(x_i) - \gamma(x_{i-1})|.$$

The ith term in this sum is the distance (in R^k) between the points $\gamma(x_{i-1})$ and $\gamma(x_i)$. Hence $\Lambda(P, \gamma)$ is the length of a polygonal path with vertices at $\gamma(x_0)$, $\gamma(x_1), \ldots, \gamma(x_n)$, in this order. As our partition becomes finer and finer, this polygon approaches the range of γ more and more closely. This makes it seem reasonable to define the *length* of γ as

$$\Lambda(\gamma) = \sup \Lambda(P, \gamma),$$

where the supremum is taken over all partitions of $[a, b]$.
If $\Lambda(\gamma) < \infty$, we say that γ is *rectifiable*.
In certain cases, $\Lambda(\gamma)$ is given by a Riemann integral. We shall prove this for *continuously differentiable* curves, i.e., for curves γ whose derivative γ' is continuous.

6.27 Theorem *If γ' is continuous on $[a, b]$, then γ is rectifiable, and*
$$\Lambda(\gamma) = \int_a^b |\gamma'(t)|\, dt.$$

Proof If $a \leq x_{i-1} < x_i \leq b$, then
$$|\gamma(x_i) - \gamma(x_{i-1})| = \left| \int_{x_{i-1}}^{x_i} \gamma'(t)\, dt \right| \leq \int_{x_{i-1}}^{x_i} |\gamma'(t)|\, dt.$$

Hence
$$\Lambda(P, \gamma) \leq \int_a^b |\gamma'(t)|\, dt$$

for every partition P of $[a, b]$. Consequently,
$$\Lambda(\gamma) \leq \int_a^b |\gamma'(t)|\, dt.$$

To prove the opposite inequality, let $\varepsilon > 0$ be given. Since γ' is uniformly continuous on $[a, b]$, there exists $\delta > 0$ such that
$$|\gamma'(s) - \gamma'(t)| < \varepsilon \quad \text{if } |s - t| < \delta.$$

Let $P = \{x_0, \ldots, x_n\}$ be a partition of $[a, b]$, with $\Delta x_i < \delta$ for all i. If $x_{i-1} \leq t \leq x_i$, it follows that
$$|\gamma'(t)| \leq |\gamma'(x_i)| + \varepsilon.$$

Hence
$$\int_{x_{i-1}}^{x_i} |\gamma'(t)|\, dt \leq |\gamma'(x_i)|\, \Delta x_i + \varepsilon\, \Delta x_i$$
$$= \left| \int_{x_{i-1}}^{x_i} [\gamma'(t) + \gamma'(x_i) - \gamma'(t)]\, dt \right| + \varepsilon\, \Delta x_i$$
$$\leq \left| \int_{x_{i-1}}^{x_i} \gamma'(t)\, dt \right| + \left| \int_{x_{i-1}}^{x_i} [\gamma'(x_i) - \gamma'(t)]\, dt \right| + \varepsilon\, \Delta x_i$$
$$\leq |\gamma(x_i) - \gamma(x_{i-1})| + 2\varepsilon\, \Delta x_i.$$

If we add these inequalities, we obtain
$$\int_a^b |\gamma'(t)|\, dt \leq \Lambda(P, \gamma) + 2\varepsilon(b - a)$$
$$\leq \Lambda(\gamma) + 2\varepsilon(b - a).$$

Since ε was arbitrary,
$$\int_a^b |\gamma'(t)|\, dt \leq \Lambda(\gamma).$$

This completes the proof.

EXERCISES

1. Suppose α increases on $[a, b]$, $a \leq x_0 \leq b$, α is continuous at x_0, $f(x_0) = 1$, and $f(x) = 0$ if $x \neq x_0$. Prove that $f \in \mathscr{R}(\alpha)$ and that $\int f \, d\alpha = 0$.

2. Suppose $f \geq 0$, f is continuous on $[a, b]$, and $\int_a^b f(x) \, dx = 0$. Prove that $f(x) = 0$ for all $x \in [a, b]$. (Compare this with Exercise 1.)

3. Define three functions $\beta_1, \beta_2, \beta_3$ as follows: $\beta_j(x) = 0$ if $x < 0$, $\beta_j(x) = 1$ if $x > 0$ for $j = 1, 2, 3$; and $\beta_1(0) = 0$, $\beta_2(0) = 1$, $\beta_3(0) = \frac{1}{2}$. Let f be a bounded function on $[-1, 1]$.

    (a) Prove that $f \in \mathscr{R}(\beta_1)$ if and only if $f(0+) = f(0)$ and that then
    $$\int f \, d\beta_1 = f(0).$$

    (b) State and prove a similar result for β_2.

    (c) Prove that $f \in \mathscr{R}(\beta_3)$ if and only if f is continuous at 0.

    (d) If f is continuous at 0 prove that
    $$\int f \, d\beta_1 = \int f \, d\beta_2 = \int f \, d\beta_3 = f(0).$$

4. If $f(x) = 0$ for all irrational x, $f(x) = 1$ for all rational x, prove that $f \notin \mathscr{R}$ on $[a, b]$ for any $a < b$.

5. Suppose f is a bounded real function on $[a, b]$, and $f^2 \in \mathscr{R}$ on $[a, b]$. Does it follow that $f \in \mathscr{R}$? Does the answer change if we assume that $f^3 \in \mathscr{R}$?

6. Let P be the Cantor set constructed in Sec. 2.44. Let f be a bounded real function on $[0, 1]$ which is continuous at every point outside P. Prove that $f \in \mathscr{R}$ on $[0, 1]$. *Hint:* P can be covered by finitely many segments whose total length can be made as small as desired. Proceed as in Theorem 6.10.

7. Suppose f is a real function on $(0, 1]$ and $f \in \mathscr{R}$ on $[c, 1]$ for every $c > 0$. Define
    $$\int_0^1 f(x) \, dx = \lim_{c \to 0} \int_c^1 f(x) \, dx$$
    if this limit exists (and is finite).

    (a) If $f \in \mathscr{R}$ on $[0, 1]$, show that this definition of the integral agrees with the old one.

    (b) Construct a function f such that the above limit exists, although it fails to exist with $|f|$ in place of f.

8. Suppose $f \in \mathscr{R}$ on $[a, b]$ for every $b > a$ where a is fixed. Define
    $$\int_a^\infty f(x) \, dx = \lim_{b \to \infty} \int_a^b f(x) \, dx$$
    if this limit exists (and is finite). In that case, we say that the integral on the left *converges*. If it also converges after f has been replaced by $|f|$, it is said to converge *absolutely*.

Assume that $f(x) \geq 0$ and that f decreases monotonically on $[1, \infty)$. Prove that

$$\int_1^\infty f(x)\, dx$$

converges if and only if

$$\sum_{n=1}^\infty f(n)$$

converges. (This is the so-called "integral test" for convergence of series.)

9. Show that integration by parts can sometimes be applied to the "improper" integrals defined in Exercises 7 and 8. (State appropriate hypotheses, formulate a theorem, and prove it.) For instance show that

$$\int_0^\infty \frac{\cos x}{1+x}\, dx = \int_0^\infty \frac{\sin x}{(1+x)^2}\, dx.$$

Show that one of these integrals converges *absolutely*, but that the other does not.

10. Let p and q be positive real numbers such that

$$\frac{1}{p} + \frac{1}{q} = 1.$$

Prove the following statements.

(a) If $u \geq 0$ and $v \geq 0$, then

$$uv \leq \frac{u^p}{p} + \frac{v^q}{q}.$$

Equality holds if and only if $u^p = v^q$.

(b) If $f \in \mathscr{R}(\alpha)$, $g \in \mathscr{R}(\alpha)$, $f \geq 0$, $g \geq 0$, and

$$\int_a^b f^p\, d\alpha = 1 = \int_a^b g^q\, d\alpha,$$

then

$$\int_a^b fg\, d\alpha \leq 1.$$

(c) If f and g are complex functions in $\mathscr{R}(\alpha)$, then

$$\left| \int_a^b fg\, d\alpha \right| \leq \left\{ \int_a^b |f|^p\, d\alpha \right\}^{1/p} \left\{ \int_a^b |g|^q\, d\alpha \right\}^{1/q}.$$

This is *Hölder's inequality*. When $p = q = 2$ it is usually called the Schwarz inequality. (Note that Theorem 1.35 is a very special case of this.)

(d) Show that Hölder's inequality is also true for the "improper" integrals described in Exercises 7 and 8.

11. Let α be a fixed increasing function on $[a, b]$. For $u \in \mathscr{R}(\alpha)$, define
$$\|u\|_2 = \left\{ \int_a^b |u|^2 \, d\alpha \right\}^{1/2}.$$
Suppose $f, g, h \in \mathscr{R}(\alpha)$, and prove the triangle inequality
$$\|f - h\|_2 \leq \|f - g\|_2 + \|g - h\|_2$$
as a consequence of the Schwarz inequality, as in the proof of Theorem 1.37.

12. With the notations of Exercise 11, suppose $f \in \mathscr{R}(\alpha)$ and $\varepsilon > 0$. Prove that there exists a continuous function g on $[a, b]$ such that $\|f - g\|_2 < \varepsilon$.

Hint: Let $P = \{x_0, \ldots, x_n\}$ be a suitable partition of $[a, b]$, define
$$g(t) = \frac{x_i - t}{\Delta x_i} f(x_{i-1}) + \frac{t - x_{i-1}}{\Delta x_i} f(x_i)$$
if $x_{i-1} \leq t \leq x_i$.

13. Define
$$f(x) = \int_x^{x+1} \sin(t^2) \, dt.$$

(a) Prove that $|f(x)| < 1/x$ if $x > 0$.

Hint: Put $t^2 = u$ and integrate by parts, to show that $f(x)$ is equal to
$$\frac{\cos(x^2)}{2x} - \frac{\cos[(x+1)^2]}{2(x+1)} - \int_{x^2}^{(x+1)^2} \frac{\cos u}{4u^{3/2}} \, du.$$
Replace $\cos u$ by -1.

(b) Prove that
$$2xf(x) = \cos(x^2) - \cos[(x+1)^2] + r(x)$$
where $|r(x)| < c/x$ and c is a constant.

(c) Find the upper and lower limits of $xf(x)$, as $x \to \infty$.

(d) Does $\int_0^\infty \sin(t^2) \, dt$ converge?

14. Deal similarly with
$$f(x) = \int_x^{x+1} \sin(e^t) \, dt.$$
Show that
$$e^x |f(x)| < 2$$
and that
$$e^x f(x) = \cos(e^x) - e^{-1} \cos(e^{x+1}) + r(x),$$
where $|r(x)| < Ce^{-x}$, for some constant C.

15. Suppose f is a real, continuously differentiable function on $[a, b]$, $f(a) = f(b) = 0$, and
$$\int_a^b f^2(x)\, dx = 1.$$
Prove that
$$\int_a^b xf(x)f'(x)\, dx = -\tfrac{1}{2}$$
and that
$$\int_a^b [f'(x)]^2\, dx \cdot \int_a^b x^2 f^2(x)\, dx > \tfrac{1}{4}.$$

16. For $1 < s < \infty$, define
$$\zeta(s) = \sum_{n=1}^{\infty} \frac{1}{n^s}.$$
(This is Riemann's zeta function, of great importance in the study of the distribution of prime numbers.) Prove that

(a) $\zeta(s) = s \displaystyle\int_1^{\infty} \frac{[x]}{x^{s+1}}\, dx$

and that

(b) $\zeta(s) = \dfrac{s}{s-1} - s \displaystyle\int_1^{\infty} \frac{x - [x]}{x^{s+1}}\, dx,$

where $[x]$ denotes the greatest integer $\leq x$.

Prove that the integral in (b) converges for all $s > 0$.

Hint: To prove (a), compute the difference between the integral over $[1, N]$ and the Nth partial sum of the series that defines $\zeta(s)$.

17. Suppose α increases monotonically on $[a, b]$, g is continuous, and $g(x) = G'(x)$ for $a \leq x \leq b$. Prove that
$$\int_a^b \alpha(x) g(x)\, dx = G(b)\alpha(b) - G(a)\alpha(a) - \int_a^b G\, d\alpha.$$

Hint: Take g real, without loss of generality. Given $P = \{x_0, x_1, \ldots, x_n\}$, choose $t_i \in (x_{i-1}, x_i)$ so that $g(t_i)\Delta x_i = G(x_i) - G(x_{i-1})$. Show that
$$\sum_{i=1}^n \alpha(x_i) g(t_i)\, \Delta x_i = G(b)\alpha(b) - G(a)\alpha(a) - \sum_{i=1}^n G(x_{i-1})\, \Delta \alpha_i.$$

18. Let $\gamma_1, \gamma_2, \gamma_3$ be curves in the complex plane, defined on $[0, 2\pi]$ by
$$\gamma_1(t) = e^{it}, \qquad \gamma_2(t) = e^{2it}, \qquad \gamma_3(t) = e^{2\pi i t \sin(1/t)}.$$
Show that these three curves have the same range, that γ_1 and γ_2 are rectifiable, that the length of γ_1 is 2π, that the length of γ_2 is 4π, and that γ_3 is not rectifiable.

19. Let γ_1 be a curve in R^k, defined on $[a, b]$; let ϕ be a continuous 1-1 mapping of $[c, d]$ onto $[a, b]$, such that $\phi(c) = a$; and define $\gamma_2(s) = \gamma_1(\phi(s))$. Prove that γ_2 is an arc, a closed curve, or a rectifiable curve if and only if the same is true of γ_1. Prove that γ_2 and γ_1 have the same length.

7
SEQUENCES AND SERIES OF FUNCTIONS

In the present chapter we confine our attention to complex-valued functions (including the real-valued ones, of course), although many of the theorems and proofs which follow extend without difficulty to vector-valued functions, and even to mappings into general metric spaces. We choose to stay within this simple framework in order to focus attention on the most important aspects of the problems that arise when limit processes are interchanged.

DISCUSSION OF MAIN PROBLEM

7.1 Definition Suppose $\{f_n\}$, $n = 1, 2, 3, \ldots$, is a sequence of functions defined on a set E, and suppose that the sequence of numbers $\{f_n(x)\}$ converges for every $x \in E$. We can then define a function f by

(1) $$f(x) = \lim_{n \to \infty} f_n(x) \qquad (x \in E).$$

Under these circumstances we say that $\{f_n\}$ converges on E and that f is the *limit*, or the *limit function*, of $\{f_n\}$. Sometimes we shall use a more descriptive terminology and shall say that "$\{f_n\}$ converges to f pointwise on E" if (1) holds. Similarly, if $\Sigma f_n(x)$ converges for every $x \in E$, and if we define

$$(2) \qquad f(x) = \sum_{n=1}^{\infty} f_n(x) \qquad (x \in E),$$

the function f is called the *sum* of the series Σf_n.

The main problem which arises is to determine whether important properties of functions are preserved under the limit operations (1) and (2). For instance, if the functions f_n are continuous, or differentiable, or integrable, is the same true of the limit function? What are the relations between f_n' and f', say, or between the integrals of f_n and that of f?

To say that f is continuous at a limit point x means

$$\lim_{t \to x} f(t) = f(x).$$

Hence, to ask whether the limit of a sequence of continuous functions is continuous is the same as to ask whether

$$(3) \qquad \lim_{t \to x} \lim_{n \to \infty} f_n(t) = \lim_{n \to \infty} \lim_{t \to x} f_n(t),$$

i.e., whether the order in which limit processes are carried out is immaterial. On the left side of (3), we first let $n \to \infty$, then $t \to x$; on the right side, $t \to x$ first, then $n \to \infty$.

We shall now show by means of several examples that limit processes cannot in general be interchanged without affecting the result. Afterward, we shall prove that under certain conditions the order in which limit operations are carried out is immaterial.

Our first example, and the simplest one, concerns a "double sequence."

7.2 Example For $m = 1, 2, 3, \ldots, n = 1, 2, 3, \ldots$, let

$$s_{m,n} = \frac{m}{m+n}.$$

Then, for every fixed n,

$$\lim_{m \to \infty} s_{m,n} = 1,$$

so that

$$(4) \qquad \lim_{n \to \infty} \lim_{m \to \infty} s_{m,n} = 1.$$

On the other hand, for every fixed m,

$$\lim_{n \to \infty} s_{m,n} = 0,$$

so that

(5) $$\lim_{m \to \infty} \lim_{n \to \infty} s_{m,n} = 0.$$

7.3 Example Let

$$f_n(x) = \frac{x^2}{(1+x^2)^n} \qquad (x \text{ real}; n = 0, 1, 2, \ldots),$$

and consider

(6) $$f(x) = \sum_{n=0}^{\infty} f_n(x) = \sum_{n=0}^{\infty} \frac{x^2}{(1+x^2)^n}.$$

Since $f_n(0) = 0$, we have $f(0) = 0$. For $x \neq 0$, the last series in (6) is a convergent geometric series with sum $1 + x^2$ (Theorem 3.26). Hence

(7) $$f(x) = \begin{cases} 0 & (x = 0), \\ 1 + x^2 & (x \neq 0), \end{cases}$$

so that a convergent series of continuous functions may have a discontinuous sum.

7.4 Example For $m = 1, 2, 3, \ldots$, put

$$f_m(x) = \lim_{n \to \infty} (\cos m!\pi x)^{2n}.$$

When $m!x$ is an integer, $f_m(x) = 1$. For all other values of x, $f_m(x) = 0$. Now let

$$f(x) = \lim_{m \to \infty} f_m(x).$$

For irrational x, $f_m(x) = 0$ for every m; hence $f(x) = 0$. For rational x, say $x = p/q$, where p and q are integers, we see that $m!x$ is an integer if $m \geq q$, so that $f(x) = 1$. Hence

(8) $$\lim_{m \to \infty} \lim_{n \to \infty} (\cos m!\pi x)^{2n} = \begin{cases} 0 & (x \text{ irrational}), \\ 1 & (x \text{ rational}). \end{cases}$$

We have thus obtained an everywhere discontinuous limit function, which is not Riemann-integrable (Exercise 4, Chap. 6).

7.5 Example Let

(9) $$f_n(x) = \frac{\sin nx}{\sqrt{n}} \qquad (x \text{ real}, n = 1, 2, 3, \ldots),$$

and
$$f(x) = \lim_{n \to \infty} f_n(x) = 0.$$

Then $f'(x) = 0$, and
$$f_n'(x) = \sqrt{n} \cos nx,$$

so that $\{f_n'\}$ does not converge to f'. For instance,
$$f_n'(0) = \sqrt{n} \to +\infty$$

as $n \to \infty$, whereas $f'(0) = 0$.

7.6 Example Let

(10) $$f_n(x) = n^2 x(1 - x^2)^n \qquad (0 \le x \le 1, n = 1, 2, 3, \ldots).$$

For $0 < x \le 1$, we have
$$\lim_{n \to \infty} f_n(x) = 0,$$

by Theorem 3.20(d). Since $f_n(0) = 0$, we see that

(11) $$\lim_{n \to \infty} f_n(x) = 0 \qquad (0 \le x \le 1).$$

A simple calculation shows that
$$\int_0^1 x(1 - x^2)^n \, dx = \frac{1}{2n + 2}.$$

Thus, in spite of (11),
$$\int_0^1 f_n(x) \, dx = \frac{n^2}{2n + 2} \to +\infty$$

as $n \to \infty$.

If, in (10), we replace n^2 by n, (11) still holds, but we now have
$$\lim_{n \to \infty} \int_0^1 f_n(x) \, dx = \lim_{n \to \infty} \frac{n}{2n + 2} = \frac{1}{2},$$

whereas
$$\int_0^1 \left[\lim_{n \to \infty} f_n(x) \right] dx = 0.$$

Thus the limit of the integral need not be equal to the integral of the limit, even if both are finite.

After these examples, which show what can go wrong if limit processes are interchanged carelessly, we now define a new mode of convergence, stronger than pointwise convergence as defined in Definition 7.1, which will enable us to arrive at positive results.

UNIFORM CONVERGENCE

7.7 Definition We say that a sequence of functions $\{f_n\}$, $n = 1, 2, 3, \ldots$, converges *uniformly* on E to a function f if for every $\varepsilon > 0$ there is an integer N such that $n \geq N$ implies
$$|f_n(x) - f(x)| \leq \varepsilon \tag{12}$$
for all $x \in E$.

It is clear that every uniformly convergent sequence is pointwise convergent. Quite explicitly, the difference between the two concepts is this: If $\{f_n\}$ converges pointwise on E, then there exists a function f such that, for every $\varepsilon > 0$, and for every $x \in E$, there is an integer N, depending on ε *and* on x, such that (12) holds if $n \geq N$; if $\{f_n\}$ converges uniformly on E, it is possible, for each $\varepsilon > 0$, to find *one* integer N which will do for *all* $x \in E$.

We say that the series $\Sigma f_n(x)$ converges uniformly on E if the sequence $\{s_n\}$ of partial sums defined by
$$\sum_{i=1}^{n} f_i(x) = s_n(x)$$
converges uniformly on E.

The Cauchy criterion for uniform convergence is as follows.

7.8 Theorem *The sequence of functions* $\{f_n\}$, *defined on E, converges uniformly on E if and only if for every $\varepsilon > 0$ there exists an integer N such that $m \geq N$, $n \geq N$, $x \in E$ implies*
$$|f_n(x) - f_m(x)| \leq \varepsilon. \tag{13}$$

Proof Suppose $\{f_n\}$ converges uniformly on E, and let f be the limit function. Then there is an integer N such that $n \geq N$, $x \in E$ implies
$$|f_n(x) - f(x)| \leq \frac{\varepsilon}{2},$$
so that
$$|f_n(x) - f_m(x)| \leq |f_n(x) - f(x)| + |f(x) - f_m(x)| \leq \varepsilon$$
if $n \geq N$, $m \geq N$, $x \in E$.

Conversely, suppose the Cauchy condition holds. By Theorem 3.11, the sequence $\{f_n(x)\}$ converges, for every x, to a limit which we may call $f(x)$. Thus the sequence $\{f_n\}$ converges on E, to f. We have to prove that the convergence is uniform.

Let $\varepsilon > 0$ be given, and choose N such that (13) holds. Fix n, and let $m \to \infty$ in (13). Since $f_m(x) \to f(x)$ as $m \to \infty$, this gives

$$|f_n(x) - f(x)| \le \varepsilon \tag{14}$$

for every $n \ge N$ and every $x \in E$, which completes the proof.

The following criterion is sometimes useful.

7.9 Theorem *Suppose*

$$\lim_{n \to \infty} f_n(x) = f(x) \qquad (x \in E).$$

Put

$$M_n = \sup_{x \in E} |f_n(x) - f(x)|.$$

Then $f_n \to f$ uniformly on E if and only if $M_n \to 0$ as $n \to \infty$.

Since this is an immediate consequence of Definition 7.7, we omit the details of the proof.

For series, there is a very convenient test for uniform convergence, due to Weierstrass.

7.10 Theorem *Suppose $\{f_n\}$ is a sequence of functions defined on E, and suppose*

$$|f_n(x)| \le M_n \qquad (x \in E, n = 1, 2, 3, \ldots).$$

Then Σf_n converges uniformly on E if ΣM_n converges.

Note that the converse is not asserted (and is, in fact, not true).

Proof If ΣM_n converges, then, for arbitrary $\varepsilon > 0$,

$$\left| \sum_{i=n}^{m} f_i(x) \right| \le \sum_{i=n}^{m} M_i \le \varepsilon \qquad (x \in E),$$

provided m and n are large enough. Uniform convergence now follows from Theorem 7.8.

UNIFORM CONVERGENCE AND CONTINUITY

7.11 Theorem *Suppose $f_n \to f$ uniformly on a set E in a metric space. Let x be a limit point of E, and suppose that*

(15) $$\lim_{t \to x} f_n(t) = A_n \qquad (n = 1, 2, 3, \ldots).$$

Then $\{A_n\}$ converges, and

(16) $$\lim_{t \to x} f(t) = \lim_{n \to \infty} A_n.$$

In other words, the conclusion is that

(17) $$\lim_{t \to x} \lim_{n \to \infty} f_n(t) = \lim_{n \to \infty} \lim_{t \to x} f_n(t).$$

Proof Let $\varepsilon > 0$ be given. By the uniform convergence of $\{f_n\}$, there exists N such that $n \geq N$, $m \geq N$, $t \in E$ imply

(18) $$|f_n(t) - f_m(t)| \leq \varepsilon.$$

Letting $t \to x$ in (18), we obtain

$$|A_n - A_m| \leq \varepsilon$$

for $n \geq N$, $m \geq N$, so that $\{A_n\}$ is a Cauchy sequence and therefore converges, say to A.

Next,

(19) $$|f(t) - A| \leq |f(t) - f_n(t)| + |f_n(t) - A_n| + |A_n - A|.$$

We first choose n such that

(20) $$|f(t) - f_n(t)| \leq \frac{\varepsilon}{3}$$

for all $t \in E$ (this is possible by the uniform convergence), and such that

(21) $$|A_n - A| \leq \frac{\varepsilon}{3}.$$

Then, for this n, we choose a neighborhood V of x such that

(22) $$|f_n(t) - A_n| \leq \frac{\varepsilon}{3}$$

if $t \in V \cap E$, $t \neq x$.

Substituting the inequalities (20) to (22) into (19), we see that

$$|f(t) - A| \leq \varepsilon,$$

provided $t \in V \cap E$, $t \neq x$. This is equivalent to (16).

7.12 Theorem *If $\{f_n\}$ is a sequence of continuous functions on E, and if $f_n \to f$ uniformly on E, then f is continuous on E.*

This very important result is an immediate corollary of Theorem 7.11.

The converse is not true; that is, a sequence of continuous functions may converge to a continuous function, although the convergence is not uniform. Example 7.6 is of this kind (to see this, apply Theorem 7.9). But there is a case in which we can assert the converse.

7.13 Theorem *Suppose K is compact, and*

(a) $\{f_n\}$ is a sequence of continuous functions on K,
(b) $\{f_n\}$ converges pointwise to a continuous function f on K,
(c) $f_n(x) \geq f_{n+1}(x)$ for all $x \in K$, $n = 1, 2, 3, \ldots$.

Then $f_n \to f$ uniformly on K.

Proof Put $g_n = f_n - f$. Then g_n is continuous, $g_n \to 0$ pointwise, and $g_n \geq g_{n+1}$. We have to prove that $g_n \to 0$ uniformly on K.

Let $\varepsilon > 0$ be given. Let K_n be the set of all $x \in K$ with $g_n(x) \geq \varepsilon$. Since g_n is continuous, K_n is closed (Theorem 4.8), hence compact (Theorem 2.35). Since $g_n \geq g_{n+1}$, we have $K_n \supset K_{n+1}$. Fix $x \in K$. Since $g_n(x) \to 0$, we see that $x \notin K_n$ if n is sufficiently large. Thus $x \notin \bigcap K_n$. In other words, $\bigcap K_n$ is empty. Hence K_N is empty for some N (Theorem 2.36). It follows that $0 \leq g_n(x) < \varepsilon$ for all $x \in K$ and for all $n \geq N$. This proves the theorem.

Let us note that compactness is really needed here. For instance, if

$$f_n(x) = \frac{1}{nx + 1} \qquad (0 < x < 1; n = 1, 2, 3, \ldots)$$

then $f_n(x) \to 0$ monotonically in $(0, 1)$, but the convergence is not uniform.

7.14 Definition If X is a metric space, $\mathscr{C}(X)$ will denote the set of all complex-valued, continuous, bounded functions with domain X.

[Note that boundedness is redundant if X is compact (Theorem 4.15). Thus $\mathscr{C}(X)$ consists of all complex continuous functions on X if X is compact.]

We associate with each $f \in \mathscr{C}(X)$ its *supremum norm*

$$\|f\| = \sup_{x \in X} |f(x)|.$$

Since f is assumed to be bounded, $\|f\| < \infty$. It is obvious that $\|f\| = 0$ only if $f(x) = 0$ for every $x \in X$, that is, only if $f = 0$. If $h = f + g$, then

$$|h(x)| \leq |f(x)| + |g(x)| \leq \|f\| + \|g\|$$

for all $x \in X$; hence

$$\|f + g\| \leq \|f\| + \|g\|.$$

If we define the distance between $f \in \mathscr{C}(X)$ and $g \in \mathscr{C}(X)$ to be $\|f - g\|$, it follows that Axioms 2.15 for a metric are satisfied.

We have thus made $\mathscr{C}(X)$ into a metric space.

Theorem 7.9 can be rephrased as follows:

A sequence $\{f_n\}$ converges to f with respect to the metric of $\mathscr{C}(X)$ if and only if $f_n \to f$ uniformly on X.

Accordingly, closed subsets of $\mathscr{C}(X)$ are sometimes called *uniformly closed*, the closure of a set $\mathscr{A} \subset \mathscr{C}(X)$ is called its *uniform closure*, and so on.

7.15 Theorem *The above metric makes $\mathscr{C}(X)$ into a complete metric space.*

Proof Let $\{f_n\}$ be a Cauchy sequence in $\mathscr{C}(X)$. This means that to each $\varepsilon > 0$ corresponds an N such that $\|f_n - f_m\| < \varepsilon$ if $n \geq N$ and $m \geq N$. It follows (by Theorem 7.8) that there is a function f with domain X to which $\{f_n\}$ converges uniformly. By Theorem 7.12, f is continuous. Moreover, f is bounded, since there is an n such that $|f(x) - f_n(x)| < 1$ for all $x \in X$, and f_n is bounded.

Thus $f \in \mathscr{C}(X)$, and since $f_n \to f$ uniformly on X, we have $\|f - f_n\| \to 0$ as $n \to \infty$.

UNIFORM CONVERGENCE AND INTEGRATION

7.16 Theorem *Let α be monotonically increasing on $[a, b]$. Suppose $f_n \in \mathscr{R}(\alpha)$ on $[a, b]$, for $n = 1, 2, 3, \ldots$, and suppose $f_n \to f$ uniformly on $[a, b]$. Then $f \in \mathscr{R}(\alpha)$ on $[a, b]$, and*

$$(23) \qquad \int_a^b f \, d\alpha = \lim_{n \to \infty} \int_a^b f_n \, d\alpha.$$

(The existence of the limit is part of the conclusion.)

Proof It suffices to prove this for real f_n. Put

$$(24) \qquad \varepsilon_n = \sup |f_n(x) - f(x)|,$$

the supremum being taken over $a \leq x \leq b$. Then

$$f_n - \varepsilon_n \leq f \leq f_n + \varepsilon_n,$$

so that the upper and lower integrals of f (see Definition 6.2) satisfy

$$(25) \qquad \int_a^b (f_n - \varepsilon_n) \, d\alpha \leq \underline{\int} f \, d\alpha \leq \overline{\int} f \, d\alpha \leq \int_a^b (f_n + \varepsilon_n) \, d\alpha.$$

Hence

$$0 \leq \overline{\int} f \, d\alpha - \underline{\int} f \, d\alpha \leq 2\varepsilon_n [\alpha(b) - \alpha(a)].$$

Since $\varepsilon_n \to 0$ as $n \to \infty$ (Theorem 7.9), the upper and lower integrals of f are equal.

Thus $f \in \mathscr{R}(\alpha)$. Another application of (25) now yields

(26) $$\left| \int_a^b f\, d\alpha - \int_a^b f_n\, d\alpha \right| \leq \varepsilon_n [\alpha(b) - \alpha(a)].$$

This implies (23).

Corollary *If $f_n \in \mathscr{R}(\alpha)$ on $[a, b]$ and if*

$$f(x) = \sum_{n=1}^{\infty} f_n(x) \qquad (a \leq x \leq b),$$

the series converging uniformly on $[a, b]$, then

$$\int_a^b f\, d\alpha = \sum_{n=1}^{\infty} \int_a^b f_n\, d\alpha.$$

In other words, the series may be integrated term by term.

UNIFORM CONVERGENCE AND DIFFERENTIATION

We have already seen, in Example 7.5, that uniform convergence of $\{f_n\}$ implies nothing about the sequence $\{f_n'\}$. Thus stronger hypotheses are required for the assertion that $f_n' \to f'$ if $f_n \to f$.

7.17 Theorem *Suppose $\{f_n\}$ is a sequence of functions, differentiable on $[a, b]$ and such that $\{f_n(x_0)\}$ converges for some point x_0 on $[a, b]$. If $\{f_n'\}$ converges uniformly on $[a, b]$, then $\{f_n\}$ converges uniformly on $[a, b]$, to a function f, and*

(27) $$f'(x) = \lim_{n \to \infty} f_n'(x) \qquad (a \leq x \leq b).$$

Proof Let $\varepsilon > 0$ be given. Choose N such that $n \geq N$, $m \geq N$, implies

(28) $$|f_n(x_0) - f_m(x_0)| < \frac{\varepsilon}{2}$$

and

(29) $$|f_n'(t) - f_m'(t)| < \frac{\varepsilon}{2(b - a)} \qquad (a \leq t \leq b).$$

If we apply the mean value theorem 5.19 to the function $f_n - f_m$, (29) shows that

(30) $$|f_n(x) - f_m(x) - f_n(t) + f_m(t)| \leq \frac{|x-t|\varepsilon}{2(b-a)} \leq \frac{\varepsilon}{2}$$

for any x and t on $[a, b]$, if $n \geq N$, $m \geq N$. The inequality

$$|f_n(x) - f_m(x)| \leq |f_n(x) - f_m(x) - f_n(x_0) + f_m(x_0)| + |f_n(x_0) - f_m(x_0)|$$

implies, by (28) and (30), that

$$|f_n(x) - f_m(x)| < \varepsilon \qquad (a \leq x \leq b, n \geq N, m \geq N),$$

so that $\{f_n\}$ converges uniformly on $[a, b]$. Let

$$f(x) = \lim_{n \to \infty} f_n(x) \qquad (a \leq x \leq b).$$

Let us now fix a point x on $[a, b]$ and define

(31) $$\phi_n(t) = \frac{f_n(t) - f_n(x)}{t - x}, \qquad \phi(t) = \frac{f(t) - f(x)}{t - x}$$

for $a \leq t \leq b$, $t \neq x$. Then

(32) $$\lim_{t \to x} \phi_n(t) = f_n'(x) \qquad (n = 1, 2, 3, \ldots).$$

The first inequality in (30) shows that

$$|\phi_n(t) - \phi_m(t)| \leq \frac{\varepsilon}{2(b-a)} \qquad (n \geq N, m \geq N),$$

so that $\{\phi_n\}$ converges uniformly, for $t \neq x$. Since $\{f_n\}$ converges to f, we conclude from (31) that

(33) $$\lim_{n \to \infty} \phi_n(t) = \phi(t)$$

uniformly for $a \leq t \leq b$, $t \neq x$.

If we now apply Theorem 7.11 to $\{\phi_n\}$, (32) and (33) show that

$$\lim_{t \to x} \phi(t) = \lim_{n \to \infty} f_n'(x);$$

and this is (27), by the definition of $\phi(t)$.

Remark: If the continuity of the functions f_n' is assumed in addition to the above hypotheses, then a much shorter proof of (27) can be based on Theorem 7.16 and the fundamental theorem of calculus.

7.18 Theorem *There exists a real continuous function on the real line which is nowhere differentiable.*

Proof Define

(34) $$\varphi(x) = |x| \quad (-1 \leq x \leq 1)$$

and extend the definition of $\varphi(x)$ to all real x by requiring that

(35) $$\varphi(x + 2) = \varphi(x).$$

Then, for all s and t,

(36) $$|\varphi(s) - \varphi(t)| \leq |s - t|.$$

In particular, φ is continuous on R^1. Define

(37) $$f(x) = \sum_{n=0}^{\infty} \left(\tfrac{3}{4}\right)^n \varphi(4^n x).$$

Since $0 \leq \varphi \leq 1$, Theorem 7.10 shows that the series (37) converges uniformly on R^1. By Theorem 7.12, f is continuous on R^1.

Now fix a real number x and a positive integer m. Put

(38) $$\delta_m = \pm \tfrac{1}{2} \cdot 4^{-m}$$

where the sign is so chosen that no integer lies between $4^m x$ and $4^m(x + \delta_m)$. This can be done, since $4^m |\delta_m| = \tfrac{1}{2}$. Define

(39) $$\gamma_n = \frac{\varphi(4^n(x + \delta_m)) - \varphi(4^n x)}{\delta_m}.$$

When $n > m$, then $4^n \delta_m$ is an even integer, so that $\gamma_n = 0$. When $0 \leq n \leq m$, (36) implies that $|\gamma_n| \leq 4^n$.

Since $|\gamma_m| = 4^m$, we conclude that

$$\left| \frac{f(x + \delta_m) - f(x)}{\delta_m} \right| = \left| \sum_{n=0}^{m} \left(\tfrac{3}{4}\right)^n \gamma_n \right|$$

$$\geq 3^m - \sum_{n=0}^{m-1} 3^n$$

$$= \tfrac{1}{2}(3^m + 1).$$

As $m \to \infty$, $\delta_m \to 0$. It follows that f is not differentiable at x.

EQUICONTINUOUS FAMILIES OF FUNCTIONS

In Theorem 3.6 we saw that every bounded sequence of complex numbers contains a convergent subsequence, and the question arises whether something similar is true for sequences of functions. To make the question more precise, we shall define two kinds of boundedness.

7.19 Definition Let $\{f_n\}$ be a sequence of functions defined on a set E.

We say that $\{f_n\}$ is *pointwise bounded* on E if the sequence $\{f_n(x)\}$ is bounded for every $x \in E$, that is, if there exists a finite-valued function ϕ defined on E such that

$$|f_n(x)| < \phi(x) \qquad (x \in E, n = 1, 2, 3, \ldots).$$

We say that $\{f_n\}$ is *uniformly bounded* on E if there exists a number M such that

$$|f_n(x)| < M \qquad (x \in E, n = 1, 2, 3, \ldots).$$

Now if $\{f_n\}$ is pointwise bounded on E and E_1 is a countable subset of E, it is always possible to find a subsequence $\{f_{n_k}\}$ such that $\{f_{n_k}(x)\}$ converges for every $x \in E_1$. This can be done by the diagonal process which is used in the proof of Theorem 7.23.

However, even if $\{f_n\}$ is a uniformly bounded sequence of continuous functions on a compact set E, there need not exist a subsequence which converges pointwise on E. In the following example, this would be quite troublesome to prove with the equipment which we have at hand so far, but the proof is quite simple if we appeal to a theorem from Chap. 11.

7.20 Example Let

$$f_n(x) = \sin nx \qquad (0 \le x \le 2\pi, n = 1, 2, 3, \ldots).$$

Suppose there exists a sequence $\{n_k\}$ such that $\{\sin n_k x\}$ converges, for every $x \in [0, 2\pi]$. In that case we must have

$$\lim_{k \to \infty} (\sin n_k x - \sin n_{k+1} x) = 0 \qquad (0 \le x \le 2\pi);$$

hence

(40) $$\lim_{k \to \infty} (\sin n_k x - \sin n_{k+1} x)^2 = 0 \qquad (0 \le x \le 2\pi).$$

By Lebesgue's theorem concerning integration of boundedly convergent sequences (Theorem 11.32), (40) implies

(41) $$\lim_{k \to \infty} \int_0^{2\pi} (\sin n_k x - \sin n_{k+1} x)^2 \, dx = 0.$$

But a simple calculation shows that

$$\int_0^{2\pi} (\sin n_k x - \sin n_{k+1} x)^2 \, dx = 2\pi,$$

which contradicts (41).

Another question is whether every convergent sequence contains a uniformly convergent subsequence. Our next example will show that this need not be so, even if the sequence is uniformly bounded on a compact set. (Example 7.6 shows that a sequence of bounded functions may converge without being uniformly bounded; but it is trivial to see that uniform convergence of a sequence of bounded functions implies uniform boundedness.)

7.21 Example Let

$$f_n(x) = \frac{x^2}{x^2 + (1 - nx)^2} \qquad (0 \leq x \leq 1, n = 1, 2, 3, \ldots).$$

Then $|f_n(x)| \leq 1$, so that $\{f_n\}$ is uniformly bounded on $[0, 1]$. Also

$$\lim_{n \to \infty} f_n(x) = 0 \qquad (0 \leq x \leq 1),$$

but

$$f_n\left(\frac{1}{n}\right) = 1 \qquad (n = 1, 2, 3, \ldots),$$

so that no subsequence can converge uniformly on $[0, 1]$.

The concept which is needed in this connection is that of equicontinuity; it is given in the following definition.

7.22 Definition A family $\mathscr{F}$ of complex functions f defined on a set E in a metric space X is said to be *equicontinuous* on E if for every $\varepsilon > 0$ there exists a $\delta > 0$ such that

$$|f(x) - f(y)| < \varepsilon$$

whenever $d(x, y) < \delta$, $x \in E$, $y \in E$, and $f \in \mathscr{F}$. Here d denotes the metric of X.

It is clear that every member of an equicontinuous family is uniformly continuous.

The sequence of Example 7.21 is not equicontinuous.

Theorems 7.24 and 7.25 will show that there is a very close relation between equicontinuity, on the one hand, and uniform convergence of sequences of continuous functions, on the other. But first we describe a selection process which has nothing to do with continuity.

7.23 Theorem *If $\{f_n\}$ is a pointwise bounded sequence of complex functions on a countable set E, then $\{f_n\}$ has a subsequence $\{f_{n_k}\}$ such that $\{f_{n_k}(x)\}$ converges for every $x \in E$.*

Proof Let $\{x_i\}$, $i = 1, 2, 3, \ldots$, be the points of E, arranged in a sequence. Since $\{f_n(x_1)\}$ is bounded, there exists a subsequence, which we shall denote by $\{f_{1,k}\}$, such that $\{f_{1,k}(x_1)\}$ converges as $k \to \infty$.

Let us now consider sequences $S_1, S_2, S_3, \ldots$, which we represent by the array

$$\begin{array}{llllll} S_1: & f_{1,1} & f_{1,2} & f_{1,3} & f_{1,4} & \cdots \\ S_2: & f_{2,1} & f_{2,2} & f_{2,3} & f_{2,4} & \cdots \\ S_3: & f_{3,1} & f_{3,2} & f_{3,3} & f_{3,4} & \cdots \\ \cdots & \cdots & \cdots & \cdots & \cdots & \end{array}$$

and which have the following properties:

(a) S_n is a subsequence of S_{n-1}, for $n = 2, 3, 4, \ldots$.
(b) $\{f_{n,k}(x_n)\}$ converges, as $k \to \infty$ (the boundedness of $\{f_n(x_n)\}$ makes it possible to choose S_n in this way);
(c) The order in which the functions appear is the same in each sequence; i.e., if one function precedes another in S_1, they are in the same relation in every S_n, until one or the other is deleted. Hence, when going from one row in the above array to the next below, functions may move to the left but never to the right.

We now go down the diagonal of the array; i.e., we consider the sequence

$$S: \quad f_{1,1} \quad f_{2,2} \quad f_{3,3} \quad f_{4,4} \cdots.$$

By (c), the sequence S (except possibly its first $n - 1$ terms) is a subsequence of S_n, for $n = 1, 2, 3, \ldots$. Hence (b) implies that $\{f_{n,n}(x_i)\}$ converges, as $n \to \infty$, for every $x_i \in E$.

7.24 Theorem *If K is a compact metric space, if $f_n \in \mathscr{C}(K)$ for $n = 1, 2, 3, \ldots$, and if $\{f_n\}$ converges uniformly on K, then $\{f_n\}$ is equicontinuous on K.*

Proof Let $\varepsilon > 0$ be given. Since $\{f_n\}$ converges uniformly, there is an integer N such that

(42) $$\|f_n - f_N\| < \varepsilon \qquad (n > N).$$

(See Definition 7.14.) Since continuous functions are uniformly continuous on compact sets, there is a $\delta > 0$ such that

(43) $$|f_i(x) - f_i(y)| < \varepsilon$$

if $1 \leq i \leq N$ and $d(x, y) < \delta$.

If $n > N$ and $d(x, y) < \delta$, it follows that

$$|f_n(x) - f_n(y)| \leq |f_n(x) - f_N(x)| + |f_N(x) - f_N(y)| + |f_N(y) - f_n(y)| < 3\varepsilon.$$

In conjunction with (43), this proves the theorem.

7.25 Theorem *If K is compact, if $f_n \in \mathscr{C}(K)$ for $n = 1, 2, 3, \ldots$, and if $\{f_n\}$ is pointwise bounded and equicontinuous on K, then*

(a) *$\{f_n\}$ is uniformly bounded on K,*
(b) *$\{f_n\}$ contains a uniformly convergent subsequence.*

Proof

(a) Let $\varepsilon > 0$ be given and choose $\delta > 0$, in accordance with Definition 7.22, so that

(44) $$|f_n(x) - f_n(y)| < \varepsilon$$

for all n, provided that $d(x, y) < \delta$.

Since K is compact, there are finitely many points $p_1, \ldots, p_r$ in K such that to every $x \in K$ corresponds at least one p_i with $d(x, p_i) < \delta$. Since $\{f_n\}$ is pointwise bounded, there exist $M_i < \infty$ such that $|f_n(p_i)| < M_i$ for all n. If $M = \max(M_1, \ldots, M_r)$, then $|f_n(x)| < M + \varepsilon$ for every $x \in K$. This proves (a).

(b) Let E be a countable dense subset of K. (For the existence of such a set E, see Exercise 25, Chap. 2.) Theorem 7.23 shows that $\{f_n\}$ has a subsequence $\{f_{n_i}\}$ such that $\{f_{n_i}(x)\}$ converges for every $x \in E$.

Put $f_{n_i} = g_i$, to simplify the notation. We shall prove that $\{g_i\}$ converges uniformly on K.

Let $\varepsilon > 0$, and pick $\delta > 0$ as in the beginning of this proof. Let $V(x, \delta)$ be the set of all $y \in K$ with $d(x, y) < \delta$. Since E is dense in K, and K is compact, there are finitely many points $x_1, \ldots, x_m$ in E such that

(45) $$K \subset V(x_1, \delta) \cup \cdots \cup V(x_m, \delta).$$

Since $\{g_i(x)\}$ converges for every $x \in E$, there is an integer N such that

(46) $$|g_i(x_s) - g_j(x_s)| < \varepsilon$$

whenever $i \geq N, j \geq N, 1 \leq s \leq m$.

If $x \in K$, (45) shows that $x \in V(x_s, \delta)$ for some s, so that

$$|g_i(x) - g_i(x_s)| < \varepsilon$$

for every i. If $i \geq N$ and $j \geq N$, it follows from (46) that

$$|g_i(x) - g_j(x)| \leq |g_i(x) - g_i(x_s)| + |g_i(x_s) - g_j(x_s)| + |g_j(x_s) - g_j(x)|$$
$$< 3\varepsilon.$$

This completes the proof.

THE STONE-WEIERSTRASS THEOREM

7.26 Theorem *If f is a continuous complex function on $[a, b]$, there exists a sequence of polynomials P_n such that*
$$\lim_{n \to \infty} P_n(x) = f(x)$$
uniformly on $[a, b]$. If f is real, the P_n may be taken real.

This is the form in which the theorem was originally discovered by Weierstrass.

Proof We may assume, without loss of generality, that $[a, b] = [0, 1]$. We may also assume that $f(0) = f(1) = 0$. For if the theorem is proved for this case, consider
$$g(x) = f(x) - f(0) - x[f(1) - f(0)] \qquad (0 \leq x \leq 1).$$
Here $g(0) = g(1) = 0$, and if g can be obtained as the limit of a uniformly convergent sequence of polynomials, it is clear that the same is true for f, since $f - g$ is a polynomial.

Furthermore, we define $f(x)$ to be zero for x outside $[0, 1]$. Then f is uniformly continuous on the whole line.

We put

(47) $$Q_n(x) = c_n(1 - x^2)^n \qquad (n = 1, 2, 3, \ldots),$$

where c_n is chosen so that

(48) $$\int_{-1}^{1} Q_n(x) \, dx = 1 \qquad (n = 1, 2, 3, \ldots).$$

We need some information about the order of magnitude of c_n. Since
$$\int_{-1}^{1} (1 - x^2)^n \, dx = 2 \int_0^1 (1 - x^2)^n \, dx \geq 2 \int_0^{1/\sqrt{n}} (1 - x^2)^n \, dx$$
$$\geq 2 \int_0^{1/\sqrt{n}} (1 - nx^2) \, dx$$
$$= \frac{4}{3\sqrt{n}}$$
$$> \frac{1}{\sqrt{n}},$$

it follows from (48) that

(49) $$c_n < \sqrt{n}.$$

The inequality $(1 - x^2)^n \geq 1 - nx^2$ which we used above is easily shown to be true by considering the function

$$(1 - x^2)^n - 1 + nx^2$$

which is zero at $x = 0$ and whose derivative is positive in $(0, 1)$.

For any $\delta > 0$, (49) implies

(50) $$Q_n(x) \leq \sqrt{n}(1 - \delta^2)^n \qquad (\delta \leq |x| \leq 1),$$

so that $Q_n \to 0$ uniformly in $\delta \leq |x| \leq 1$.

Now set

(51) $$P_n(x) = \int_{-1}^{1} f(x + t)Q_n(t)\,dt \qquad (0 \leq x \leq 1).$$

Our assumptions about f show, by a simple change of variable, that

$$P_n(x) = \int_{-x}^{1-x} f(x + t)Q_n(t)\,dt = \int_{0}^{1} f(t)Q_n(t - x)\,dt,$$

and the last integral is clearly a polynomial in x. Thus $\{P_n\}$ is a sequence of polynomials, which are real if f is real.

Given $\varepsilon > 0$, we choose $\delta > 0$ such that $|y - x| < \delta$ implies

$$|f(y) - f(x)| < \frac{\varepsilon}{2}.$$

Let $M = \sup |f(x)|$. Using (48), (50), and the fact that $Q_n(x) \geq 0$, we see that for $0 \leq x \leq 1$,

$$|P_n(x) - f(x)| = \left|\int_{-1}^{1} [f(x + t) - f(x)]Q_n(t)\,dt\right|$$

$$\leq \int_{-1}^{1} |f(x + t) - f(x)|Q_n(t)\,dt$$

$$\leq 2M \int_{-1}^{-\delta} Q_n(t)\,dt + \frac{\varepsilon}{2}\int_{-\delta}^{\delta} Q_n(t)\,dt + 2M \int_{\delta}^{1} Q_n(t)\,dt$$

$$\leq 4M\sqrt{n}(1 - \delta^2)^n + \frac{\varepsilon}{2}$$

$$< \varepsilon$$

for all large enough n, which proves the theorem.

It is instructive to sketch the graphs of Q_n for a few values of n; also, note that we needed uniform continuity of f to deduce uniform convergence of $\{P_n\}$.

In the proof of Theorem 7.32 we shall not need the full strength of Theorem 7.26, but only the following special case, which we state as a corollary.

7.27 Corollary *For every interval* $[-a, a]$ *there is a sequence of real polynomials* P_n *such that* $P_n(0) = 0$ *and such that*
$$\lim_{n \to \infty} P_n(x) = |x|$$
uniformly on $[-a, a]$.

Proof By Theorem 7.26, there exists a sequence $\{P_n^*\}$ of real polynomials which converges to $|x|$ uniformly on $[-a, a]$. In particular, $P_n^*(0) \to 0$ as $n \to \infty$. The polynomials
$$P_n(x) = P_n^*(x) - P_n^*(0) \qquad (n = 1, 2, 3, \ldots)$$
have desired properties.

We shall now isolate those properties of the polynomials which make the Weierstrass theorem possible.

7.28 Definition A family $\mathscr{A}$ of complex functions defined on a set E is said to be an *algebra* if (i) $f + g \in \mathscr{A}$, (ii) $fg \in \mathscr{A}$, and (iii) $cf \in \mathscr{A}$ for all $f \in \mathscr{A}$, $g \in \mathscr{A}$ and for all complex constants c, that is, if $\mathscr{A}$ is closed under addition, multiplication, and scalar multiplication. We shall also have to consider algebras of real functions; in this case, (iii) is of course only required to hold for all real c.

If $\mathscr{A}$ has the property that $f \in \mathscr{A}$ whenever $f_n \in \mathscr{A}$ $(n = 1, 2, 3, \ldots)$ and $f_n \to f$ uniformly on E, then $\mathscr{A}$ is said to be *uniformly closed*.

Let $\mathscr{B}$ be the set of all functions which are limits of uniformly convergent sequences of members of $\mathscr{A}$. Then $\mathscr{B}$ is called the *uniform closure* of $\mathscr{A}$. (See Definition 7.14.)

For example, the set of all polynomials is an algebra, and the Weierstrass theorem may be stated by saying that the set of continuous functions on $[a, b]$ is the uniform closure of the set of polynomials on $[a, b]$.

7.29 Theorem *Let $\mathscr{B}$ be the uniform closure of an algebra $\mathscr{A}$ of bounded functions. Then $\mathscr{B}$ is a uniformly closed algebra.*

Proof If $f \in \mathscr{B}$ and $g \in \mathscr{B}$, there exist uniformly convergent sequences $\{f_n\}, \{g_n\}$ such that $f_n \to f$, $g_n \to g$ and $f_n \in \mathscr{A}$, $g_n \in \mathscr{A}$. Since we are dealing with bounded functions, it is easy to show that
$$f_n + g_n \to f + g, \qquad f_n g_n \to fg, \qquad cf_n \to cf,$$
where c is any constant, the convergence being uniform in each case.

Hence $f + g \in \mathscr{B}$, $fg \in \mathscr{B}$, and $cf \in \mathscr{B}$, so that $\mathscr{B}$ is an algebra.
By Theorem 2.27, $\mathscr{B}$ is (uniformly) closed.

7.30 Definition Let $\mathscr{A}$ be a family of functions on a set E. Then $\mathscr{A}$ is said to *separate points* on E if to every pair of distinct points $x_1, x_2 \in E$ there corresponds a function $f \in \mathscr{A}$ such that $f(x_1) \neq f(x_2)$.

If to each $x \in E$ there corresponds a function $g \in \mathscr{A}$ such that $g(x) \neq 0$, we say that $\mathscr{A}$ *vanishes at no point of* E.

The algebra of all polynomials in one variable clearly has these properties on R^1. An example of an algebra which does not separate points is the set of all even polynomials, say on $[-1, 1]$, since $f(-x) = f(x)$ for every even function f.

The following theorem will illustrate these concepts further.

7.31 Theorem *Suppose $\mathscr{A}$ is an algebra of functions on a set E, $\mathscr{A}$ separates points on E, and $\mathscr{A}$ vanishes at no point of E. Suppose x_1, x_2 are distinct points of E, and c_1, c_2 are constants (real if $\mathscr{A}$ is a real algebra). Then $\mathscr{A}$ contains a function f such that*

$$f(x_1) = c_1, \quad f(x_2) = c_2.$$

Proof The assumptions show that $\mathscr{A}$ contains functions $g, h,$ and k such that

$$g(x_1) \neq g(x_2), \quad h(x_1) \neq 0, \quad k(x_2) \neq 0.$$

Put

$$u = gk - g(x_1)k, \quad v = gh - g(x_2)h.$$

Then $u \in \mathscr{A}$, $v \in \mathscr{A}$, $u(x_1) = v(x_2) = 0$, $u(x_2) \neq 0$, and $v(x_1) \neq 0$. Therefore

$$f = \frac{c_1 v}{v(x_1)} + \frac{c_2 u}{u(x_2)}$$

has the desired properties.

We now have all the material needed for Stone's generalization of the Weierstrass theorem.

7.32 Theorem *Let $\mathscr{A}$ be an algebra of real continuous functions on a compact set K. If $\mathscr{A}$ separates points on K and if $\mathscr{A}$ vanishes at no point of K, then the uniform closure $\mathscr{B}$ of $\mathscr{A}$ consists of all real continuous functions on K.*

We shall divide the proof into four steps.

STEP 1 *If $f \in \mathscr{B}$, then $|f| \in \mathscr{B}$.*

Proof Let

(52) $$a = \sup |f(x)| \quad (x \in K)$$

and let $\varepsilon > 0$ be given. By Corollary 7.27 there exist real numbers $c_1, \ldots, c_n$ such that

(53) $$\left| \sum_{i=1}^{n} c_i y^i - |y| \right| < \varepsilon \qquad (-a \le y \le a).$$

Since $\mathscr{B}$ is an algebra, the function

$$g = \sum_{i=1}^{n} c_i f^i$$

is a member of $\mathscr{B}$. By (52) and (53), we have

$$\big| g(x) - |f(x)| \big| < \varepsilon \qquad (x \in K).$$

Since $\mathscr{B}$ is uniformly closed, this shows that $|f| \in \mathscr{B}$.

STEP 2 *If $f \in \mathscr{B}$ and $g \in \mathscr{B}$, then $\max(f, g) \in \mathscr{B}$ and $\min(f, g) \in \mathscr{B}$.*

By $\max(f, g)$ we mean the function h defined by

$$h(x) = \begin{cases} f(x) & \text{if } f(x) \ge g(x), \\ g(x) & \text{if } f(x) < g(x), \end{cases}$$

and $\min(f, g)$ is defined likewise.

Proof Step 2 follows from step 1 and the identities

$$\max(f, g) = \frac{f + g}{2} + \frac{|f - g|}{2},$$

$$\min(f, g) = \frac{f + g}{2} - \frac{|f - g|}{2}.$$

By iteration, the result can of course be extended to any finite set of functions: If $f_1, \ldots, f_n \in \mathscr{B}$, then $\max(f_1, \ldots, f_n) \in \mathscr{B}$, and

$$\min(f_1, \ldots, f_n) \in \mathscr{B}.$$

STEP 3 *Given a real function f, continuous on K, a point $x \in K$, and $\varepsilon > 0$, there exists a function $g_x \in \mathscr{B}$ such that $g_x(x) = f(x)$ and*

(54) $$g_x(t) > f(t) - \varepsilon \qquad (t \in K).$$

Proof Since $\mathscr{A} \subset \mathscr{B}$ and $\mathscr{A}$ satisfies the hypotheses of Theorem 7.31 so does $\mathscr{B}$. Hence, for every $y \in K$, we can find a function $h_y \in \mathscr{B}$ such that

(55) $$h_y(x) = f(x), \qquad h_y(y) = f(y).$$

By the continuity of h_y there exists an open set J_y, containing y, such that

(56) $$h_y(t) > f(t) - \varepsilon \qquad (t \in J_y).$$

Since K is compact, there is a finite set of points $y_1, \ldots, y_n$ such that

(57) $$K \subset J_{y_1} \cup \cdots \cup J_{y_n}.$$

Put

$$g_x = \max(h_{y_1}, \ldots, h_{y_n}).$$

By step 2, $g_x \in \mathscr{B}$, and the relations (55) to (57) show that g_x has the other required properties.

STEP 4 *Given a real function f, continuous on K, and $\varepsilon > 0$, there exists a function $h \in \mathscr{B}$ such that*

(58) $$|h(x) - f(x)| < \varepsilon \qquad (x \in K).$$

Since $\mathscr{B}$ is uniformly closed, this statement is equivalent to the conclusion of the theorem.

Proof Let us consider the functions g_x, for each $x \in K$, constructed in step 3. By the continuity of g_x, there exist open sets V_x containing x, such that

(59) $$g_x(t) < f(t) + \varepsilon \qquad (t \in V_x).$$

Since K is compact, there exists a finite set of points $x_1, \ldots, x_m$ such that

(60) $$K \subset V_{x_1} \cup \cdots \cup V_{x_m}.$$

Put

$$h = \min(g_{x_1}, \ldots, g_{x_m}).$$

By step 2, $h \in \mathscr{B}$, and (54) implies

(61) $$h(t) > f(t) - \varepsilon \qquad (t \in K),$$

whereas (59) and (60) imply

(62) $$h(t) < f(t) + \varepsilon \qquad (t \in K).$$

Finally, (58) follows from (61) and (62).

Theorem 7.32 does not hold for complex algebras. A counterexample is given in Exercise 21. However, the conclusion of the theorem does hold, even for complex algebras, if an extra condition is imposed on $\mathscr{A}$, namely, that $\mathscr{A}$ be *self-adjoint*. This means that for every $f \in \mathscr{A}$ its complex conjugate $\bar{f}$ must also belong to $\mathscr{A}$; $\bar{f}$ is defined by $\bar{f}(x) = \overline{f(x)}$.

7.33 Theorem *Suppose $\mathscr{A}$ is a self-adjoint algebra of complex continuous functions on a compact set K, $\mathscr{A}$ separates points on K, and $\mathscr{A}$ vanishes at no point of K. Then the uniform closure $\mathscr{B}$ of $\mathscr{A}$ consists of all complex continuous functions on K. In other words, $\mathscr{A}$ is dense $\mathscr{C}(K)$.*

Proof Let $\mathscr{A}_R$ be the set of all real functions on K which belong to $\mathscr{A}$.

If $f \in \mathscr{A}$ and $f = u + iv$, with u, v real, then $2u = f + \bar{f}$, and since $\mathscr{A}$ is self-adjoint, we see that $u \in \mathscr{A}_R$. If $x_1 \neq x_2$, there exists $f \in \mathscr{A}$ such that $f(x_1) = 1, f(x_2) = 0$; hence $0 = u(x_2) \neq u(x_1) = 1$, which shows that $\mathscr{A}_R$ separates points on K. If $x \in K$, then $g(x) \neq 0$ for some $g \in \mathscr{A}$, and there is a complex number λ such that $\lambda g(x) > 0$; if $f = \lambda g, f = u + iv$, it follows that $u(x) > 0$; hence $\mathscr{A}_R$ vanishes at no point of K.

Thus $\mathscr{A}_R$ satisfies the hypotheses of Theorem 7.32. It follows that every real continuous function on K lies in the uniform closure of $\mathscr{A}_R$, hence lies in $\mathscr{B}$. If f is a complex continuous function on K, $f = u + iv$, then $u \in \mathscr{B}$, $v \in \mathscr{B}$, hence $f \in \mathscr{B}$. This completes the proof.

EXERCISES

1. Prove that every uniformly convergent sequence of bounded functions is uniformly bounded.
2. If $\{f_n\}$ and $\{g_n\}$ converge uniformly on a set E, prove that $\{f_n + g_n\}$ converges uniformly on E. If, in addition, $\{f_n\}$ and $\{g_n\}$ are sequences of bounded functions, prove that $\{f_n g_n\}$ converges uniformly on E.
3. Construct sequences $\{f_n\}$, $\{g_n\}$ which converge uniformly on some set E, but such that $\{f_n g_n\}$ does not converge uniformly on E (of course, $\{f_n g_n\}$ must converge on E).
4. Consider

$$f(x) = \sum_{n=1}^{\infty} \frac{1}{1 + n^2 x}.$$

For what values of x does the series converge absolutely? On what intervals does it converge uniformly? On what intervals does it fail to converge uniformly? Is f continuous wherever the series converges? Is f bounded?

5. Let

$$f_n(x) = \begin{cases} 0 & \left(x < \dfrac{1}{n+1}\right), \\ \sin^2 \dfrac{\pi}{x} & \left(\dfrac{1}{n+1} \leq x \leq \dfrac{1}{n}\right), \\ 0 & \left(\dfrac{1}{n} < x\right). \end{cases}$$

Show that $\{f_n\}$ converges to a continuous function, but not uniformly. Use the series Σf_n to show that absolute convergence, even for all x, does not imply uniform convergence.

6. Prove that the series

$$\sum_{n=1}^{\infty} (-1)^n \frac{x^2 + n}{n^2}$$

converges uniformly in every bounded interval, but does not converge absolutely for any value of x.

7. For $n = 1, 2, 3, \ldots$, x real, put

$$f_n(x) = \frac{x}{1 + nx^2}.$$

Show that $\{f_n\}$ converges uniformly to a function f, and that the equation

$$f'(x) = \lim_{n \to \infty} f_n'(x)$$

is correct if $x \neq 0$, but false if $x = 0$.

8. If

$$I(x) = \begin{cases} 0 & (x \leq 0), \\ 1 & (x > 0), \end{cases}$$

if $\{x_n\}$ is a sequence of distinct points of (a, b), and if $\Sigma |c_n|$ converges, prove that the series

$$f(x) = \sum_{n=1}^{\infty} c_n I(x - x_n) \qquad (a \leq x \leq b)$$

converges uniformly, and that f is continuous for every $x \neq x_n$.

9. Let $\{f_n\}$ be a sequence of continuous functions which converges uniformly to a function f on a set E. Prove that

$$\lim_{n \to \infty} f_n(x_n) = f(x)$$

for every sequence of points $x_n \in E$ such that $x_n \to x$, and $x \in E$. Is the converse of this true?

10. Letting (x) denote the fractional part of the real number x (see Exercise 16, Chap. 4, for the definition), consider the function

$$f(x) = \sum_{n=1}^{\infty} \frac{(nx)}{n^2} \qquad (x \text{ real}).$$

Find all discontinuities of f, and show that they form a countable dense set. Show that f is nevertheless Riemann-integrable on every bounded interval.

11. Suppose $\{f_n\}$, $\{g_n\}$ are defined on E, and
  (a) Σf_n has uniformly bounded partial sums;
  (b) $g_n \to 0$ uniformly on E;
  (c) $g_1(x) \geq g_2(x) \geq g_3(x) \geq \cdots$ for every $x \in E$.
  Prove that $\Sigma f_n g_n$ converges uniformly on E. *Hint*: Compare with Theorem 3.42.

12. Suppose g and f_n ($n = 1, 2, 3, \ldots$) are defined on $(0, \infty)$, are Riemann-integrable on $[t, T]$ whenever $0 < t < T < \infty$, $|f_n| \leq g$, $f_n \to f$ uniformly on every compact subset of $(0, \infty)$, and

$$\int_0^\infty g(x)\, dx < \infty.$$

Prove that

$$\lim_{n \to \infty} \int_0^\infty f_n(x)\, dx = \int_0^\infty f(x)\, dx.$$

(See Exercises 7 and 8 of Chap. 6 for the relevant definitions.)

This is a rather weak form of Lebesgue's dominated convergence theorem (Theorem 11.32). Even in the context of the Riemann integral, uniform convergence can be replaced by pointwise convergence if it is assumed that $f \in \mathscr{R}$. (See the articles by F. Cunningham in *Math. Mag.*, vol. 40, 1967, pp. 179–186, and by H. Kestelman in *Amer. Math. Monthly*, vol. 77, 1970, pp. 182–187.)

13. Assume that $\{f_n\}$ is a sequence of monotonically increasing functions on R^1 with $0 \leq f_n(x) \leq 1$ for all x and all n.
  (a) Prove that there is a function f and a sequence $\{n_k\}$ such that

$$f(x) = \lim_{k \to \infty} f_{n_k}(x)$$

for every $x \in R^1$. (The existence of such a pointwise convergent subsequence is usually called *Helly's selection theorem*.)
  (b) If, moreover, f is continuous, prove that $f_{n_k} \to f$ uniformly on compact sets.
  Hint: (i) Some subsequence $\{f_{n_i}\}$ converges at all rational points r, say, to $f(r)$. (ii) Define $f(x)$, for any $x \in R^1$, to be sup $f(r)$, the sup being taken over all $r \leq x$. (iii) Show that $f_{n_i}(x) \to f(x)$ at every x at which f is continuous. (This is where monotonicity is strongly used.) (iv) A subsequence of $\{f_{n_i}\}$ converges at every point of discontinuity of f since there are at most countably many such points. This proves (a). To prove (b), modify your proof of (iii) appropriately.

14. Let f be a continuous real function on R^1 with the following properties: $0 \le f(t) \le 1$, $f(t+2) = f(t)$ for every t, and

$$f(t) = \begin{cases} 0 & (0 \le t \le \tfrac{1}{3}) \\ 1 & (\tfrac{2}{3} \le t \le 1). \end{cases}$$

Put $\Phi(t) = (x(t), y(t))$, where

$$x(t) = \sum_{n=1}^{\infty} 2^{-n} f(3^{2n-1} t), \qquad y(t) = \sum_{n=1}^{\infty} 2^{-n} f(3^{2n} t).$$

Prove that Φ is *continuous* and that Φ maps $I = [0, 1]$ *onto* the unit square $I^2 \subset R^2$. If fact, show that Φ maps the Cantor set onto I^2.

Hint: Each $(x_0, y_0) \in I^2$ has the form

$$x_0 = \sum_{n=1}^{\infty} 2^{-n} a_{2n-1}, \qquad y_0 = \sum_{n=1}^{\infty} 2^{-n} a_{2n}$$

where each a_i is 0 or 1. If

$$t_0 = \sum_{i=1}^{\infty} 3^{-i-1}(2a_i)$$

show that $f(3^k t_0) = a_k$, and hence that $x(t_0) = x_0$, $y(t_0) = y_0$.

(This simple example of a so-called "space-filling curve" is due to I. J. Schoenberg, *Bull. A.M.S.*, vol. 44, 1938, pp. 519.)

15. Suppose f is a real continuous function on R^1, $f_n(t) = f(nt)$ for $n = 1, 2, 3, \ldots$, and $\{f_n\}$ is equicontinuous on $[0, 1]$. What conclusion can you draw about f?

16. Suppose $\{f_n\}$ is an equicontinuous sequence of functions on a compact set K, and $\{f_n\}$ converges pointwise on K. Prove that $\{f_n\}$ converges uniformly on K.

17. Define the notions of uniform convergence and equicontinuity for mappings into any metric space. Show that Theorems 7.9 and 7.12 are valid for mappings into any metric space, that Theorems 7.8 and 7.11 are valid for mappings into any complete metric space, and that Theorems 7.10, 7.16, 7.17, 7.24, and 7.25 hold for vector-valued functions, that is, for mappings into any R^k.

18. Let $\{f_n\}$ be a uniformly bounded sequence of functions which are Riemann-integrable on $[a, b]$, and put

$$F_n(x) = \int_a^x f_n(t)\, dt \qquad (a \le x \le b).$$

Prove that there exists a subsequence $\{F_{n_k}\}$ which converges uniformly on $[a, b]$.

19. Let K be a compact metric space, let S be a subset of $\mathscr{C}(K)$. Prove that S is compact (with respect to the metric defined in Section 7.14) if and only if S is uniformly closed, pointwise bounded, and equicontinuous. (If S is not equicontinuous, then S contains a sequence which has no equicontinuous subsequence, hence has no subsequence that converges uniformly on K.)

20. If f is continuous on $[0, 1]$ and if

$$\int_0^1 f(x)x^n \, dx = 0 \quad (n = 0, 1, 2, \ldots),$$

prove that $f(x) = 0$ on $[0, 1]$. *Hint:* The integral of the product of f with any polynomial is zero. Use the Weierstrass theorem to show that $\int_0^1 f^2(x) \, dx = 0$.

21. Let K be the unit circle in the complex plane (i.e., the set of all z with $|z| = 1$), and let $\mathscr{A}$ be the algebra of all functions of the form

$$f(e^{i\theta}) = \sum_{n=0}^N c_n e^{in\theta} \quad (\theta \text{ real}).$$

Then $\mathscr{A}$ separates points on K and $\mathscr{A}$ vanishes at no point of K, but nevertheless there are continuous functions on K which are not in the uniform closure of $\mathscr{A}$. *Hint:* For every $f \in \mathscr{A}$

$$\int_0^{2\pi} f(e^{i\theta}) e^{i\theta} \, d\theta = 0,$$

and this is also true for every f in the closure of $\mathscr{A}$.

22. Assume $f \in \mathscr{R}(\alpha)$ on $[a, b]$, and prove that there are polynomials P_n such that

$$\lim_{n \to \infty} \int_a^b |f - P_n|^2 \, d\alpha = 0.$$

(Compare with Exercise 12, Chap. 6.)

23. Put $P_0 = 0$, and define, for $n = 0, 1, 2, \ldots$,

$$P_{n+1}(x) = P_n(x) + \frac{x^2 - P_n^2(x)}{2}.$$

Prove that

$$\lim_{n \to \infty} P_n(x) = |x|,$$

uniformly on $[-1, 1]$.

(This makes it possible to prove the Stone-Weierstrass theorem without first proving Theorem 7.26.)

Hint: Use the identity

$$|x| - P_{n+1}(x) = [|x| - P_n(x)]\left[1 - \frac{|x| + P_n(x)}{2}\right]$$

to prove that $0 \leq P_n(x) \leq P_{n+1}(x) \leq |x|$ if $|x| \leq 1$, and that

$$|x| - P_n(x) \leq |x|\left(1 - \frac{|x|}{2}\right)^n < \frac{2}{n+1}$$

if $|x| \leq 1$.

24. Let X be a metric space, with metric d. Fix a point $a \in X$. Assign to each $p \in X$ the function f_p defined by

$$f_p(x) = d(x, p) - d(x, a) \qquad (x \in X).$$

Prove that $|f_p(x)| \leq d(a, p)$ for all $x \in X$, and that therefore $f_p \in \mathscr{C}(X)$.
Prove that

$$\|f_p - f_q\| = d(p, q)$$

for all $p, q \in X$.

If $\Phi(p) = f_p$, it follows that Φ is an *isometry* (a distance-preserving mapping) of X onto $\Phi(X) \subset \mathscr{C}(X)$.

Let Y be the closure of $\Phi(X)$ in $\mathscr{C}(X)$. Show that Y is complete.

Conclusion: X is isometric to a dense subset of a complete metric space Y. (Exercise 24, Chap. 3 contains a different proof of this.)

25. Suppose ϕ is a continuous bounded real function in the strip defined by $0 \leq x \leq 1$, $-\infty < y < \infty$. Prove that the initial-value problem

$$y' = \phi(x, y), \qquad y(0) = c$$

has a solution. (Note that the hypotheses of this existence theorem are less stringent than those of the corresponding uniqueness theorem; see Exercise 27, Chap. 5.)

Hint: Fix n. For $i = 0, \ldots, n$, put $x_i = i/n$. Let f_n be a continuous function on $[0, 1]$ such that $f_n(0) = c$,

$$f_n'(t) = \phi(x_i, f_n(x_i)) \qquad \text{if } x_i < t < x_{i+1},$$

and put

$$\Delta_n(t) = f_n'(t) - \phi(t, f_n(t)),$$

except at the points x_i, where $\Delta_n(t) = 0$. Then

$$f_n(x) = c + \int_0^x [\phi(t, f_n(t)) + \Delta_n(t)] \, dt.$$

Choose $M < \infty$ so that $|\phi| \leq M$. Verify the following assertions.

(a) $|f_n'| \leq M$, $|\Delta_n| \leq 2M$, $\Delta_n \in \mathscr{R}$, and $|f_n| \leq |c| + M = M_1$, say, on $[0, 1]$, for all n.
(b) $\{f_n\}$ is equicontinuous on $[0, 1]$, since $|f_n'| \leq M$.
(c) Some $\{f_{n_k}\}$ converges to some f, uniformly on $[0, 1]$.
(d) Since ϕ is uniformly continuous on the rectangle $0 \leq x \leq 1$, $|y| \leq M_1$,

$$\phi(t, f_{n_k}(t)) \to \phi(t, f(t))$$

uniformly on $[0, 1]$.
(e) $\Delta_n(t) \to 0$ uniformly on $[0, 1]$, since

$$\Delta_n(t) = \phi(x_i, f_n(x_i)) - \phi(t, f_n(t))$$

in (x_i, x_{i+1}).

(f) Hence
$$f(x) = c + \int_0^x \phi(t, f(t))\, dt.$$
This f is a solution of the given problem.

26. Prove an analogous existence theorem for the initial-value problem
$$\mathbf{y}' = \mathbf{\Phi}(x, \mathbf{y}), \qquad \mathbf{y}(0) = \mathbf{c},$$
where now $\mathbf{c} \in R^k$, $\mathbf{y} \in R^k$, and $\mathbf{\Phi}$ is a continuous bounded mapping of the part of R^{k+1} defined by $0 \leq x \leq 1$, $\mathbf{y} \in R^k$ into R^k. (Compare Exercise 28, Chap. 5.) *Hint:* Use the vector-valued version of Theorem 7.25.

8

SOME SPECIAL FUNCTIONS

POWER SERIES

In this section we shall derive some properties of functions which are represented by power series, i.e., functions of the form

(1) $$f(x) = \sum_{n=0}^{\infty} c_n x^n$$

or, more generally,

(2) $$f(x) = \sum_{n=0}^{\infty} c_n (x - a)^n.$$

These are called *analytic functions*.

We shall restrict ourselves to real values of x. Instead of circles of convergence (see Theorem 3.39) we shall therefore encounter intervals of convergence.

If (1) converges for all x in $(-R, R)$, for some $R > 0$ (R may be $+\infty$), we say that f is expanded in a power series about the point $x = 0$. Similarly, if (2) converges for $|x - a| < R$, f is said to be expanded in a power series about the point $x = a$. As a matter of convenience, we shall often take $a = 0$ without any loss of generality.

8.1 Theorem *Suppose the series*

(3) $$\sum_{n=0}^{\infty} c_n x^n$$

converges for $|x| < R$, *and define*

(4) $$f(x) = \sum_{n=0}^{\infty} c_n x^n \qquad (|x| < R).$$

Then (3) *converges uniformly on* $[-R + \varepsilon, R - \varepsilon]$, *no matter which* $\varepsilon > 0$ *is chosen. The function f is continuous and differentiable in* $(-R, R)$, *and*

(5) $$f'(x) = \sum_{n=1}^{\infty} n c_n x^{n-1} \qquad (|x| < R).$$

Proof Let $\varepsilon > 0$ be given. For $|x| \leq R - \varepsilon$, we have

$$|c_n x^n| \leq |c_n (R - \varepsilon)^n|;$$

and since

$$\Sigma c_n (R - \varepsilon)^n$$

converges absolutely (every power series converges absolutely in the interior of its interval of convergence, by the root test), Theorem 7.10 shows the uniform convergence of (3) on $[-R + \varepsilon, R - \varepsilon]$.

Since $\sqrt[n]{n} \to 1$ as $n \to \infty$, we have

$$\limsup_{n \to \infty} \sqrt[n]{n|c_n|} = \limsup_{n \to \infty} \sqrt[n]{|c_n|},$$

so that the series (4) and (5) have the same interval of convergence.

Since (5) is a power series, it converges uniformly in $[-R + \varepsilon, R - \varepsilon]$, for every $\varepsilon > 0$, and we can apply Theorem 7.17 (for series instead of sequences). It follows that (5) holds if $|x| \leq R - \varepsilon$.

But, given any x such that $|x| < R$, we can find an $\varepsilon > 0$ such that $|x| < R - \varepsilon$. This shows that (5) holds for $|x| < R$.

Continuity of f follows from the existence of f' (Theorem 5.2).

Corollary *Under the hypotheses of Theorem 8.1, f has derivatives of all orders in* $(-R, R)$, *which are given by*

(6) $$f^{(k)}(x) = \sum_{n=k}^{\infty} n(n-1) \cdots (n-k+1) c_n x^{n-k}.$$

In particular,

(7) $$f^{(k)}(0) = k! c_k \qquad (k = 0, 1, 2, \ldots).$$

(Here $f^{(0)}$ means f, and $f^{(k)}$ is the kth derivative of f, for $k = 1, 2, 3, \ldots$).

Proof Equation (6) follows if we apply Theorem 8.1 successively to f, f', f'', ... Putting $x = 0$ in (6), we obtain (7).

Formula (7) is very interesting. It shows, on the one hand, that the coefficients of the power series development of f are determined by the values of f and of its derivatives at a single point. On the other hand, if the coefficients are given, the values of the derivatives of f at the center of the interval of convergence can be read off immediately from the power series.

Note, however, that although a function f may have derivatives of all orders, the series $\Sigma c_n x^n$, where c_n is computed by (7), need not converge to $f(x)$ for any $x \neq 0$. In this case, f cannot be expanded in a power series about $x = 0$. For if we had $f(x) = \Sigma a_n x^n$, we should have

$$n! a_n = f^{(n)}(0);$$

hence $a_n = c_n$. An example of this situation is given in Exercise 1.

If the series (3) converges at an endpoint, say at $x = R$, then f is continuous not only in $(-R, R)$, but also at $x = R$. This follows from Abel's theorem (for simplicity of notation, we take $R = 1$):

8.2 Theorem *Suppose Σc_n converges. Put*

$$f(x) = \sum_{n=0}^{\infty} c_n x^n \qquad (-1 < x < 1).$$

Then

(8) $$\lim_{x \to 1} f(x) = \sum_{n=0}^{\infty} c_n.$$

Proof Let $s_n = c_0 + \cdots + c_n$, $s_{-1} = 0$. Then

$$\sum_{n=0}^{m} c_n x^n = \sum_{n=0}^{m} (s_n - s_{n-1}) x^n = (1-x) \sum_{n=0}^{m-1} s_n x^n + s_m x^m.$$

For $|x| < 1$, we let $m \to \infty$ and obtain

(9) $$f(x) = (1-x) \sum_{n=0}^{\infty} s_n x^n.$$

Suppose $s = \lim_{n \to \infty} s_n$. Let $\varepsilon > 0$ be given. Choose N so that $n > N$ implies

$$|s - s_n| < \frac{\varepsilon}{2}.$$

Then, since

$$(1-x)\sum_{n=0}^{\infty} x^n = 1 \qquad (|x| < 1),$$

we obtain from (9)

$$|f(x) - s| = \left|(1-x)\sum_{n=0}^{\infty}(s_n - s)x^n\right| \le (1-x)\sum_{n=0}^{N}|s_n - s||x|^n + \frac{\varepsilon}{2} \le \varepsilon$$

if $x > 1 - \delta$, for some suitably chosen $\delta > 0$. This implies (8).

As an application, let us prove Theorem 3.51, which asserts: *If Σa_n, Σb_n, Σc_n, converge to A, B, C, and if $c_n = a_0 b_n + \cdots + a_n b_0$, then $C = AB$.* We let

$$f(x) = \sum_{n=0}^{\infty} a_n x^n, \qquad g(x) = \sum_{n=0}^{\infty} b_n x^n, \qquad h(x) = \sum_{n=0}^{\infty} c_n x^n,$$

for $0 \le x \le 1$. For $x < 1$, these series converge absolutely and hence may be multiplied according to Definition 3.48; when the multiplication is carried out, we see that

(10) $$f(x) \cdot g(x) = h(x) \qquad (0 \le x < 1).$$

By Theorem 8.2,

(11) $$f(x) \to A, \qquad g(x) \to B, \qquad h(x) \to C$$

as $x \to 1$. Equations (10) and (11) imply $AB = C$.

We now require a theorem concerning an inversion in the order of summation. (See Exercises 2 and 3.)

8.3 Theorem *Given a double sequence $\{a_{ij}\}$, $i = 1, 2, 3, \ldots$, $j = 1, 2, 3, \ldots$, suppose that*

(12) $$\sum_{j=1}^{\infty} |a_{ij}| = b_i \qquad (i = 1, 2, 3, \ldots)$$

and Σb_i converges. Then

(13) $$\sum_{i=1}^{\infty} \sum_{j=1}^{\infty} a_{ij} = \sum_{j=1}^{\infty} \sum_{i=1}^{\infty} a_{ij}.$$

Proof We could establish (13) by a direct procedure similar to (although more involved than) the one used in Theorem 3.55. However, the following method seems more interesting.

Let E be a countable set, consisting of the points $x_0, x_1, x_2, \ldots$, and suppose $x_n \to x_0$ as $n \to \infty$. Define

(14) $$f_i(x_0) = \sum_{j=1}^{\infty} a_{ij} \qquad (i = 1, 2, 3, \ldots),$$

(15) $$f_i(x_n) = \sum_{j=1}^{n} a_{ij} \qquad (i, n = 1, 2, 3, \ldots),$$

(16) $$g(x) = \sum_{i=1}^{\infty} f_i(x) \qquad (x \in E).$$

Now, (14) and (15), together with (12), show that each f_i is continuous at x_0. Since $|f_i(x)| \le b_i$ for $x \in E$, (16) converges uniformly, so that g is continuous at x_0 (Theorem 7.11). It follows that

$$\sum_{i=1}^{\infty} \sum_{j=1}^{\infty} a_{ij} = \sum_{i=1}^{\infty} f_i(x_0) = g(x_0) = \lim_{n \to \infty} g(x_n)$$
$$= \lim_{n \to \infty} \sum_{i=1}^{\infty} f_i(x_n) = \lim_{n \to \infty} \sum_{i=1}^{\infty} \sum_{j=1}^{n} a_{ij}$$
$$= \lim_{n \to \infty} \sum_{j=1}^{n} \sum_{i=1}^{\infty} a_{ij} = \sum_{j=1}^{\infty} \sum_{i=1}^{\infty} a_{ij}.$$

8.4 Theorem *Suppose*

$$f(x) = \sum_{n=0}^{\infty} c_n x^n,$$

the series converging in $|x| < R$. *If* $-R < a < R$, *then* f *can be expanded in a power series about the point* $x = a$ *which converges in* $|x - a| < R - |a|$, *and*

(17) $$f(x) = \sum_{n=0}^{\infty} \frac{f^{(n)}(a)}{n!} (x - a)^n \qquad (|x - a| < R - |a|).$$

This is an extension of Theorem 5.15 and is also known as *Taylor's theorem*.

Proof We have

$$f(x) = \sum_{n=0}^{\infty} c_n [(x - a) + a]^n$$
$$= \sum_{n=0}^{\infty} c_n \sum_{m=0}^{n} \binom{n}{m} a^{n-m} (x - a)^m$$
$$= \sum_{m=0}^{\infty} \left[\sum_{n=m}^{\infty} \binom{n}{m} c_n a^{n-m} \right] (x - a)^m.$$

This is the desired expansion about the point $x = a$. To prove its validity, we have to justify the change which was made in the order of summation. Theorem 8.3 shows that this is permissible if

(18) $$\sum_{n=0}^{\infty} \sum_{m=0}^{n} \left| c_n \binom{n}{m} a^{n-m}(x-a)^m \right|$$

converges. But (18) is the same as

(19) $$\sum_{n=0}^{\infty} |c_n| \cdot (|x-a| + |a|)^n,$$

and (19) converges if $|x - a| + |a| < R$.

Finally, the form of the coefficients in (17) follows from (7).

It should be noted that (17) may actually converge in a larger interval than the one given by $|x - a| < R - |a|$.

If two power series converge to the same function in $(-R, R)$, (7) shows that the two series must be identical, i.e., they must have the same coefficients. It is interesting that the same conclusion can be deduced from much weaker hypotheses:

8.5 Theorem *Suppose the series $\Sigma a_n x^n$ and $\Sigma b_n x^n$ converge in the segment $S = (-R, R)$. Let E be the set of all $x \in S$ at which*

(20) $$\sum_{n=0}^{\infty} a_n x^n = \sum_{n=0}^{\infty} b_n x^n.$$

If E has a limit point in S, then $a_n = b_n$ for $n = 0, 1, 2, \ldots$. Hence (20) holds for all $x \in S$.

Proof Put $c_n = a_n - b_n$ and

(21) $$f(x) = \sum_{n=0}^{\infty} c_n x^n \qquad (x \in S).$$

Then $f(x) = 0$ on E.

Let A be the set of all limit points of E in S, and let B consist of all other points of S. It is clear from the definition of "limit point" that B is open. Suppose we can prove that A is open. Then A and B are disjoint open sets. Hence they are separated (Definition 2.45). Since $S = A \cup B$, and S is connected, one of A and B must be empty. By hypothesis, A is not empty. Hence B is empty, and $A = S$. Since f is continuous in S, $A \subset E$. Thus $E = S$, and (7) shows that $c_n = 0$ for $n = 0, 1, 2, \ldots$, which is the desired conclusion.

Thus we have to prove that A is open. If $x_0 \in A$, Theorem 8.4 shows that

(22) $$f(x) = \sum_{n=0}^{\infty} d_n(x - x_0)^n \qquad (|x - x_0| < R - |x_0|).$$

We claim that $d_n = 0$ for all n. Otherwise, let k be the smallest non-negative integer such that $d_k \neq 0$. Then

(23) $$f(x) = (x - x_0)^k g(x) \qquad (|x - x_0| < R - |x_0|),$$

where

(24) $$g(x) = \sum_{m=0}^{\infty} d_{k+m}(x - x_0)^m.$$

Since g is continuous at x_0 and

$$g(x_0) = d_k \neq 0,$$

there exists a $\delta > 0$ such that $g(x) \neq 0$ if $|x - x_0| < \delta$. It follows from (23) that $f(x) \neq 0$ if $0 < |x - x_0| < \delta$. But this contradicts the fact that x_0 is a limit point of E.

Thus $d_n = 0$ for all n, so that $f(x) = 0$ for all x for which (22) holds, i.e., in a neighborhood of x_0. This shows that A is open, and completes the proof.

THE EXPONENTIAL AND LOGARITHMIC FUNCTIONS

We define

(25) $$E(z) = \sum_{n=0}^{\infty} \frac{z^n}{n!}$$

The ratio test shows that this series converges for every complex z. Applying Theorem 3.50 on multiplication of absolutely convergent series, we obtain

$$E(z)E(w) = \sum_{n=0}^{\infty} \frac{z^n}{n!} \sum_{m=0}^{\infty} \frac{w^m}{m!} = \sum_{n=0}^{\infty} \sum_{k=0}^{n} \frac{z^k w^{n-k}}{k!(n-k)!}$$

$$= \sum_{n=0}^{\infty} \frac{1}{n!} \sum_{k=0}^{n} \binom{n}{k} z^k w^{n-k} = \sum_{n=0}^{\infty} \frac{(z+w)^n}{n!},$$

which gives us the important addition formula

(26) $$E(z + w) = E(z)E(w) \qquad (z, w \text{ complex}).$$

One consequence is that

(27) $$E(z)E(-z) = E(z - z) = E(0) = 1 \qquad (z \text{ complex}).$$

This shows that $E(z) \neq 0$ for all z. By (25), $E(x) > 0$ if $x > 0$; hence (27) shows that $E(x) > 0$ for all real x. By (25), $E(x) \to +\infty$ as $x \to +\infty$; hence (27) shows that $E(x) \to 0$ as $x \to -\infty$ along the real axis. By (25), $0 < x < y$ implies that $E(x) < E(y)$; by (27), it follows that $E(-y) < E(-x)$; hence E is strictly increasing on the whole real axis.

The addition formula also shows that

$$(28) \qquad \lim_{h=0} \frac{E(z+h) - E(z)}{h} = E(z) \lim_{h=0} \frac{E(h) - 1}{h} = E(z);$$

the last equality follows directly from (25).

Iteration of (26) gives

$$(29) \qquad E(z_1 + \cdots + z_n) = E(z_1) \cdots E(z_n).$$

Let us take $z_1 = \cdots = z_n = 1$. Since $E(1) = e$, where e is the number defined in Definition 3.30, we obtain

$$(30) \qquad E(n) = e^n \qquad (n = 1, 2, 3, \ldots).$$

If $p = n/m$, where n, m are positive integers, then

$$(31) \qquad [E(p)]^m = E(mp) = E(n) = e^n,$$

so that

$$(32) \qquad E(p) = e^p \qquad (p > 0, p \text{ rational}).$$

It follows from (27) that $E(-p) = e^{-p}$ if p is positive and rational. Thus (32) holds for all rational p.

In Exercise 6, Chap. 1, we suggested the definition

$$(33) \qquad x^y = \sup x^p,$$

where the sup is taken over all rational p such that $p < y$, for any real y, and $x > 1$. If we thus define, for any real x,

$$(34) \qquad e^x = \sup e^p \qquad (p < x, p \text{ rational}),$$

the continuity and monotonicity properties of E, together with (32), show that

$$(35) \qquad E(x) = e^x$$

for all real x. Equation (35) explains why E is called the exponential function.

The notation $\exp(x)$ is often used in place of e^x, expecially when x is a complicated expression.

Actually one may very well use (35) instead of (34) as the definition of e^x; (35) is a much more convenient starting point for the investigation of the properties of e^x. We shall see presently that (33) may also be replaced by a more convenient definition [see (43)].

We now revert to the customary notation, e^x, in place of $E(x)$, and summarize what we have proved so far.

8.6 Theorem *Let e^x be defined on R^1 by (35) and (25). Then*
  (a) *e^x is continuous and differentiable for all x;*
  (b) *$(e^x)' = e^x$;*
  (c) *e^x is a strictly increasing function of x, and $e^x > 0$;*
  (d) *$e^{x+y} = e^x e^y$;*
  (e) *$e^x \to +\infty$ as $x \to +\infty$, $e^x \to 0$ as $x \to -\infty$;*
  (f) *$\lim_{x \to +\infty} x^n e^{-x} = 0$, for every n.*

Proof We have already proved (a) to (e); (25) shows that

$$e^x > \frac{x^{n+1}}{(n+1)!}$$

for $x > 0$, so that

$$x^n e^{-x} < \frac{(n+1)!}{x},$$

and (f) follows. Part (f) shows that e^x tends to $+\infty$ "faster" than any power of x, as $x \to +\infty$.

Since E is strictly increasing and differentiable on R^1, it has an inverse function L which is also strictly increasing and differentiable and whose domain is $E(R^1)$, that is, the set of all positive numbers. L is defined by

(36) $$E(L(y)) = y \quad (y > 0),$$

or, equivalently, by

(37) $$L(E(x)) = x \quad (x \text{ real}).$$

Differentiating (37), we get (compare Theorem 5.5)

$$L'(E(x)) \cdot E(x) = 1.$$

Writing $y = E(x)$, this gives us

(38) $$L'(y) = \frac{1}{y} \quad (y > 0).$$

Taking $x = 0$ in (37), we see that $L(1) = 0$. Hence (38) implies

(39) $$L(y) = \int_1^y \frac{dx}{x}.$$

Quite frequently, (39) is taken as the starting point of the theory of the logarithm and the exponential function. Writing $u = E(x)$, $v = E(y)$, (26) gives

$$L(uv) = L(E(x) \cdot E(y)) = L(E(x + y)) = x + y,$$

so that

(40) $$L(uv) = L(u) + L(v) \qquad (u > 0, v > 0).$$

This shows that L has the familiar property which makes logarithms useful tools for computation. The customary notation for $L(x)$ is of course $\log x$.

As to the behavior of $\log x$ as $x \to +\infty$ and as $x \to 0$, Theorem 8.6(e) shows that

$$\log x \to +\infty \qquad \text{as } x \to +\infty,$$
$$\log x \to -\infty \qquad \text{as } x \to 0.$$

It is easily seen that

(41) $$x^n = E(nL(x))$$

if $x > 0$ and n is an integer. Similarly, if m is a positive integer, we have

(42) $$x^{1/m} = E\left(\frac{1}{m} L(x)\right),$$

since each term of (42), when raised to the mth power, yields the corresponding term of (36). Combining (41) and (42), we obtain

(43) $$x^\alpha = E(\alpha L(x)) = e^{\alpha \log x}$$

for any rational α.

We now define x^α, for any real α and any $x > 0$, by (43). The continuity and monotonicity of E and L show that this definition leads to the same result as the previously suggested one. The facts stated in Exercise 6 of Chap. 1, are trivial consequences of (43).

If we differentiate (43), we obtain, by Theorem 5.5,

(44) $$(x^\alpha)' = E(\alpha L(x)) \cdot \frac{\alpha}{x} = \alpha x^{\alpha - 1}.$$

Note that we have previously used (44) only for integral values of α, in which case (44) follows easily from Theorem 5.3(b). To prove (44) directly from the definition of the derivative, if x^α is defined by (33) and α is irrational, is quite troublesome.

The well-known integration formula for x^α follows from (44) if $\alpha \neq -1$, and from (38) if $\alpha = -1$. We wish to demonstrate one more property of $\log x$, namely,

(45) $$\lim_{x \to +\infty} x^{-\alpha} \log x = 0$$

for every $\alpha > 0$. That is, $\log x \to +\infty$ "slower" than any positive power of x, as $x \to +\infty$.

For if $0 < \varepsilon < \alpha$, and $x > 1$, then

$$x^{-\alpha} \log x = x^{-\alpha} \int_1^x t^{-1}\, dt < x^{-\alpha} \int_1^x t^{\varepsilon-1}\, dt$$

$$= x^{-\alpha} \cdot \frac{x^\varepsilon - 1}{\varepsilon} < \frac{x^{\varepsilon-\alpha}}{\varepsilon},$$

and (45) follows. We could also have used Theorem 8.6(f) to derive (45).

THE TRIGONOMETRIC FUNCTIONS

Let us define

(46) $\qquad C(x) = \frac{1}{2}[E(ix) + E(-ix)], \qquad S(x) = \frac{1}{2i}[E(ix) - E(-ix)].$

We shall show that $C(x)$ and $S(x)$ coincide with the functions $\cos x$ and $\sin x$, whose definition is usually based on geometric considerations. By (25), $E(\bar{z}) = \overline{E(z)}$. Hence (46) shows that $C(x)$ and $S(x)$ are real for real x. Also,

(47) $\qquad E(ix) = C(x) + iS(x).$

Thus $C(x)$ and $S(x)$ are the real and imaginary parts, respectively, of $E(ix)$, if x is real. By (27),

$$|E(ix)|^2 = E(ix)\overline{E(ix)} = E(ix)E(-ix) = 1,$$

so that

(48) $\qquad |E(ix)| = 1 \qquad (x \text{ real}).$

From (46) we can read off that $C(0) = 1$, $S(0) = 0$, and (28) shows that

(49) $\qquad C'(x) = -S(x), \qquad S'(x) = C(x).$

We assert that there exist positive numbers x such that $C(x) = 0$. For suppose this is not so. Since $C(0) = 1$, it then follows that $C(x) > 0$ for all $x > 0$, hence $S'(x) > 0$, by (49), hence S is strictly increasing; and since $S(0) = 0$, we have $S(x) > 0$ if $x > 0$. Hence if $0 < x < y$, we have

(50) $\qquad S(x)(y - x) < \int_x^y S(t)\, dt = C(x) - C(y) \leq 2.$

The last inequality follows from (48) and (47). Since $S(x) > 0$, (50) cannot be true for large y, and we have a contradiction.

Let x_0 be the smallest positive number such that $C(x_0) = 0$. This exists, since the set of zeros of a continuous function is closed, and $C(0) \neq 0$. We define the number π by

(51) $$\pi = 2x_0.$$

Then $C(\pi/2) = 0$, and (48) shows that $S(\pi/2) = \pm 1$. Since $C(x) > 0$ in $(0, \pi/2)$, S is increasing in $(0, \pi/2)$; hence $S(\pi/2) = 1$. Thus

$$E\left(\frac{\pi i}{2}\right) = i,$$

and the addition formula gives

(52) $$E(\pi i) = -1, \quad E(2\pi i) = 1;$$

hence

(53) $$E(z + 2\pi i) = E(z) \quad (z \text{ complex}).$$

8.7 Theorem

(a) *The function E is periodic, with period $2\pi i$.*
(b) *The functions C and S are periodic, with period 2π.*
(c) *If $0 < t < 2\pi$, then $E(it) \neq 1$.*
(d) *If z is a complex number with $|z| = 1$, there is a unique t in $[0, 2\pi)$ such that $E(it) = z$.*

Proof By (53), (a) holds; and (b) follows from (a) and (46).

Suppose $0 < t < \pi/2$ and $E(it) = x + iy$, with x, y real. Our preceding work shows that $0 < x < 1$, $0 < y < 1$. Note that

$$E(4it) = (x + iy)^4 = x^4 - 6x^2 y^2 + y^4 + 4ixy(x^2 - y^2).$$

If $E(4it)$ is real, it follows that $x^2 - y^2 = 0$; since $x^2 + y^2 = 1$, by (48), we have $x^2 = y^2 = \frac{1}{2}$, hence $E(4it) = -1$. This proves (c).

If $0 \leq t_1 < t_2 < 2\pi$, then

$$E(it_2)[E(it_1)]^{-1} = E(it_2 - it_1) \neq 1,$$

by (c). This establishes the uniqueness assertion in (d).

To prove the existence assertion in (d), fix z so that $|z| = 1$. Write $z = x + iy$, with x and y real. Suppose first that $x \geq 0$ and $y \geq 0$. On $[0, \pi/2]$, C decreases from 1 to 0. Hence $C(t) = x$ for some $t \in [0, \pi/2]$. Since $C^2 + S^2 = 1$ and $S \geq 0$ on $[0, \pi/2]$, it follows that $z = E(it)$.

If $x < 0$ and $y \geq 0$, the preceding conditions are satisfied by $-iz$. Hence $-iz = E(it)$ for some $t \in [0, \pi/2]$, and since $i = E(\pi i/2)$, we obtain $z = E(i(t + \pi/2))$. Finally, if $y < 0$, the preceding two cases show that

$-z = E(it)$ for some $t \in (0, \pi)$. Hence $z = -E(it) = E(i(t + \pi))$.
This proves (d), and hence the theorem.

It follows from (d) and (48) that the curve γ defined by
$$\gamma(t) = E(it) \qquad (0 \le t \le 2\pi) \tag{54}$$
is a simple closed curve whose range is the unit circle in the plane. Since $\gamma'(t) = iE(it)$, the length of γ is
$$\int_0^{2\pi} |\gamma'(t)|\, dt = 2\pi,$$
by Theorem 6.27. This is of course the expected result for the circumference of a circle of radius 1. It shows that π, defined by (51), has the usual geometric significance.

In the same way we see that the point $\gamma(t)$ describes a circular arc of length t_0 as t increases from 0 to t_0. Consideration of the triangle whose vertices are
$$z_1 = 0, \qquad z_2 = \gamma(t_0), \qquad z_3 = C(t_0)$$
shows that $C(t)$ and $S(t)$ are indeed identical with $\cos t$ and $\sin t$, if the latter are defined in the usual way as ratios of the sides of a right triangle.

It should be stressed that we derived the basic properties of the trigonometric functions from (46) and (25), without any appeal to the geometric notion of angle. There are other nongeometric approaches to these functions. The papers by W. F. Eberlein (*Amer. Math. Monthly*, vol. 74, 1967, pp. 1223–1225) and by G. B. Robison (*Math. Mag.*, vol. 41, 1968, pp. 66–70) deal with these topics.

THE ALGEBRAIC COMPLETENESS OF THE COMPLEX FIELD

We are now in a position to give a simple proof of the fact that the complex field is algebraically complete, that is to say, that every nonconstant polynomial with complex coefficients has a complex root.

8.8 Theorem *Suppose $a_0, \ldots, a_n$ are complex numbers, $n \ge 1$, $a_n \ne 0$,*
$$P(z) = \sum_0^n a_k z^k.$$
Then $P(z) = 0$ for some complex number z.

Proof Without loss of generality, assume $a_n = 1$. Put
$$\mu = \inf |P(z)| \qquad (z \text{ complex}) \tag{55}$$
If $|z| = R$, then
$$|P(z)| \ge R^n [1 - |a_{n-1}|R^{-1} - \cdots - |a_0|R^{-n}]. \tag{56}$$

The right side of (56) tends to ∞ as $R \to \infty$. Hence there exists R_0 such that $|P(z)| > \mu$ if $|z| > R_0$. Since $|P|$ is continuous on the closed disc with center at 0 and radius R_0, Theorem 4.16 shows that $|P(z_0)| = \mu$ for some z_0.

We claim that $\mu = 0$.

If not, put $Q(z) = P(z + z_0)/P(z_0)$. Then Q is a nonconstant polynomial, $Q(0) = 1$, and $|Q(z)| \geq 1$ for all z. There is a smallest integer k, $1 \leq k \leq n$, such that

(57) $$Q(z) = 1 + b_k z^k + \cdots + b_n z^n, \quad b_k \neq 0.$$

By Theorem 8.7(d) there is a real θ such that

(58) $$e^{ik\theta} b_k = -|b_k|.$$

If $r > 0$ and $r^k |b_k| < 1$, (58) implies

$$|1 + b_k r^k e^{ik\theta}| = 1 - r^k |b_k|,$$

so that

$$|Q(re^{i\theta})| \leq 1 - r^k \{|b_k| - r|b_{k+1}| - \cdots - r^{n-k}|b_n|\}.$$

For sufficiently small r, the expression in braces is positive; hence $|Q(re^{i\theta})| < 1$, a contradiction.

Thus $\mu = 0$, that is, $P(z_0) = 0$.

Exercise 27 contains a more general result.

FOURIER SERIES

8.9 Definition A *trigonometric polynomial* is a finite sum of the form

(59) $$f(x) = a_0 + \sum_{n=1}^{N} (a_n \cos nx + b_n \sin nx) \quad (x \text{ real}),$$

where $a_0, \ldots, a_N, b_1, \ldots, b_N$ are complex numbers. On account of the identities (46), (59) can also be written in the form

(60) $$f(x) = \sum_{-N}^{N} c_n e^{inx} \quad (x \text{ real}),$$

which is more convenient for most purposes. It is clear that every trigonometric polynomial is periodic, with period 2π.

If n is a nonzero integer, e^{inx} is the derivative of e^{inx}/in, which also has period 2π. Hence

(61) $$\frac{1}{2\pi} \int_{-\pi}^{\pi} e^{inx} \, dx = \begin{cases} 1 & (\text{if } n = 0), \\ 0 & (\text{if } n = \pm 1, \pm 2, \ldots). \end{cases}$$

Let us multiply (60) by e^{-imx}, where m is an integer; if we integrate the product, (61) shows that

$$(62) \qquad c_m = \frac{1}{2\pi} \int_{-\pi}^{\pi} f(x) e^{-imx} \, dx$$

for $|m| \le N$. If $|m| > N$, the integral in (62) is 0.

The following observation can be read off from (60) and (62): The trigonometric polynomial f, given by (60), is *real* if and only if $c_{-n} = \overline{c_n}$ for $n = 0, \ldots, N$.

In agreement with (60), we define a *trigonometric series* to be a series of the form

$$(63) \qquad \sum_{-\infty}^{\infty} c_n e^{inx} \qquad (x \text{ real});$$

the Nth partial sum of (63) is defined to be the right side of (60).

If f is an integrable function on $[-\pi, \pi]$, the numbers c_m defined by (62) for all integers m are called the *Fourier coefficients* of f, and the series (63) formed with these coefficients is called the *Fourier series* of f.

The natural question which now arises is whether the Fourier series of f converges to f, or, more generally, whether f is determined by its Fourier series. That is to say, if we know the Fourier coefficients of a function, can we find the function, and if so, how?

The study of such series, and, in particular, the problem of representing a given function by a trigonometric series, originated in physical problems such as the theory of oscillations and the theory of heat conduction (Fourier's "Théorie analytique de la chaleur" was published in 1822). The many difficult and delicate problems which arose during this study caused a thorough revision and reformulation of the whole theory of functions of a real variable. Among many prominent names, those of Riemann, Cantor, and Lebesgue are intimately connected with this field, which nowadays, with all its generalizations and ramifications, may well be said to occupy a central position in the whole of analysis.

We shall be content to derive some basic theorems which are easily accessible by the methods developed in the preceding chapters. For more thorough investigations, the Lebesgue integral is a natural and indispensable tool.

We shall first study more general systems of functions which share a property analogous to (61).

8.10 Definition Let $\{\phi_n\}$ ($n = 1, 2, 3, \ldots$) be a sequence of complex functions on $[a, b]$, such that

$$(64) \qquad \int_a^b \phi_n(x) \overline{\phi_m(x)} \, dx = 0 \qquad (n \ne m).$$

Then $\{\phi_n\}$ is said to be an *orthogonal system of functions* on $[a, b]$. If, in addition,

(65) $$\int_a^b |\phi_n(x)|^2 \, dx = 1$$

for all n, $\{\phi_n\}$ is said to be *orthonormal*.

For example, the functions $(2\pi)^{-\frac{1}{2}} e^{inx}$ form an orthonormal system on $[-\pi, \pi]$. So do the real functions

$$\frac{1}{\sqrt{2\pi}}, \frac{\cos x}{\sqrt{\pi}}, \frac{\sin x}{\sqrt{\pi}}, \frac{\cos 2x}{\sqrt{\pi}}, \frac{\sin 2x}{\sqrt{\pi}}, \ldots.$$

If $\{\phi_n\}$ is orthonormal on $[a, b]$ and if

(66) $$c_n = \int_a^b f(t) \overline{\phi_n(t)} \, dt \qquad (n = 1, 2, 3, \ldots),$$

we call c_n the nth Fourier coefficient of f relative to $\{\phi_n\}$. We write

(67) $$f(x) \sim \sum_1^\infty c_n \phi_n(x)$$

and call this series the Fourier series of f (relative to $\{\phi_n\}$).

Note that the symbol $\sim$ used in (67) implies nothing about the convergence of the series; it merely says that the coefficients are given by (66).

The following theorems show that the partial sums of the Fourier series of f have a certain minimum property. We shall assume here and in the rest of this chapter that $f \in \mathcal{R}$, although this hypothesis can be weakened.

8.11 Theorem *Let $\{\phi_n\}$ be orthonormal on $[a, b]$. Let*

(68) $$s_n(x) = \sum_{m=1}^n c_m \phi_m(x)$$

be the nth partial sum of the Fourier series of f, and suppose

(69) $$t_n(x) = \sum_{m=1}^n \gamma_m \phi_m(x).$$

Then

(70) $$\int_a^b |f - s_n|^2 \, dx \leq \int_a^b |f - t_n|^2 \, dx,$$

and equality holds if and only if

(71) $$\gamma_m = c_m \qquad (m = 1, \ldots, n).$$

That is to say, among all functions t_n, s_n gives the best possible mean square approximation to f.

Proof Let $\int$ denote the integral over $[a, b]$, Σ the sum from 1 to n. Then
$$\int f \bar{t}_n = \int f \sum \bar{\gamma}_m \bar{\phi}_m = \sum c_m \bar{\gamma}_m$$
by the definition of $\{c_m\}$,
$$\int |t_n|^2 = \int t_n \bar{t}_n = \int \sum \gamma_m \phi_m \sum \bar{\gamma}_k \bar{\phi}_k = \sum |\gamma_m|^2$$
since $\{\phi_m\}$ is orthonormal, and so
$$\int |f - t_n|^2 = \int |f|^2 - \int f \bar{t}_n - \int \bar{f} t_n + \int |t_n|^2$$
$$= \int |f|^2 - \sum c_m \bar{\gamma}_m - \sum \bar{c}_m \gamma_m + \sum \gamma_m \bar{\gamma}_m$$
$$= \int |f|^2 - \sum |c_m|^2 + \sum |\gamma_m - c_m|^2,$$
which is evidently minimized if and only if $\gamma_m = c_m$.

Putting $\gamma_m = c_m$ in this calculation, we obtain
$$(72) \qquad \int_a^b |s_n(x)|^2 \, dx = \sum_1^n |c_m|^2 \leq \int_a^b |f(x)|^2 \, dx,$$
since $\int |f - t_n|^2 \geq 0$.

8.12 Theorem *If $\{\phi_n\}$ is orthonormal on $[a, b]$, and if*
$$f(x) \sim \sum_{n=1}^\infty c_n \phi_n(x),$$
then
$$(73) \qquad \sum_{n=1}^\infty |c_n|^2 \leq \int_a^b |f(x)|^2 \, dx.$$
In particular,
$$(74) \qquad \lim_{n \to \infty} c_n = 0.$$

Proof Letting $n \to \infty$ in (72), we obtain (73), the so-called "Bessel inequality."

8.13 Trigonometric series From now on we shall deal only with the trigonometric system. We shall consider functions f that have period 2π and that are Riemann-integrable on $[-\pi, \pi]$ (and hence on every bounded interval). The Fourier series of f is then the series (63) whose coefficients c_n are given by the integrals (62), and
$$(75) \qquad s_N(x) = s_N(f; x) = \sum_{-N}^N c_n e^{inx}$$

is the Nth partial sum of the Fourier series of f. The inequality (72) now takes the form

$$\text{(76)} \qquad \frac{1}{2\pi}\int_{-\pi}^{\pi}|s_N(x)|^2\,dx = \sum_{-N}^{N}|c_n|^2 \le \frac{1}{2\pi}\int_{-\pi}^{\pi}|f(x)|^2\,dx.$$

In order to obtain an expression for s_N that is more manageable than (75) we introduce the *Dirichlet kernel*

$$\text{(77)} \qquad D_N(x) = \sum_{n=-N}^{N} e^{inx} = \frac{\sin(N+\tfrac{1}{2})x}{\sin(x/2)}.$$

The first of these equalities is the definition of $D_N(x)$. The second follows if both sides of the identity

$$(e^{ix} - 1)D_N(x) = e^{i(N+1)x} - e^{-iNx}$$

are multiplied by $e^{-ix/2}$.

By (62) and (75), we have

$$s_N(f;x) = \sum_{-N}^{N} \frac{1}{2\pi}\int_{-\pi}^{\pi} f(t)e^{-int}\,dt\, e^{inx}$$

$$= \frac{1}{2\pi}\int_{-\pi}^{\pi} f(t) \sum_{-N}^{N} e^{in(x-t)}\,dt,$$

so that

$$\text{(78)} \qquad s_N(f;x) = \frac{1}{2\pi}\int_{-\pi}^{\pi} f(t) D_N(x-t)\,dt = \frac{1}{2\pi}\int_{-\pi}^{\pi} f(x-t) D_N(t)\,dt.$$

The periodicity of all functions involved shows that it is immaterial over which interval we integrate, as long as its length is 2π. This shows that the two integrals in (78) are equal.

We shall prove just one theorem about the pointwise convergence of Fourier series.

8.14 Theorem *If, for some x, there are constants $\delta > 0$ and $M < \infty$ such that*

$$\text{(79)} \qquad |f(x+t) - f(x)| \le M|t|$$

for all $t \in (-\delta, \delta)$, then

$$\text{(80)} \qquad \lim_{N\to\infty} s_N(f;x) = f(x).$$

Proof Define

$$\text{(81)} \qquad g(t) = \frac{f(x-t) - f(x)}{\sin(t/2)}$$

for $0 < |t| \leq \pi$, and put $g(0) = 0$. By the definition (77),
$$\frac{1}{2\pi} \int_{-\pi}^{\pi} D_N(x)\, dx = 1.$$
Hence (78) shows that
$$s_N(f; x) - f(x) = \frac{1}{2\pi} \int_{-\pi}^{\pi} g(t) \sin\left(N + \frac{1}{2}\right)t\, dt$$
$$= \frac{1}{2\pi} \int_{-\pi}^{\pi} \left[g(t) \cos \frac{t}{2}\right] \sin Nt\, dt + \frac{1}{2\pi} \int_{-\pi}^{\pi} \left[g(t) \sin \frac{t}{2}\right] \cos Nt\, dt.$$

By (79) and (81), $g(t) \cos (t/2)$ and $g(t) \sin (t/2)$ are bounded. The last two integrals thus tend to 0 as $N \to \infty$, by (74). This proves (80).

Corollary *If $f(x) = 0$ for all x in some segment J, then $\lim s_N(f; x) = 0$ for every $x \in J$.*

Here is another formulation of this corollary:

If $f(t) = g(t)$ for all t in some neighborhood of x, then
$$s_N(f; x) - s_N(g; x) = s_N(f - g; x) \to 0 \text{ as } N \to \infty.$$

This is usually called the *localization theorem*. It shows that the behavior of the sequence $\{s_N(f; x)\}$, as far as convergence is concerned, depends only on the values of f in some (arbitrarily small) neighborhood of x. Two Fourier series may thus have the same behavior in one interval, but may behave in entirely different ways in some other interval. We have here a very striking contrast between Fourier series and power series (Theorem 8.5).

We conclude with two other approximation theorems.

8.15 Theorem *If f is continuous (with period 2π) and if $\varepsilon > 0$, then there is a trigonometric polynomial P such that*
$$|P(x) - f(x)| < \varepsilon$$
for all real x.

Proof If we identify x and $x + 2\pi$, we may regard the 2π-periodic functions on R^1 as functions on the unit circle T, by means of the mapping $x \to e^{ix}$. The trigonometric polynomials, i.e., the functions of the form (60), form a self-adjoint algebra $\mathscr{A}$, which separates points on T, and which vanishes at no point of T. Since T is compact, Theorem 7.33 tells us that $\mathscr{A}$ is dense in $\mathscr{C}(T)$. This is exactly what the theorem asserts.

A more precise form of this theorem appears in Exercise 15.

8.16 Parseval's theorem *Suppose f and g are Riemann-integrable functions with period 2π, and*

(82) $$f(x) \sim \sum_{-\infty}^{\infty} c_n e^{inx}, \qquad g(x) \sim \sum_{-\infty}^{\infty} \gamma_n e^{inx}.$$

Then

(83) $$\lim_{N \to \infty} \frac{1}{2\pi} \int_{-\pi}^{\pi} |f(x) - s_N(f; x)|^2 \, dx = 0,$$

(84) $$\frac{1}{2\pi} \int_{-\pi}^{\pi} f(x) \overline{g(x)} \, dx = \sum_{-\infty}^{\infty} c_n \bar{\gamma}_n,$$

(85) $$\frac{1}{2\pi} \int_{-\pi}^{\pi} |f(x)|^2 \, dx = \sum_{-\infty}^{\infty} |c_n|^2.$$

Proof Let us use the notation

(86) $$\|h\|_2 = \left\{ \frac{1}{2\pi} \int_{-\pi}^{\pi} |h(x)|^2 \, dx \right\}^{1/2}.$$

Let $\varepsilon > 0$ be given. Since $f \in \mathscr{R}$ and $f(\pi) = f(-\pi)$, the construction described in Exercise 12 of Chap. 6 yields a continuous 2π-periodic function h with

(87) $$\|f - h\|_2 < \varepsilon.$$

By Theorem 8.15, there is a trigonometric polynomial P such that $|h(x) - P(x)| < \varepsilon$ for all x. Hence $\|h - P\|_2 < \varepsilon$. If P has degree N_0, Theorem 8.11 shows that

(88) $$\|h - s_N(h)\|_2 \leq \|h - P\|_2 < \varepsilon$$

for all $N \geq N_0$. By (72), with $h - f$ in place of f,

(89) $$\|s_N(h) - s_N(f)\|_2 = \|s_N(h - f)\|_2 \leq \|h - f\|_2 < \varepsilon.$$

Now the triangle inequality (Exercise 11, Chap. 6), combined with (87), (88), and (89), shows that

(90) $$\|f - s_N(f)\|_2 < 3\varepsilon \qquad (N \geq N_0).$$

This proves (83). Next,

(91) $$\frac{1}{2\pi} \int_{-\pi}^{\pi} s_N(f) \bar{g} \, dx = \sum_{-N}^{N} c_n \frac{1}{2\pi} \int_{-\pi}^{\pi} e^{inx} \overline{g(x)} \, dx = \sum_{-N}^{N} c_n \bar{\gamma}_n,$$

and the Schwarz inequality shows that

(92) $$\left| \int f\bar{g} - \int s_N(f) \bar{g} \right| \leq \int |f - s_N(f)||g| \leq \left\{ \int |f - s_N|^2 \int |g|^2 \right\}^{1/2},$$

which tends to 0, as $N \to \infty$, by (83). Comparison of (91) and (92) gives (84). Finally, (85) is the special case $g = f$ of (84).

A more general version of Theorem 8.16 appears in Chap. 11.

THE GAMMA FUNCTION

This function is closely related to factorials and crops up in many unexpected places in analysis. Its origin, history, and development are very well described in an interesting article by P. J. Davis (*Amer. Math. Monthly*, vol. 66, 1959, pp. 849–869). Artin's book (cited in the Bibliography) is another good elementary introduction.

Our presentation will be very condensed, with only a few comments after each theorem. This section may thus be regarded as a large exercise, and as an opportunity to apply some of the material that has been presented so far.

8.17 Definition For $0 < x < \infty$,

$$\Gamma(x) = \int_0^\infty t^{x-1} e^{-t}\, dt. \tag{93}$$

The integral converges for these x. (When $x < 1$, both 0 and ∞ have to be looked at.)

8.18 Theorem
  (a) *The functional equation*
$$\Gamma(x+1) = x\Gamma(x)$$
  holds if $0 < x < \infty$.
  (b) $\Gamma(n+1) = n!$ *for* $n = 1, 2, 3, \ldots$.
  (c) $\log \Gamma$ *is convex on* $(0, \infty)$.

Proof An integration by parts proves (a). Since $\Gamma(1) = 1$, (a) implies (b), by induction. If $1 < p < \infty$ and $(1/p) + (1/q) = 1$, apply Hölder's inequality (Exercise 10, Chap. 6) to (93), and obtain

$$\Gamma\!\left(\frac{x}{p} + \frac{y}{q}\right) \leq \Gamma(x)^{1/p} \Gamma(y)^{1/q}.$$

This is equivalent to (c).

It is a rather surprising fact, discovered by Bohr and Mollerup, that these three properties characterize Γ completely.

8.19 Theorem *If f is a positive function on $(0, \infty)$ such that*
  (a) $f(x + 1) = xf(x)$,
  (b) $f(1) = 1$,
  (c) $\log f$ is convex,
then $f(x) = \Gamma(x)$.

Proof Since Γ satisfies (a), (b), and (c), it is enough to prove that $f(x)$ is uniquely determined by (a), (b), (c), for all $x > 0$. By (a), it is enough to do this for $x \in (0, 1)$.

Put $\varphi = \log f$. Then

(94) $$\varphi(x + 1) = \varphi(x) + \log x \quad (0 < x < \infty),$$

$\varphi(1) = 0$, and φ is convex. Suppose $0 < x < 1$, and n is a positive integer. By (94), $\varphi(n + 1) = \log(n!)$. Consider the difference quotients of φ on the intervals $[n, n + 1]$, $[n + 1, n + 1 + x]$, $[n + 1, n + 2]$. Since φ is convex

$$\log n \leq \frac{\varphi(n + 1 + x) - \varphi(n + 1)}{x} \leq \log(n + 1).$$

Repeated application of (94) gives

$$\varphi(n + 1 + x) = \varphi(x) + \log[x(x + 1) \cdots (x + n)].$$

Thus

$$0 \leq \varphi(x) - \log\left[\frac{n!n^x}{x(x + 1) \cdots (x + n)}\right] \leq x \log\left(1 + \frac{1}{n}\right).$$

The last expression tends to 0 as $n \to \infty$. Hence $\varphi(x)$ is determined, and the proof is complete.

As a by-product we obtain the relation

(95) $$\Gamma(x) = \lim_{n \to \infty} \frac{n!n^x}{x(x + 1) \cdots (x + n)}$$

at least when $0 < x < 1$; from this one can deduce that (95) holds for all $x > 0$, since $\Gamma(x + 1) = x\Gamma(x)$.

8.20 Theorem *If $x > 0$ and $y > 0$, then*

(96) $$\int_0^1 t^{x-1}(1 - t)^{y-1} dt = \frac{\Gamma(x)\Gamma(y)}{\Gamma(x + y)}.$$

This integral is the so-called *beta function* $B(x, y)$.

Proof Note that $B(1, y) = 1/y$, that $\log B(x, y)$ is a convex function of x, for each fixed y, by Hölder's inequality, as in Theorem 8.18, and that

(97) $$B(x + 1, y) = \frac{x}{x + y} B(x, y).$$

To prove (97), perform an integration by parts on

$$B(x + 1, y) = \int_0^1 \left(\frac{t}{1-t}\right)^x (1 - t)^{x+y-1} \, dt.$$

These three properties of $B(x, y)$ show, for each y, that Theorem 8.19 applies to the function f defined by

$$f(x) = \frac{\Gamma(x + y)}{\Gamma(y)} B(x, y).$$

Hence $f(x) = \Gamma(x)$.

8.21 Some consequences The substitution $t = \sin^2 \theta$ turns (96) into

(98) $$2 \int_0^{\pi/2} (\sin \theta)^{2x-1} (\cos \theta)^{2y-1} \, d\theta = \frac{\Gamma(x)\Gamma(y)}{\Gamma(x + y)}.$$

The special case $x = y = \frac{1}{2}$ gives

(99) $$\Gamma(\tfrac{1}{2}) = \sqrt{\pi}.$$

The substitution $t = s^2$ turns (93) into

(100) $$\Gamma(x) = 2 \int_0^\infty s^{2x-1} e^{-s^2} \, ds \qquad (0 < x < \infty).$$

The special case $x = \frac{1}{2}$ gives

(101) $$\int_{-\infty}^\infty e^{-s^2} \, ds = \sqrt{\pi}.$$

By (99), the identity

(102) $$\Gamma(x) = \frac{2^{x-1}}{\sqrt{\pi}} \Gamma\left(\frac{x}{2}\right) \Gamma\left(\frac{x+1}{2}\right)$$

follows directly from Theorem 8.19.

8.22 Stirling's formula This provides a simple approximate expression for $\Gamma(x + 1)$ when x is large (hence for $n!$ when n is large). The formula is

(103) $$\lim_{x \to \infty} \frac{\Gamma(x + 1)}{(x/e)^x \sqrt{2\pi x}} = 1.$$

Here is a proof. Put $t = x(1 + u)$ in (93). This gives

(104) $$\Gamma(x + 1) = x^{x+1} e^{-x} \int_{-1}^{\infty} [(1 + u)e^{-u}]^x \, du.$$

Determine $h(u)$ so that $h(0) = 1$ and

(105) $$(1 + u)e^{-u} = \exp\left[-\frac{u^2}{2} h(u)\right]$$

if $-1 < u < \infty$, $u \neq 0$. Then

(106) $$h(u) = \frac{2}{u^2} [u - \log(1 + u)].$$

It follows that h is continuous, and that $h(u)$ decreases monotonically from ∞ to 0 as u increases from -1 to ∞.

The substitution $u = s\sqrt{2/x}$ turns (104) into

(107) $$\Gamma(x + 1) = x^x e^{-x} \sqrt{2x} \int_{-\infty}^{\infty} \psi_x(s) \, ds$$

where

$$\psi_x(s) = \begin{cases} \exp[-s^2 h(s\sqrt{2/x})] & (-\sqrt{x/2} < s < \infty), \\ 0 & (s \leq -\sqrt{x/2}). \end{cases}$$

Note the following facts about $\psi_x(s)$:

(a) For every s, $\psi_x(s) \to e^{-s^2}$ as $x \to \infty$.
(b) The convergence in (a) is uniform on $[-A, A]$, for every $A < \infty$.
(c) When $s < 0$, then $0 < \psi_x(s) < e^{-s^2}$.
(d) When $s > 0$ and $x > 1$, then $0 < \psi_x(s) < \psi_1(s)$.
(e) $\int_0^{\infty} \psi_1(s) \, ds < \infty$.

The convergence theorem stated in Exercise 12 of Chap. 7 can therefore be applied to the integral (107), and shows that this integral converges to $\sqrt{\pi}$ as $x \to \infty$, by (101). This proves (103).

A more detailed version of this proof may be found in R. C. Buck's "Advanced Calculus," pp. 216–218. For two other, entirely different, proofs, see W. Feller's article in *Amer. Math. Monthly*, vol. 74, 1967, pp. 1223–1225 (with a correction in vol. 75, 1968, p. 518) and pp. 20–24 of Artin's book.

Exercise 20 gives a simpler proof of a less precise result.

EXERCISES

1. Define

$$f(x) = \begin{cases} e^{-1/x^2} & (x \neq 0), \\ 0 & (x = 0). \end{cases}$$

Prove that f has derivatives of all orders at $x = 0$, and that $f^{(n)}(0) = 0$ for $n = 1, 2, 3, \ldots$.

2. Let a_{ij} be the number in the ith row and jth column of the array

$$\begin{matrix} -1 & 0 & 0 & 0 & \cdots \\ \frac{1}{2} & -1 & 0 & 0 & \cdots \\ \frac{1}{4} & \frac{1}{2} & -1 & 0 & \cdots \\ \frac{1}{8} & \frac{1}{4} & \frac{1}{2} & -1 & \cdots \\ \cdots & \cdots & \cdots & \cdots & \cdots \end{matrix}$$

so that

$$a_{ij} = \begin{cases} 0 & (i < j), \\ -1 & (i = j), \\ 2^{j-i} & (i > j). \end{cases}$$

Prove that

$$\sum_i \sum_j a_{ij} = -2, \qquad \sum_j \sum_i a_{ij} = 0.$$

3. Prove that

$$\sum_i \sum_j a_{ij} = \sum_j \sum_i a_{ij}$$

if $a_{ij} \geq 0$ for all i and j (the case $+\infty = +\infty$ may occur).

4. Prove the following limit relations:

(a) $\lim\limits_{x \to 0} \dfrac{b^x - 1}{x} = \log b \quad (b > 0)$.

(b) $\lim\limits_{x \to 0} \dfrac{\log(1 + x)}{x} = 1$.

(c) $\lim\limits_{x \to 0} (1 + x)^{1/x} = e$.

(d) $\lim\limits_{n \to \infty} \left(1 + \dfrac{x}{n}\right)^n = e^x$.

5. Find the following limits

(a) $\lim\limits_{x \to 0} \dfrac{e - (1+x)^{1/x}}{x}$.

(b) $\lim\limits_{n \to \infty} \dfrac{n}{\log n} [n^{1/n} - 1]$.

(c) $\lim\limits_{x \to 0} \dfrac{\tan x - x}{x(1 - \cos x)}$.

(d) $\lim\limits_{x \to 0} \dfrac{x - \sin x}{\tan x - x}$.

6. Suppose $f(x)f(y) = f(x+y)$ for all real x and y.
   (a) Assuming that f is differentiable and not zero, prove that
   $$f(x) = e^{cx}$$
   where c is a constant.
   (b) Prove the same thing, assuming only that f is continuous.

7. If $0 < x < \dfrac{\pi}{2}$, prove that
$$\dfrac{2}{\pi} < \dfrac{\sin x}{x} < 1.$$

8. For $n = 0, 1, 2, \ldots$, and x real, prove that
$$|\sin nx| \le n|\sin x|.$$
   Note that this inequality may be false for other values of n. For instance,
   $$|\sin \tfrac{1}{2}\pi| > \tfrac{1}{2}|\sin \pi|.$$

9. (a) Put $s_N = 1 + (\tfrac{1}{2}) + \cdots + (1/N)$. Prove that
$$\lim_{N \to \infty} (s_N - \log N)$$
exists. (The limit, often denoted by γ, is called Euler's constant. Its numerical value is $0.5772\ldots$. It is not known whether γ is rational or not.)
   (b) Roughly how large must m be so that $N = 10^m$ satisfies $s_N > 100$?

10. Prove that $\sum 1/p$ diverges; the sum extends over all primes.
    (This shows that the primes form a fairly substantial subset of the positive integers.)

Hint: Given N, let $p_1, \ldots, p_k$ be those primes that divide at least one integer $\leq N$. Then

$$\sum_{n=1}^{N} \frac{1}{n} \leq \prod_{j=1}^{k} \left(1 + \frac{1}{p_j} + \frac{1}{p_j^2} + \cdots\right)$$

$$= \prod_{j=1}^{k} \left(1 - \frac{1}{p_j}\right)^{-1}$$

$$\leq \exp \sum_{j=1}^{k} \frac{2}{p_j}.$$

The last inequality holds because

$$(1-x)^{-1} \leq e^{2x}$$

if $0 \leq x \leq \frac{1}{2}$.

(There are many proofs of this result. See, for instance, the article by I. Niven in *Amer. Math. Monthly*, vol. 78, 1971, pp. 272–273, and the one by R. Bellman in *Amer. Math. Monthly*, vol. 50, 1943, pp. 318–319.)

11. Suppose $f \in \mathscr{R}$ on $[0, A]$ for all $A < \infty$, and $f(x) \to 1$ as $x \to +\infty$. Prove that

$$\lim_{t \to 0} t \int_0^{\infty} e^{-tx} f(x) \, dx = 1 \qquad (t > 0).$$

12. Suppose $0 < \delta < \pi$, $f(x) = 1$ if $|x| \leq \delta$, $f(x) = 0$ if $\delta < |x| \leq \pi$, and $f(x + 2\pi) = f(x)$ for all x.
    (a) Compute the Fourier coefficients of f.
    (b) Conclude that

$$\sum_{n=1}^{\infty} \frac{\sin(n\delta)}{n} = \frac{\pi - \delta}{2} \qquad (0 < \delta < \pi).$$

(c) Deduce from Parseval's theorem that

$$\sum_{n=1}^{\infty} \frac{\sin^2(n\delta)}{n^2 \delta} = \frac{\pi - \delta}{2}.$$

(d) Let $\delta \to 0$ and prove that

$$\int_0^{\infty} \left(\frac{\sin x}{x}\right)^2 dx = \frac{\pi}{2}.$$

(e) Put $\delta = \pi/2$ in (c). What do you get?

13. Put $f(x) = x$ if $0 \leq x < 2\pi$, and apply Parseval's theorem to conclude that

$$\sum_{n=1}^{\infty} \frac{1}{n^2} = \frac{\pi^2}{6}.$$

14. If $f(x) = (\pi - |x|)^2$ on $[-\pi, \pi]$, prove that
$$f(x) = \frac{\pi^2}{3} + \sum_{n=1}^{\infty} \frac{4}{n^2} \cos nx$$
and deduce that
$$\sum_{n=1}^{\infty} \frac{1}{n^2} = \frac{\pi^2}{6}, \qquad \sum_{n=1}^{\infty} \frac{1}{n^4} = \frac{\pi^4}{90}.$$

(A recent article by E. L. Stark contains many references to series of the form $\sum n^{-s}$, where s is a positive integer. See *Math. Mag.*, vol. 47, 1974, pp. 197–202.)

15. With D_n as defined in (77), put
$$K_N(x) = \frac{1}{N+1} \sum_{n=0}^{N} D_n(x).$$

Prove that
$$K_N(x) = \frac{1}{N+1} \cdot \frac{1 - \cos(N+1)x}{1 - \cos x}$$

and that
(a) $K_N \geq 0$,
(b) $\dfrac{1}{2\pi} \displaystyle\int_{-\pi}^{\pi} K_N(x)\, dx = 1$,
(c) $K_N(x) \leq \dfrac{1}{N+1} \cdot \dfrac{2}{1 - \cos \delta}$ if $0 < \delta \leq |x| \leq \pi$.

If $s_N = s_N(f; x)$ is the Nth partial sum of the Fourier series of f, consider the arithmetic means
$$\sigma_N = \frac{s_0 + s_1 + \cdots + s_N}{N+1}.$$

Prove that
$$\sigma_N(f; x) = \frac{1}{2\pi} \int_{-\pi}^{\pi} f(x - t) K_N(t)\, dt,$$

and hence prove Fejér's theorem:

If f is continuous, with period 2π, then $\sigma_N(f; x) \to f(x)$ uniformly on $[-\pi, \pi]$.

Hint: Use properties (a), (b), (c) to proceed as in Theorem 7.26.

16. Prove a pointwise version of Fejér's theorem:

If $f \in \mathscr{R}$ and $f(x+)$, $f(x-)$ exist for some x, then
$$\lim_{N \to \infty} \sigma_N(f; x) = \tfrac{1}{2}[f(x+) + f(x-)].$$

17. Assume f is bounded and monotonic on $[-\pi, \pi)$, with Fourier coefficients c_n, as given by (62).

    (a) Use Exercise 17 of Chap. 6 to prove that $\{nc_n\}$ is a bounded sequence.

    (b) Combine (a) with Exercise 16 and with Exercise 14(e) of Chap. 3, to conclude that
    $$\lim_{N \to \infty} s_N(f; x) = \tfrac{1}{2}[f(x+) + f(x-)]$$
    for every x.

    (c) Assume only that $f \in \mathscr{R}$ on $[-\pi, \pi]$ and that f is monotonic in some segment $(\alpha, \beta) \subset [-\pi, \pi]$. Prove that the conclusion of (b) holds for every $x \in (\alpha, \beta)$.

    (This is an application of the localization theorem.)

18. Define
    $$f(x) = x^3 - \sin^2 x \tan x$$
    $$g(x) = 2x^2 - \sin^2 x - x \tan x.$$

    Find out, for each of these two functions, whether it is positive or negative for all $x \in (0, \pi/2)$, or whether it changes sign. Prove your answer.

19. Suppose f is a continuous function on R^1, $f(x + 2\pi) = f(x)$, and α/π is irrational. Prove that
    $$\lim_{N \to \infty} \frac{1}{N} \sum_{n=1}^{N} f(x + n\alpha) = \frac{1}{2\pi} \int_{-\pi}^{\pi} f(t)\, dt$$
    for every x. *Hint*: Do it first for $f(x) = e^{ikx}$.

20. The following simple computation yields a good approximation to Stirling's formula.

    For $m = 1, 2, 3, \ldots$, define
    $$f(x) = (m + 1 - x) \log m + (x - m) \log (m + 1)$$
    if $m \leq x \leq m + 1$, and define
    $$g(x) = \frac{x}{m} - 1 + \log m$$
    if $m - \tfrac{1}{2} \leq x < m + \tfrac{1}{2}$. Draw the graphs of f and g. Note that $f(x) \leq \log x \leq g(x)$ if $x \geq 1$ and that
    $$\int_1^n f(x)\, dx = \log (n!) - \tfrac{1}{2} \log n > -\tfrac{1}{8} + \int_1^n g(x)\, dx.$$

    Integrate $\log x$ over $[1, n]$. Conclude that
    $$\tfrac{7}{8} < \log (n!) - (n + \tfrac{1}{2}) \log n + n < 1$$
    for $n = 2, 3, 4, \ldots$. (*Note*: $\log \sqrt{2\pi} \sim 0.918\ldots$.) Thus
    $$e^{7/8} < \frac{n!}{(n/e)^n \sqrt{n}} < e.$$

21. Let
$$L_n = \frac{1}{2\pi}\int_{-\pi}^{\pi} |D_n(t)|\, dt \qquad (n=1,2,3,\ldots).$$
Prove that there exists a constant $C>0$ such that
$$L_n > C \log n \qquad (n=1,2,3,\ldots),$$
or, more precisely, that the sequence
$$\left\{L_n - \frac{4}{\pi^2}\log n\right\}$$
is bounded.

22. If α is real and $-1 < x < 1$, prove Newton's binomial theorem
$$(1+x)^\alpha = 1 + \sum_{n=1}^{\infty} \frac{\alpha(\alpha-1)\cdots(\alpha-n+1)}{n!} x^n.$$

Hint: Denote the right side by $f(x)$. Prove that the series converges. Prove that
$$(1+x)f'(x) = \alpha f(x)$$
and solve this differential equation.
  Show also that
$$(1-x)^{-\alpha} = \sum_{n=0}^{\infty} \frac{\Gamma(n+\alpha)}{n!\,\Gamma(\alpha)} x^n$$
if $-1 < x < 1$ and $\alpha > 0$.

23. Let γ be a continuously differentiable *closed* curve in the complex plane, with parameter interval $[a, b]$, and assume that $\gamma(t) \neq 0$ for every $t \in [a, b]$. Define the *index* of γ to be
$$\mathrm{Ind}\,(\gamma) = \frac{1}{2\pi i}\int_a^b \frac{\gamma'(t)}{\gamma(t)}\, dt.$$
Prove that $\mathrm{Ind}\,(\gamma)$ is always an integer.

Hint: There exists φ on $[a, b]$ with $\varphi' = \gamma'/\gamma$, $\varphi(a) = 0$. Hence $\gamma\exp(-\varphi)$ is constant. Since $\gamma(a) = \gamma(b)$ it follows that $\exp\varphi(b) = \exp\varphi(a) = 1$. Note that $\varphi(b) = 2\pi i\,\mathrm{Ind}\,(\gamma)$.
  Compute $\mathrm{Ind}\,(\gamma)$ when $\gamma(t) = e^{int}$, $a = 0$, $b = 2\pi$.
  Explain why $\mathrm{Ind}\,(\gamma)$ is often called the *winding number* of γ around 0.

24. Let γ be as in Exercise 23, and assume in addition that the range of γ does not intersect the negative real axis. Prove that $\mathrm{Ind}\,(\gamma) = 0$. *Hint:* For $0 \leq c < \infty$, $\mathrm{Ind}\,(\gamma + c)$ is a continuous integer-valued function of c. Also, $\mathrm{Ind}\,(\gamma + c) \to 0$ as $c \to \infty$.

25. Suppose γ_1 and γ_2 are curves as in Exercise 23, and
$$|\gamma_1(t) - \gamma_2(t)| < |\gamma_1(t)| \qquad (a \le t \le b).$$
Prove that Ind $(\gamma_1) =$ Ind (γ_2).

Hint: Put $\gamma = \gamma_2/\gamma_1$. Then $|1 - \gamma| < 1$, hence Ind $(\gamma) = 0$, by Exercise 24. Also,
$$\frac{\gamma'}{\gamma} = \frac{\gamma_2'}{\gamma_2} - \frac{\gamma_1'}{\gamma_1}.$$

26. Let γ be a *closed* curve in the complex plane (not necessarily differentiable) with parameter interval $[0, 2\pi]$, such that $\gamma(t) \ne 0$ for every $t \in [0, 2\pi]$.

Choose $\delta > 0$ so that $|\gamma(t)| > \delta$ for all $t \in [0, 2\pi]$. If P_1 and P_2 are trigonometric polynomials such that $|P_j(t) - \gamma(t)| < \delta/4$ for all $t \in [0, 2\pi]$ (their existence is assured by Theorem 8.15), prove that
$$\text{Ind } (P_1) = \text{Ind } (P_2)$$
by applying Exercise 25.

Define this common value to be Ind (γ).

Prove that the statements of Exercises 24 and 25 hold without any differentiability assumption.

27. Let f be a continuous complex function defined in the complex plane. Suppose there is a positive integer n and a complex number $c \ne 0$ such that
$$\lim_{|z| \to \infty} z^{-n} f(z) = c.$$

Prove that $f(z) = 0$ for at least one complex number z.

Note that this is a generalization of Theorem 8.8.

Hint: Assume $f(z) \ne 0$ for all z, define
$$\gamma_r(t) = f(re^{it})$$
for $0 \le r < \infty$, $0 \le t \le 2\pi$, and prove the following statements about the curves γ_r:

(a) Ind $(\gamma_0) = 0$.
(b) Ind $(\gamma_r) = n$ for all sufficiently large r.
(c) Ind (γ_r) is a continuous function of r, on $[0, \infty)$.

[In (b) and (c), use the last part of Exercise 26.]

Show that (a), (b), and (c) are contradictory, since $n > 0$.

28. Let $\bar{D}$ be the closed unit disc in the complex plane. (Thus $z \in \bar{D}$ if and only if $|z| \le 1$.) Let g be a continuous mapping of $\bar{D}$ into the unit circle T. (Thus, $|g(z)| = 1$ for every $z \in \bar{D}$.)

Prove that $g(z) = -z$ for at least one $z \in T$.

Hint: For $0 \le r \le 1$, $0 \le t \le 2\pi$, put
$$\gamma_r(t) = g(re^{it}),$$
and put $\psi(t) = e^{-it}\gamma_1(t)$. If $g(z) \ne -z$ for every $z \in T$, then $\psi(t) \ne -1$ for every $t \in [0, 2\pi]$. Hence Ind $(\psi) = 0$, by Exercises 24 and 26. It follows that Ind $(\gamma_1) = 1$. But Ind $(\gamma_0) = 0$. Derive a contradiction, as in Exercise 27.

29. Prove that every continuous mapping f of $\bar{D}$ into $\bar{D}$ has a fixed point in $\bar{D}$.
    (This is the 2-dimensional case of Brouwer's fixed-point theorem.)

    Hint: Assume $f(z) \neq z$ for every $z \in \bar{D}$. Associate to each $z \in \bar{D}$ the point $g(z) \in T$ which lies on the ray that starts at $f(z)$ and passes through z. Then g maps $\bar{D}$ into T, $g(z) = z$ if $z \in T$, and g is continuous, because
    $$g(z) = z - s(z)[f(z) - z],$$
    where $s(z)$ is the unique nonnegative root of a certain quadratic equation whose coefficients are continuous functions of f and z. Apply Exercise 28.

30. Use Stirling's formula to prove that
    $$\lim_{x \to \infty} \frac{\Gamma(x+c)}{x^c \Gamma(x)} = 1$$
    for every real constant c.

31. In the proof of Theorem 7.26 it was shown that
    $$\int_{-1}^{1} (1-x^2)^n \, dx \geq \frac{4}{3\sqrt{n}}$$
    for $n = 1, 2, 3, \ldots$. Use Theorem 8.20 and Exercise 30 to show the more precise result
    $$\lim_{n \to \infty} \sqrt{n} \int_{-1}^{1} (1-x^2)^n \, dx = \sqrt{\pi}.$$

9
FUNCTIONS OF SEVERAL VARIABLES

LINEAR TRANSFORMATIONS

We begin this chapter with a discussion of sets of vectors in euclidean n-space R^n. The algebraic facts presented here extend without change to finite-dimensional vector spaces over any field of scalars. However, for our purposes it is quite sufficient to stay within the familiar framework provided by the euclidean spaces.

9.1 Definitions

(a) A nonempty set $X \subset R^n$ is a *vector space* if $\mathbf{x} + \mathbf{y} \in X$ and $c\mathbf{x} \in X$ for all $\mathbf{x} \in X$, $\mathbf{y} \in X$, and for all scalars c.

(b) If $\mathbf{x}_1, \ldots, \mathbf{x}_k \in R^n$ and $c_1, \ldots, c_k$ are scalars, the vector

$$c_1 \mathbf{x}_1 + \cdots + c_k \mathbf{x}_k$$

is called a *linear combination* of $\mathbf{x}_1, \ldots, \mathbf{x}_k$. If $S \subset R^n$ and if E is the set of all linear combinations of elements of S, we say that S *spans* E, or that E is the span of S.

Observe that every span is a vector space.

(c) A set consisting of vectors $\mathbf{x}_1, \ldots, \mathbf{x}_k$ (we shall use the notation $\{\mathbf{x}_1, \ldots, \mathbf{x}_k\}$ for such a set) is said to be *independent* if the relation $c_1 \mathbf{x}_1 + \cdots + c_k \mathbf{x}_k = \mathbf{0}$ implies that $c_1 = \cdots = c_k = 0$. Otherwise $\{\mathbf{x}_1, \ldots, \mathbf{x}_k\}$ is said to be *dependent*.

Observe that no independent set contains the null vector.

(d) If a vector space X contains an independent set of r vectors but contains no independent set of $r + 1$ vectors, we say that X has *dimension r*, and write: dim $X = r$.

The set consisting of $\mathbf{0}$ alone is a vector space; its dimension is 0.

(e) An independent subset of a vector space X which spans X is called a *basis* of X.

Observe that if $B = \{\mathbf{x}_1, \ldots, \mathbf{x}_r\}$ is a basis of X, then every $\mathbf{x} \in X$ has a unique representation of the form $\mathbf{x} = \Sigma c_j \mathbf{x}_j$. Such a representation exists since B spans X, and it is unique since B is independent. The numbers $c_1, \ldots, c_r$ are called the *coordinates of* $\mathbf{x}$ with respect to the basis B.

The most familiar example of a basis is the set $\{\mathbf{e}_1, \ldots, \mathbf{e}_n\}$, where $\mathbf{e}_j$ is the vector in R^n whose jth coordinate is 1 and whose other coordinates are all 0. If $\mathbf{x} \in R^n$, $\mathbf{x} = (x_1, \ldots, x_n)$, then $\mathbf{x} = \Sigma x_j \mathbf{e}_j$. We shall call

$$\{\mathbf{e}_1, \ldots, \mathbf{e}_n\}$$

the *standard basis* of R^n.

9.2 Theorem *Let r be a positive integer. If a vector space X is spanned by a set of r vectors, then* dim $X \leq r$.

Proof If this is false, there is a vector space X which contains an independent set $Q = \{\mathbf{y}_1, \ldots, \mathbf{y}_{r+1}\}$ and which is spanned by a set S_0 consisting of r vectors.

Suppose $0 \leq i < r$, and suppose a set S_i has been constructed which spans X and which consists of all $\mathbf{y}_j$ with $1 \leq j \leq i$ plus a certain collection of $r - i$ members of S_0, say $\mathbf{x}_1, \ldots, \mathbf{x}_{r-i}$. (In other words, S_i is obtained from S_0 by replacing i of its elements by members of Q, without altering the span.) Since S_i spans X, $\mathbf{y}_{i+1}$ is in the span of S_i; hence there are scalars $a_1, \ldots, a_{i+1}, b_1, \ldots, b_{r-i}$, with $a_{i+1} = 1$, such that

$$\sum_{j=1}^{i+1} a_j \mathbf{y}_j + \sum_{k=1}^{r-i} b_k \mathbf{x}_k = \mathbf{0}.$$

If all b_k's were 0, the independence of Q would force all a_j's to be 0, a contradiction. It follows that some $\mathbf{x}_k \in S_i$ is a linear combination of the other members of $T_i = S_i \cup \{\mathbf{y}_{i+1}\}$. Remove this $\mathbf{x}_k$ from T_i and call the remaining set S_{i+1}. Then S_{i+1} spans the same set as T_i, namely X, so that S_{i+1} has the properties postulated for S_i with $i + 1$ in place of i.

Starting with S_0, we thus construct sets $S_1, \ldots, S_r$. The last of these consists of $\mathbf{y}_1, \ldots, \mathbf{y}_r$, and our construction shows that it spans X. But Q is independent; hence $\mathbf{y}_{r+1}$ is not in the span of S_r. This contradiction establishes the theorem.

Corollary dim $R^n = n$.

Proof Since $\{\mathbf{e}_1, \ldots, \mathbf{e}_n\}$ spans R^n, the theorem shows that dim $R^n \leq n$. Since $\{\mathbf{e}_1, \ldots, \mathbf{e}_n\}$ is independent, dim $R^n \geq n$.

9.3 Theorem *Suppose X is a vector space, and* dim $X = n$.

(a) *A set E of n vectors in X spans X if and only if E is independent.*
(b) *X has a basis, and every basis consists of n vectors.*
(c) *If $1 \leq r \leq n$ and $\{\mathbf{y}_1, \ldots, \mathbf{y}_r\}$ is an independent set in X, then X has a basis containing $\{\mathbf{y}_1, \ldots, \mathbf{y}_r\}$.*

Proof Suppose $E = \{\mathbf{x}_1, \ldots, \mathbf{x}_n\}$. Since dim $X = n$, the set $\{\mathbf{x}_1, \ldots, \mathbf{x}_n, \mathbf{y}\}$ is dependent, for every $\mathbf{y} \in X$. If E is independent, it follows that $\mathbf{y}$ is in the span of E; hence E spans X. Conversely, if E is dependent, one of its members can be removed without changing the span of E. Hence E cannot span X, by Theorem 9.2. This proves (a).

Since dim $X = n$, X contains an independent set of n vectors, and (a) shows that every such set is a basis of X; (b) now follows from 9.1(d) and 9.2.

To prove (c), let $\{\mathbf{x}_1, \ldots, \mathbf{x}_n\}$ be a basis of X. The set

$$S = \{\mathbf{y}_1, \ldots, \mathbf{y}_r, \mathbf{x}_1, \ldots, \mathbf{x}_n\}$$

spans X and is dependent, since it contains more than n vectors. The argument used in the proof of Theorem 9.2 shows that one of the $\mathbf{x}_i$'s is a linear combination of the other members of S. If we remove this $\mathbf{x}_i$ from S, the remaining set still spans X. This process can be repeated r times and leads to a basis of X which contains $\{\mathbf{y}_1, \ldots, \mathbf{y}_r\}$, by (a).

9.4 Definitions A mapping A of a vector space X into a vector space Y is said to be a *linear transformation* if

$$A(\mathbf{x}_1 + \mathbf{x}_2) = A\mathbf{x}_1 + A\mathbf{x}_2, \quad A(c\mathbf{x}) = cA\mathbf{x}$$

for all $\mathbf{x}, \mathbf{x}_1, \mathbf{x}_2 \in X$ and all scalars c. Note that one often writes $A\mathbf{x}$ instead of $A(\mathbf{x})$ if A is linear.

Observe that $A\mathbf{0} = \mathbf{0}$ if A is linear. Observe also that a linear transformation A of X into Y is completely determined by its action on any basis: If

$\{\mathbf{x}_1, \ldots, \mathbf{x}_n\}$ is a basis of X, then every $\mathbf{x} \in X$ has a unique representation of the form

$$\mathbf{x} = \sum_{i=1}^{n} c_i \mathbf{x}_i,$$

and the linearity of A allows us to compute $A\mathbf{x}$ from the vectors $A\mathbf{x}_1, \ldots, A\mathbf{x}_n$ and the coordinates $c_1, \ldots, c_n$ by the formula

$$A\mathbf{x} = \sum_{i=1}^{n} c_i A\mathbf{x}_i.$$

Linear transformations of X into X are often called *linear operators* on X. If A is a linear operator on X which (i) is one-to-one and (ii) maps X onto X, we say that A is *invertible*. In this case we can define an operator A^{-1} on X by requiring that $A^{-1}(A\mathbf{x}) = \mathbf{x}$ for all $\mathbf{x} \in X$. It is trivial to verify that we then also have $A(A^{-1}\mathbf{x}) = \mathbf{x}$, for all $\mathbf{x} \in X$, and that A^{-1} is linear.

An important fact about linear operators on finite-dimensional vector spaces is that each of the above conditions (i) and (ii) implies the other:

9.5 Theorem *A linear operator A on a finite-dimensional vector space X is one-to-one if and only if the range of A is all of X.*

Proof Let $\{\mathbf{x}_1, \ldots, \mathbf{x}_n\}$ be a basis of X. The linearity of A shows that its range $\mathscr{R}(A)$ is the span of the set $Q = \{A\mathbf{x}_1, \ldots, A\mathbf{x}_n\}$. We therefore infer from Theorem 9.3(a) that $\mathscr{R}(A) = X$ if and only if Q is independent. We have to prove that this happens if and only if A is one-to-one.

Suppose A is one-to-one and $\Sigma c_i A\mathbf{x}_i = \mathbf{0}$. Then $A(\Sigma c_i \mathbf{x}_i) = \mathbf{0}$, hence $\Sigma c_i \mathbf{x}_i = \mathbf{0}$, hence $c_1 = \cdots = c_n = 0$, and we conclude that Q is independent.

Conversely, suppose Q is independent and $A(\Sigma c_i \mathbf{x}_i) = \mathbf{0}$. Then $\Sigma c_i A\mathbf{x}_i = \mathbf{0}$, hence $c_1 = \cdots = c_n = 0$, and we conclude: $A\mathbf{x} = \mathbf{0}$ only if $\mathbf{x} = \mathbf{0}$. If now $A\mathbf{x} = A\mathbf{y}$, then $A(\mathbf{x} - \mathbf{y}) = A\mathbf{x} - A\mathbf{y} = \mathbf{0}$, so that $\mathbf{x} - \mathbf{y} = \mathbf{0}$, and this says that A is one-to-one.

9.6 Definitions

(a) Let $L(X, Y)$ be the set of all linear transformations of the vector space X into the vector space Y. Instead of $L(X, X)$, we shall simply write $L(X)$. If $A_1, A_2 \in L(X, Y)$ and if c_1, c_2 are scalars, define $c_1 A_1 + c_2 A_2$ by

$$(c_1 A_1 + c_2 A_2)\mathbf{x} = c_1 A_1 \mathbf{x} + c_2 A_2 \mathbf{x} \qquad (\mathbf{x} \in X).$$

It is then clear that $c_1 A_1 + c_2 A_2 \in L(X, Y)$.

(b) If X, Y, Z are vector spaces, and if $A \in L(X, Y)$ and $B \in L(Y, Z)$, we define their *product* BA to be the composition of A and B:

$$(BA)\mathbf{x} = B(A\mathbf{x}) \qquad (\mathbf{x} \in X).$$

Then $BA \in L(X, Z)$.

Note that BA need not be the same as AB, even if $X = Y = Z$.

(c) For $A \in L(R^n, R^m)$, define the *norm* $\|A\|$ of A to be the sup of all numbers $|A\mathbf{x}|$, where $\mathbf{x}$ ranges over all vectors in R^n with $|\mathbf{x}| \le 1$.

Observe that the inequality

$$|A\mathbf{x}| \le \|A\| \, |\mathbf{x}|$$

holds for all $\mathbf{x} \in R^n$. Also, if λ is such that $|A\mathbf{x}| \le \lambda |\mathbf{x}|$ for all $\mathbf{x} \in R^n$, then $\|A\| \le \lambda$.

9.7 Theorem

(a) If $A \in L(R^n, R^m)$, then $\|A\| < \infty$ and A is a uniformly continuous mapping of R^n into R^m.

(b) If $A, B \in L(R^n, R^m)$ and c is a scalar, then

$$\|A + B\| \le \|A\| + \|B\|, \qquad \|cA\| = |c| \, \|A\|.$$

With the distance between A and B defined as $\|A - B\|$, $L(R^n, R^m)$ is a metric space.

(c) If $A \in L(R^n, R^m)$ and $B \in L(R^m, R^k)$, then

$$\|BA\| \le \|B\| \, \|A\|.$$

Proof

(a) Let $\{\mathbf{e}_1, \ldots, \mathbf{e}_n\}$ be the standard basis in R^n and suppose $\mathbf{x} = \Sigma c_i \mathbf{e}_i$, $|\mathbf{x}| \le 1$, so that $|c_i| \le 1$ for $i = 1, \ldots, n$. Then

$$|A\mathbf{x}| = \left| \sum c_i A\mathbf{e}_i \right| \le \sum |c_i| \, |A\mathbf{e}_i| \le \sum |A\mathbf{e}_i|$$

so that

$$\|A\| \le \sum_{i=1}^{n} |A\mathbf{e}_i| < \infty.$$

Since $|A\mathbf{x} - A\mathbf{y}| \le \|A\| \, |\mathbf{x} - \mathbf{y}|$ if $\mathbf{x}, \mathbf{y} \in R^n$, we see that A is uniformly continuous.

(b) The inequality in (b) follows from

$$|(A + B)\mathbf{x}| = |A\mathbf{x} + B\mathbf{x}| \le |A\mathbf{x}| + |B\mathbf{x}| \le (\|A\| + \|B\|) |\mathbf{x}|.$$

The second part of (b) is proved in the same manner. If

$$A, B, C \in L(R^n, R^m),$$

we have the triangle inequality

$$\|A - C\| = \|(A - B) + (B - C)\| \le \|A - B\| + \|B - C\|,$$

and it is easily verified that $\|A - B\|$ has the other properties of a metric (Definition 2.15).

(c) Finally, (c) follows from
$$|(BA)\mathbf{x}| = |B(A\mathbf{x})| \leq \|B\| \, |A\mathbf{x}| \leq \|B\| \, \|A\| \, |\mathbf{x}|.$$

Since we now have metrics in the spaces $L(R^n, R^m)$, the concepts of open set, continuity, etc., make sense for these spaces. Our next theorem utilizes these concepts.

9.8 Theorem *Let Ω be the set of all invertible linear operators on R^n.*

(a) *If $A \in \Omega$, $B \in L(R^n)$, and*
$$\|B - A\| \cdot \|A^{-1}\| < 1,$$
then $B \in \Omega$.

(b) *Ω is an open subset of $L(R^n)$, and the mapping $A \to A^{-1}$ is continuous on Ω.*

(This mapping is also obviously a $1 - 1$ mapping of Ω onto Ω, which is its own inverse.)

Proof

(a) Put $\|A^{-1}\| = 1/\alpha$, put $\|B - A\| = \beta$. Then $\beta < \alpha$. For every $\mathbf{x} \in R^n$,
$$\alpha|\mathbf{x}| = \alpha|A^{-1}A\mathbf{x}| \leq \alpha\|A^{-1}\| \cdot |A\mathbf{x}|$$
$$= |A\mathbf{x}| \leq |(A - B)\mathbf{x}| + |B\mathbf{x}| \leq \beta|\mathbf{x}| + |B\mathbf{x}|,$$
so that

(1) $$(\alpha - \beta)|\mathbf{x}| \leq |B\mathbf{x}| \quad (\mathbf{x} \in R^n).$$

Since $\alpha - \beta > 0$, (1) shows that $B\mathbf{x} \neq 0$ if $\mathbf{x} \neq 0$. Hence B is $1 - 1$. By Theorem 9.5, $B \in \Omega$. This holds for all B with $\|B - A\| < \alpha$. Thus we have (a) and the fact that Ω is open.

(b) Next, replace $\mathbf{x}$ by $B^{-1}\mathbf{y}$ in (1). The resulting inequality

(2) $$(\alpha - \beta)|B^{-1}\mathbf{y}| \leq |BB^{-1}\mathbf{y}| = |\mathbf{y}| \quad (\mathbf{y} \in R^n)$$

shows that $\|B^{-1}\| \leq (\alpha - \beta)^{-1}$. The identity
$$B^{-1} - A^{-1} = B^{-1}(A - B)A^{-1},$$
combined with Theorem 9.7(c), implies therefore that
$$\|B^{-1} - A^{-1}\| \leq \|B^{-1}\| \, \|A - B\| \, \|A^{-1}\| \leq \frac{\beta}{\alpha(\alpha - \beta)}.$$

This establishes the continuity assertion made in (b), since $\beta \to 0$ as $B \to A$.

9.9 Matrices Suppose $\{x_1, \ldots, x_n\}$ and $\{y_1, \ldots, y_m\}$ are bases of vector spaces X and Y, respectively. Then every $A \in L(X, Y)$ determines a set of numbers a_{ij} such that

$$ A x_j = \sum_{i=1}^{m} a_{ij} y_i \quad (1 \leq j \leq n). \tag{3} $$

It is convenient to visualize these numbers in a rectangular array of m rows and n columns, called an *m by n matrix*:

$$ [A] = \begin{bmatrix} a_{11} & a_{12} & \cdots & a_{1n} \\ a_{21} & a_{22} & \cdots & a_{2n} \\ \vdots & & & \vdots \\ a_{m1} & a_{m2} & \cdots & a_{mn} \end{bmatrix} $$

Observe that the coordinates a_{ij} of the vector Ax_j (with respect to the basis $\{y_1, \ldots, y_m\}$) appear in the jth column of $[A]$. The vectors Ax_j are therefore sometimes called the *column vectors* of $[A]$. With this terminology, the *range of A is spanned by the column vectors of $[A]$*.

If $x = \Sigma c_j x_j$, the linearity of A, combined with (3), shows that

$$ Ax = \sum_{i=1}^{m} \left(\sum_{j=1}^{n} a_{ij} c_j \right) y_i. \tag{4} $$

Thus the coordinates of Ax are $\Sigma_j a_{ij} c_j$. Note that in (3) the summation ranges over the first subscript of a_{ij}, but that we sum over the second subscript when computing coordinates.

Suppose next that an m by n matrix is given, with real entries a_{ij}. If A is then defined by (4), it is clear that $A \in L(X, Y)$ and that $[A]$ is the given matrix. Thus there is a natural 1-1 correspondence between $L(X, Y)$ and the set of all real m by n matrices. We emphasize, though, that $[A]$ depends not only on A but also on the choice of bases in X and Y. The same A may give rise to many different matrices if we change bases, and vice versa. We shall not pursue this observation any further, since we shall usually work with fixed bases. (Some remarks on this may be found in Sec. 9.37.)

If Z is a third vector space, with basis $\{z_1, \ldots, z_p\}$, if A is given by (3), and if

$$ By_i = \sum_k b_{ki} z_k, \quad (BA)x_j = \sum_k c_{kj} z_k, $$

then $A \in L(X, Y)$, $B \in L(Y, Z)$, $BA \in L(X, Z)$, and since

$$ B(Ax_j) = B \sum_i a_{ij} y_i = \sum_i a_{ij} By_i $$
$$ = \sum_i a_{ij} \sum_k b_{ki} z_k = \sum_k \left(\sum_i b_{ki} a_{ij} \right) z_k, $$

the independence of $\{z_1, \ldots, z_p\}$ implies that

(5) $$c_{kj} = \sum_i b_{ki} a_{ij} \qquad (1 \le k \le p, 1 \le j \le n).$$

This shows how to compute the p by n matrix $[BA]$ from $[B]$ and $[A]$. If we define the product $[B][A]$ to be $[BA]$, then (5) describes the usual rule of matrix multiplication.

Finally, suppose $\{x_1, \ldots, x_n\}$ and $\{y_1, \ldots, y_m\}$ are standard bases of R^n and R^m, and A is given by (4). The Schwarz inequality shows that

$$|A\mathbf{x}|^2 = \sum_i \left(\sum_j a_{ij} c_j\right)^2 \le \sum_i \left(\sum_j a_{ij}^2 \cdot \sum_j c_j^2\right) = \sum_{i,j} a_{ij}^2 |\mathbf{x}|^2.$$

Thus

(6) $$\|A\| \le \left\{\sum_{i,j} a_{ij}^2\right\}^{1/2}.$$

If we apply (6) to $B - A$ in place of A, where $A, B \in L(R^n, R^m)$, we see that if the matrix elements a_{ij} are continuous functions of a parameter, then the same is true of A. More precisely:

If S is a metric space, if $a_{11}, \ldots, a_{mn}$ are real continuous functions on S, and if, for each $p \in S$, A_p is the linear transformation of R^n into R^m whose matrix has entries $a_{ij}(p)$, then the mapping $p \to A_p$ is a continuous mapping of S into $L(R^n, R^m)$.

DIFFERENTIATION

9.10 Preliminaries In order to arrive at a definition of the derivative of a function whose domain is R^n (or an open subset of R^n), let us take another look at the familiar case $n = 1$, and let us see how to interpret the derivative in that case in a way which will naturally extend to $n > 1$.

If f is a real function with domain $(a, b) \subset R^1$ and if $x \in (a, b)$, then $f'(x)$ is usually defined to be the real number

(7) $$\lim_{h \to 0} \frac{f(x+h) - f(x)}{h},$$

provided, of course, that this limit exists. Thus

(8) $$f(x+h) - f(x) = f'(x)h + r(h)$$

where the "remainder" $r(h)$ is small, in the sense that

(9) $$\lim_{h \to 0} \frac{r(h)}{h} = 0.$$

Note that (8) expresses the difference $f(x+h) - f(x)$ as the sum of the *linear function* that takes h to $f'(x)h$, plus a small remainder.

We can therefore regard the derivative of f at x, not as a real number, but as the linear operator on R^1 that takes h to $f'(x)h$.

[Observe that every real number α gives rise to a linear operator on R^1; the operator in question is simply multiplication by α. Conversely, every linear function that carries R^1 to R^1 is multiplication by some real number. It is this natural 1-1 correspondence between R^1 and $L(R^1)$ which motivates the preceding statements.]

Let us next consider a function $\mathbf{f}$ that maps $(a, b) \subset R^1$ into R^m. In that case, $\mathbf{f}'(x)$ was defined to be that vector $\mathbf{y} \in R^m$ (if there is one) for which

$$(10) \qquad \lim_{h \to 0} \left\{ \frac{\mathbf{f}(x+h) - \mathbf{f}(x)}{h} - \mathbf{y} \right\} = \mathbf{0}.$$

We can again rewrite this in the form

$$(11) \qquad \mathbf{f}(x+h) - \mathbf{f}(x) = h\mathbf{y} + \mathbf{r}(h),$$

where $\mathbf{r}(h)/h \to \mathbf{0}$ as $h \to 0$. The main term on the right side of (11) is again a *linear* function of h. Every $\mathbf{y} \in R^m$ induces a linear transformation of R^1 into R^m, by associating to each $h \in R^1$ the vector $h\mathbf{y} \in R^m$. This identification of R^m with $L(R^1, R^m)$ allows us to regard $\mathbf{f}'(x)$ as a member of $L(R^1, R^m)$.

Thus, if $\mathbf{f}$ is a differentiable mapping of $(a, b) \subset R^1$ into R^m, and if $x \in (a, b)$, then $\mathbf{f}'(x)$ is the linear transformation of R^1 into R^m that satisfies

$$(12) \qquad \lim_{h \to 0} \frac{\mathbf{f}(x+h) - \mathbf{f}(x) - \mathbf{f}'(x)h}{h} = \mathbf{0},$$

or, equivalently,

$$(13) \qquad \lim_{h \to 0} \frac{|\mathbf{f}(x+h) - \mathbf{f}(x) - \mathbf{f}'(x)h|}{|h|} = 0.$$

We are now ready for the case $n > 1$.

9.11 Definition Suppose E is an open set in R^n, $\mathbf{f}$ maps E into R^m, and $\mathbf{x} \in E$. If there exists a linear transformation A of R^n into R^m such that

$$(14) \qquad \lim_{\mathbf{h} \to \mathbf{0}} \frac{|\mathbf{f}(\mathbf{x}+\mathbf{h}) - \mathbf{f}(\mathbf{x}) - A\mathbf{h}|}{|\mathbf{h}|} = 0,$$

then we say that $\mathbf{f}$ is *differentiable at* $\mathbf{x}$, and we write

$$(15) \qquad \mathbf{f}'(\mathbf{x}) = A.$$

If $\mathbf{f}$ is differentiable at every $\mathbf{x} \in E$, we say that $\mathbf{f}$ is *differentiable in E*.

It is of course understood in (14) that $\mathbf{h} \in R^n$. If $|\mathbf{h}|$ is small enough, then $\mathbf{x} + \mathbf{h} \in E$, since E is open. Thus $\mathbf{f}(\mathbf{x} + \mathbf{h})$ is defined, $\mathbf{f}(\mathbf{x} + \mathbf{h}) \in R^m$, and since $A \in L(R^n, R^m)$, $A\mathbf{h} \in R^m$. Thus

$$\mathbf{f}(\mathbf{x} + \mathbf{h}) - \mathbf{f}(\mathbf{x}) - A\mathbf{h} \in R^m.$$

The norm in the numerator of (14) is that of R^m. In the denominator we have the R^n-norm of $\mathbf{h}$.

There is an obvious uniqueness problem which has to be settled before we go any further.

9.12 Theorem *Suppose E and $\mathbf{f}$ are as in Definition 9.11, $\mathbf{x} \in E$, and (14) holds with $A = A_1$ and with $A = A_2$. Then $A_1 = A_2$.*

Proof If $B = A_1 - A_2$, the inequality

$$|B\mathbf{h}| \leq |\mathbf{f}(\mathbf{x} + \mathbf{h}) - \mathbf{f}(\mathbf{x}) - A_1\mathbf{h}| + |\mathbf{f}(\mathbf{x} + \mathbf{h}) - \mathbf{f}(\mathbf{x}) - A_2\mathbf{h}|$$

shows that $|B\mathbf{h}|/|\mathbf{h}| \to 0$ as $\mathbf{h} \to 0$. For fixed $\mathbf{h} \neq 0$, it follows that

$$(16) \qquad \frac{|B(t\mathbf{h})|}{|t\mathbf{h}|} \to 0 \quad \text{as} \quad t \to 0.$$

The linearity of B shows that the left side of (16) is independent of t. Thus $B\mathbf{h} = 0$ for every $\mathbf{h} \in R^n$. Hence $B = 0$.

9.13 Remarks

(a) The relation (14) can be rewritten in the form

$$(17) \qquad \mathbf{f}(\mathbf{x} + \mathbf{h}) - \mathbf{f}(\mathbf{x}) = \mathbf{f}'(\mathbf{x})\mathbf{h} + \mathbf{r}(\mathbf{h})$$

where the remainder $\mathbf{r}(\mathbf{h})$ satisfies

$$(18) \qquad \lim_{\mathbf{h} \to 0} \frac{|\mathbf{r}(\mathbf{h})|}{|\mathbf{h}|} = 0.$$

We may interpret (17), as in Sec. 9.10, by saying that for fixed $\mathbf{x}$ and small $\mathbf{h}$, the left side of (17) is approximately equal to $\mathbf{f}'(\mathbf{x})\mathbf{h}$, that is, to the value of a linear transformation applied to $\mathbf{h}$.

(b) Suppose $\mathbf{f}$ and E are as in Definition 9.11, and $\mathbf{f}$ is differentiable in E. For every $\mathbf{x} \in E$, $\mathbf{f}'(\mathbf{x})$ is then a function, namely, a linear transformation of R^n into R^m. But $\mathbf{f}'$ is also a function: $\mathbf{f}'$ maps E into $L(R^n, R^m)$.

(c) A glance at (17) shows that $\mathbf{f}$ is continuous at any point at which $\mathbf{f}$ is differentiable.

(d) The derivative defined by (14) or (17) is often called the *differential* of $\mathbf{f}$ at $\mathbf{x}$, or the *total derivative* of $\mathbf{f}$ at $\mathbf{x}$, to distinguish it from the partial derivatives that will occur later.

9.14 Example We have defined derivatives of functions carrying R^n to R^m to be linear transformations of R^n into R^m. What is the derivative of such a linear transformation? The answer is very simple.

If $A \in L(R^n, R^m)$ and if $\mathbf{x} \in R^n$, then

(19) $$A'(\mathbf{x}) = A.$$

Note that $\mathbf{x}$ appears on the left side of (19), but not on the right. Both sides of (19) are members of $L(R^n, R^m)$, whereas $A\mathbf{x} \in R^m$.

The proof of (19) is a triviality, since

(20) $$A(\mathbf{x} + \mathbf{h}) - A\mathbf{x} = A\mathbf{h},$$

by the linearity of A. With $\mathbf{f}(\mathbf{x}) = A\mathbf{x}$, the numerator in (14) is thus 0 for every $\mathbf{h} \in R^n$. In (17), $\mathbf{r}(\mathbf{h}) = \mathbf{0}$.

We now extend the chain rule (Theorem 5.5) to the present situation.

9.15 Theorem *Suppose E is an open set in R^n, $\mathbf{f}$ maps E into R^m, $\mathbf{f}$ is differentiable at $\mathbf{x}_0 \in E$, $\mathbf{g}$ maps an open set containing $\mathbf{f}(E)$ into R^k, and $\mathbf{g}$ is differentiable at $\mathbf{f}(\mathbf{x}_0)$. Then the mapping $\mathbf{F}$ of E into R^k defined by*

$$\mathbf{F}(\mathbf{x}) = \mathbf{g}(\mathbf{f}(\mathbf{x}))$$

is differentiable at $\mathbf{x}_0$, and

(21) $$\mathbf{F}'(\mathbf{x}_0) = \mathbf{g}'(\mathbf{f}(\mathbf{x}_0))\mathbf{f}'(\mathbf{x}_0).$$

On the right side of (21), we have the product of two linear transformations, as defined in Sec. 9.6.

Proof Put $\mathbf{y}_0 = \mathbf{f}(\mathbf{x}_0)$, $A = \mathbf{f}'(\mathbf{x}_0)$, $B = \mathbf{g}'(\mathbf{y}_0)$, and define

$$\mathbf{u}(\mathbf{h}) = \mathbf{f}(\mathbf{x}_0 + \mathbf{h}) - \mathbf{f}(\mathbf{x}_0) - A\mathbf{h},$$
$$\mathbf{v}(\mathbf{k}) = \mathbf{g}(\mathbf{y}_0 + \mathbf{k}) - \mathbf{g}(\mathbf{y}_0) - B\mathbf{k},$$

for all $\mathbf{h} \in R^n$ and $\mathbf{k} \in R^m$ for which $\mathbf{f}(\mathbf{x}_0 + \mathbf{h})$ and $\mathbf{g}(\mathbf{y}_0 + \mathbf{k})$ are defined. Then

(22) $$|\mathbf{u}(\mathbf{h})| = \varepsilon(\mathbf{h})|\mathbf{h}|, \qquad |\mathbf{v}(\mathbf{k})| = \eta(\mathbf{k})|\mathbf{k}|,$$

where $\varepsilon(\mathbf{h}) \to 0$ as $\mathbf{h} \to \mathbf{0}$ and $\eta(\mathbf{k}) \to 0$ as $\mathbf{k} \to \mathbf{0}$.

Given $\mathbf{h}$, put $\mathbf{k} = \mathbf{f}(\mathbf{x}_0 + \mathbf{h}) - \mathbf{f}(\mathbf{x}_0)$. Then

(23) $$|\mathbf{k}| = |A\mathbf{h} + \mathbf{u}(\mathbf{h})| \le [\|A\| + \varepsilon(\mathbf{h})]\,|\mathbf{h}|,$$

and

$$\mathbf{F}(\mathbf{x}_0 + \mathbf{h}) - \mathbf{F}(\mathbf{x}_0) - BA\mathbf{h} = \mathbf{g}(\mathbf{y}_0 + \mathbf{k}) - \mathbf{g}(\mathbf{y}_0) - BA\mathbf{h}$$
$$= B(\mathbf{k} - A\mathbf{h}) + \mathbf{v}(\mathbf{k})$$
$$= B\mathbf{u}(\mathbf{h}) + \mathbf{v}(\mathbf{k}).$$

Hence (22) and (23) imply, for $\mathbf{h} \neq \mathbf{0}$, that

$$\frac{|\mathbf{F}(\mathbf{x}_0 + \mathbf{h}) - \mathbf{F}(\mathbf{x}_0) - BA\mathbf{h}|}{|\mathbf{h}|} \leq \|B\| \varepsilon(\mathbf{h}) + [\|A\| + \varepsilon(\mathbf{h})]\eta(\mathbf{k}).$$

Let $\mathbf{h} \to \mathbf{0}$. Then $\varepsilon(\mathbf{h}) \to 0$. Also, $\mathbf{k} \to \mathbf{0}$, by (23), so that $\eta(\mathbf{k}) \to 0$. It follows that $\mathbf{F}'(\mathbf{x}_0) = BA$, which is what (21) asserts.

9.16 Partial derivatives We again consider a function $\mathbf{f}$ that maps an open set $E \subset R^n$ into R^m. Let $\{\mathbf{e}_1, \ldots, \mathbf{e}_n\}$ and $\{\mathbf{u}_1, \ldots, \mathbf{u}_m\}$ be the standard bases of R^n and R^m. The *components of* $\mathbf{f}$ are the real functions $f_1, \ldots, f_m$ defined by

$$(24) \qquad \mathbf{f}(\mathbf{x}) = \sum_{i=1}^{m} f_i(\mathbf{x})\mathbf{u}_i \qquad (\mathbf{x} \in E),$$

or, equivalently, by $f_i(\mathbf{x}) = \mathbf{f}(\mathbf{x}) \cdot \mathbf{u}_i$, $1 \leq i \leq m$.

For $\mathbf{x} \in E$, $1 \leq i \leq m$, $1 \leq j \leq n$, we define

$$(25) \qquad (D_j f_i)(\mathbf{x}) = \lim_{t \to 0} \frac{f_i(\mathbf{x} + t\mathbf{e}_j) - f_i(\mathbf{x})}{t},$$

provided the limit exists. Writing $f_i(x_1, \ldots, x_n)$ in place of $f_i(\mathbf{x})$, we see that $D_j f_i$ is the derivative of f_i with respect to x_j, keeping the other variables fixed. The notation

$$(26) \qquad \frac{\partial f_i}{\partial x_j}$$

is therefore often used in place of $D_j f_i$, and $D_j f_i$ is called a *partial derivative*.

In many cases where the existence of a derivative is sufficient when dealing with functions of one variable, continuity or at least boundedness of the partial derivatives is needed for functions of several variables. For example, the functions f and g described in Exercise 7, Chap. 4, are not continuous, although their partial derivatives exist at every point of R^2. Even for continuous functions, the existence of all partial derivatives does not imply differentiability in the sense of Definition 9.11; see Exercises 6 and 14, and Theorem 9.21.

However, if $\mathbf{f}$ is known to be differentiable at a point $\mathbf{x}$, then its partial derivatives exist at $\mathbf{x}$, and they determine the linear transformation $\mathbf{f}'(\mathbf{x})$ completely:

9.17 Theorem *Suppose $\mathbf{f}$ maps an open set $E \subset R^n$ into R^m, and $\mathbf{f}$ is differentiable at a point $\mathbf{x} \in E$. Then the partial derivatives $(D_j f_i)(\mathbf{x})$ exist, and*

$$(27) \qquad \mathbf{f}'(\mathbf{x})\mathbf{e}_j = \sum_{i=1}^{m} (D_j f_i)(\mathbf{x})\mathbf{u}_i \qquad (1 \leq j \leq n).$$

Here, as in Sec. 9.16, $\{e_1, \ldots, e_n\}$ and $\{u_1, \ldots, u_m\}$ are the standard bases of R^n and R^m.

Proof Fix j. Since f is differentiable at x,

$$f(x + te_j) - f(x) = f'(x)(te_j) + r(te_j)$$

where $|r(te_j)|/t \to 0$ as $t \to 0$. The linearity of $f'(x)$ shows therefore that

(28) $$\lim_{t \to 0} \frac{f(x + te_j) - f(x)}{t} = f'(x)e_j.$$

If we now represent f in terms of its components, as in (24), then (28) becomes

(29) $$\lim_{t \to 0} \sum_{i=1}^{m} \frac{f_i(x + te_j) - f_i(x)}{t} u_i = f'(x)e_j.$$

It follows that each quotient in this sum has a limit, as $t \to 0$ (see Theorem 4.10), so that each $(D_j f_i)(x)$ exists, and then (27) follows from (29).

Here are some consequences of Theorem 9.17:

Let $[f'(x)]$ be the matrix that represents $f'(x)$ with respect to our standard bases, as in Sec. 9.9.

Then $f'(x)e_j$ is the jth column vector of $[f'(x)]$, and (27) shows therefore that the number $(D_j f_i)(x)$ occupies the spot in the ith row and jth column of $[f'(x)]$. Thus

$$[f'(x)] = \begin{bmatrix} (D_1 f_1)(x) & \cdots & (D_n f_1)(x) \\ \cdots & \cdots & \cdots \\ (D_1 f_m)(x) & \cdots & (D_n f_m)(x) \end{bmatrix}.$$

If $h = \Sigma h_j e_j$ is any vector in R^n, then (27) implies that

(30) $$f'(x)h = \sum_{i=1}^{m} \left\{ \sum_{j=1}^{n} (D_j f_i)(x) h_j \right\} u_i.$$

9.18 Example Let γ be a differentiable mapping of the segment $(a, b) \subset R^1$ into an open set $E \subset R^n$, in other words, γ is a differentiable curve in E. Let f be a real-valued differentiable function with domain E. Thus f is a differentiable mapping of E into R^1. Define

(31) $$g(t) = f(\gamma(t)) \quad (a < t < b).$$

The chain rule asserts then that

(32) $$g'(t) = f'(\gamma(t))\gamma'(t) \quad (a < t < b).$$

Since $\gamma'(t) \in L(R^1, R^n)$ and $f'(\gamma(t)) \in L(R^n, R^1)$, (32) defines $g'(t)$ as a linear operator on R^1. This agrees with the fact that g maps (a, b) into R^1. However, $g'(t)$ can also be regarded as a real number. (This was discussed in Sec. 9.10.) This number can be computed in terms of the partial derivatives of f and the derivatives of the components of γ, as we shall now see.

With respect to the standard basis $\{e_1, \ldots, e_n\}$ of R^n, $[\gamma'(t)]$ is the n by 1 matrix (a "column matrix") which has $\gamma'_i(t)$ in the ith row, where $\gamma_1, \ldots, \gamma_n$ are the components of γ. For every $x \in E$, $[f'(x)]$ is the 1 by n matrix (a "row matrix") which has $(D_j f)(x)$ in the jth column. Hence $[g'(t)]$ is the 1 by 1 matrix whose only entry is the real number

$$(33) \qquad g'(t) = \sum_{i=1}^{n} (D_i f)(\gamma(t)) \gamma'_i(t).$$

This is a frequently encountered special case of the chain rule. It can be rephrased in the following manner.

Associate with each $x \in E$ a vector, the so-called "gradient" of f at x, defined by

$$(34) \qquad (\nabla f)(x) = \sum_{i=1}^{n} (D_i f)(x) e_i.$$

Since

$$(35) \qquad \gamma'(t) = \sum_{i=1}^{n} \gamma'_i(t) e_i,$$

(33) can be written in the form

$$(36) \qquad g'(t) = (\nabla f)(\gamma(t)) \cdot \gamma'(t),$$

the scalar product of the vectors $(\nabla f)(\gamma(t))$ and $\gamma'(t)$.

Let us now fix an $x \in E$, let $u \in R^n$ be a unit vector (that is, $|u| = 1$), and specialize γ so that

$$(37) \qquad \gamma(t) = x + tu \qquad (-\infty < t < \infty).$$

Then $\gamma'(t) = u$ for every t. Hence (36) shows that

$$(38) \qquad g'(0) = (\nabla f)(x) \cdot u.$$

On the other hand, (37) shows that

$$g(t) - g(0) = f(x + tu) - f(x).$$

Hence (38) gives

$$(39) \qquad \lim_{t \to 0} \frac{f(x + tu) - f(x)}{t} = (\nabla f)(x) \cdot u.$$

The limit in (39) is usually called the *directional derivative* of f at $\mathbf{x}$, in the direction of the unit vector $\mathbf{u}$, and may be denoted by $(D_\mathbf{u} f)(\mathbf{x})$.

If f and $\mathbf{x}$ are fixed, but $\mathbf{u}$ varies, then (39) shows that $(D_\mathbf{u} f)(\mathbf{x})$ attains its maximum when $\mathbf{u}$ is a positive scalar multiple of $(\nabla f)(\mathbf{x})$. [The case $(\nabla f)(\mathbf{x}) = \mathbf{0}$ should be excluded here.]

If $\mathbf{u} = \Sigma u_i \mathbf{e}_i$, then (39) shows that $(D_\mathbf{u} f)(\mathbf{x})$ can be expressed in terms of the partial derivatives of f at $\mathbf{x}$ by the formula

$$(40) \qquad (D_\mathbf{u} f)(\mathbf{x}) = \sum_{i=1}^{n} (D_i f)(\mathbf{x}) u_i.$$

Some of these ideas will play a role in the following theorem.

9.19 Theorem *Suppose* $\mathbf{f}$ *maps a convex open set* $E \subset R^n$ *into* R^m, $\mathbf{f}$ *is differentiable in* E, *and there is a real number* M *such that*

$$\|\mathbf{f}'(\mathbf{x})\| \le M$$

for every $\mathbf{x} \in E$. *Then*

$$|\mathbf{f}(\mathbf{b}) - \mathbf{f}(\mathbf{a})| \le M|\mathbf{b} - \mathbf{a}|$$

for all $\mathbf{a} \in E, \mathbf{b} \in E$.

Proof Fix $\mathbf{a} \in E$, $\mathbf{b} \in E$. Define

$$\gamma(t) = (1-t)\mathbf{a} + t\mathbf{b}$$

for all $t \in R^1$ such that $\gamma(t) \in E$. Since E is convex, $\gamma(t) \in E$ if $0 \le t \le 1$. Put

$$\mathbf{g}(t) = \mathbf{f}(\gamma(t)).$$

Then

$$\mathbf{g}'(t) = \mathbf{f}'(\gamma(t))\gamma'(t) = \mathbf{f}'(\gamma(t))(\mathbf{b} - \mathbf{a}),$$

so that

$$|\mathbf{g}'(t)| \le \|\mathbf{f}'(\gamma(t))\|\,|\mathbf{b} - \mathbf{a}| \le M|\mathbf{b} - \mathbf{a}|$$

for all $t \in [0, 1]$. By Theorem 5.19,

$$|\mathbf{g}(1) - \mathbf{g}(0)| \le M|\mathbf{b} - \mathbf{a}|.$$

But $\mathbf{g}(0) = \mathbf{f}(\mathbf{a})$ and $\mathbf{g}(1) = \mathbf{f}(\mathbf{b})$. This completes the proof.

Corollary *If, in addition,* $\mathbf{f}'(\mathbf{x}) = 0$ *for all* $\mathbf{x} \in E$, *then* $\mathbf{f}$ *is constant*.

Proof To prove this, note that the hypotheses of the theorem hold now with $M = 0$.

9.20 Definition A differentiable mapping $\mathbf{f}$ of an open set $E \subset R^n$ into R^m is said to be *continuously differentiable* in E if $\mathbf{f}'$ is a continuous mapping of E into $L(R^n, R^m)$.

More explicitly, it is required that to every $\mathbf{x} \in E$ and to every $\varepsilon > 0$ corresponds a $\delta > 0$ such that

$$\|\mathbf{f}'(\mathbf{y}) - \mathbf{f}'(\mathbf{x})\| < \varepsilon$$

if $\mathbf{y} \in E$ and $|\mathbf{x} - \mathbf{y}| < \delta$.

If this is so, we also say that $\mathbf{f}$ is a $\mathscr{C}'$-mapping, or that $\mathbf{f} \in \mathscr{C}'(E)$.

9.21 Theorem *Suppose $\mathbf{f}$ maps an open set $E \subset R^n$ into R^m. Then $\mathbf{f} \in \mathscr{C}'(E)$ if and only if the partial derivatives $D_j f_i$ exist and are continuous on E for $1 \le i \le m$, $1 \le j \le n$.*

Proof Assume first that $\mathbf{f} \in \mathscr{C}'(E)$. By (27),

$$(D_j f_i)(\mathbf{x}) = (\mathbf{f}'(\mathbf{x})\mathbf{e}_j) \cdot \mathbf{u}_i$$

for all i, j, and for all $\mathbf{x} \in E$. Hence

$$(D_j f_i)(\mathbf{y}) - (D_j f_i)(\mathbf{x}) = \{[\mathbf{f}'(\mathbf{y}) - \mathbf{f}'(\mathbf{x})]\mathbf{e}_j\} \cdot \mathbf{u}_i$$

and since $|\mathbf{u}_i| = |\mathbf{e}_j| = 1$, it follows that

$$|(D_j f_i)(\mathbf{y}) - (D_j f_i)(\mathbf{x})| \le |[\mathbf{f}'(\mathbf{y}) - \mathbf{f}'(\mathbf{x})]\mathbf{e}_j|$$
$$\le \|\mathbf{f}'(\mathbf{y}) - \mathbf{f}'(\mathbf{x})\|.$$

Hence $D_j f_i$ is continuous.

For the converse, it suffices to consider the case $m = 1$. (Why?) Fix $\mathbf{x} \in E$ and $\varepsilon > 0$. Since E is open, there is an open ball $S \subset E$, with center at $\mathbf{x}$ and radius r, and the continuity of the functions $D_j f$ shows that r can be chosen so that

(41) $$|(D_j f)(\mathbf{y}) - (D_j f)(\mathbf{x})| < \frac{\varepsilon}{n} \qquad (\mathbf{y} \in S, 1 \le j \le n).$$

Suppose $\mathbf{h} = \Sigma h_j \mathbf{e}_j$, $|\mathbf{h}| < r$, put $\mathbf{v}_0 = 0$, and $\mathbf{v}_k = h_1 \mathbf{e}_1 + \cdots + h_k \mathbf{e}_k$, for $1 \le k \le n$. Then

(42) $$f(\mathbf{x} + \mathbf{h}) - f(\mathbf{x}) = \sum_{j=1}^{n} [f(\mathbf{x} + \mathbf{v}_j) - f(\mathbf{x} + \mathbf{v}_{j-1})].$$

Since $|\mathbf{v}_k| < r$ for $1 \le k \le n$ and since S is convex, the segments with end points $\mathbf{x} + \mathbf{v}_{j-1}$ and $\mathbf{x} + \mathbf{v}_j$ lie in S. Since $\mathbf{v}_j = \mathbf{v}_{j-1} + h_j \mathbf{e}_j$, the mean value theorem (5.10) shows that the jth summand in (42) is equal to

$$h_j (D_j f)(\mathbf{x} + \mathbf{v}_{j-1} + \theta_j h_j \mathbf{e}_j)$$

for some $\theta_j \in (0, 1)$, and this differs from $h_j(D_j f)(\mathbf{x})$ by less than $|h_j|\varepsilon/n$, using (41). By (42), it follows that

$$\left| f(\mathbf{x} + \mathbf{h}) - f(\mathbf{x}) - \sum_{j=1}^{n} h_j(D_j f)(\mathbf{x}) \right| \leq \frac{1}{n} \sum_{j=1}^{n} |h_j|\varepsilon \leq |\mathbf{h}|\varepsilon$$

for all $\mathbf{h}$ such that $|\mathbf{h}| < r$.

This says that f is differentiable at $\mathbf{x}$ and that $f'(\mathbf{x})$ is the linear function which assigns the number $\Sigma h_j(D_j f)(\mathbf{x})$ to the vector $\mathbf{h} = \Sigma h_j \mathbf{e}_j$. The matrix $[f'(\mathbf{x})]$ consists of the row $(D_1 f)(\mathbf{x}), \ldots, (D_n f)(\mathbf{x})$; and since $D_1 f, \ldots, D_n f$ are continuous functions on E, the concluding remarks of Sec. 9.9 show that $f \in \mathscr{C}'(E)$.

THE CONTRACTION PRINCIPLE

We now interrupt our discussion of differentiation to insert a fixed point theorem that is valid in arbitrary complete metric spaces. It will be used in the proof of the inverse function theorem.

9.22 Definition Let X be a metric space, with metric d. If φ maps X into X and if there is a number $c < 1$ such that

(43) $$d(\varphi(x), \varphi(y)) \leq c\, d(x, y)$$

for all $x, y \in X$, then φ is said to be a *contraction* of X into X.

9.23 Theorem *If X is a complete metric space, and if φ is a contraction of X into X, then there exists one and only one $x \in X$ such that $\varphi(x) = x$.*

In other words, φ has a unique fixed point. The uniqueness is a triviality, for if $\varphi(x) = x$ and $\varphi(y) = y$, then (43) gives $d(x, y) \leq c\, d(x, y)$, which can only happen when $d(x, y) = 0$.

The *existence* of a fixed point of φ is the essential part of the theorem. The proof actually furnishes a constructive method for locating the fixed point.

Proof Pick $x_0 \in X$ arbitrarily, and define $\{x_n\}$ recursively, by setting

(44) $$x_{n+1} = \varphi(x_n) \qquad (n = 0, 1, 2, \ldots).$$

Choose $c < 1$ so that (43) holds. For $n \geq 1$ we then have

$$d(x_{n+1}, x_n) = d(\varphi(x_n), \varphi(x_{n-1})) \leq c\, d(x_n, x_{n-1}).$$

Hence induction gives

(45) $$d(x_{n+1}, x_n) \leq c^n\, d(x_1, x_0) \qquad (n = 0, 1, 2, \ldots).$$

If $n < m$, it follows that

$$d(x_n, x_m) \leq \sum_{i=n+1}^{m} d(x_i, x_{i-1})$$
$$\leq (c^n + c^{n+1} + \cdots + c^{m-1}) d(x_1, x_0)$$
$$\leq [(1-c)^{-1} d(x_1, x_0)] c^n.$$

Thus $\{x_n\}$ is a Cauchy sequence. Since X is complete, $\lim x_n = x$ for some $x \in X$.

Since φ is a contraction, φ is continuous (in fact, uniformly continuous) on X. Hence

$$\varphi(x) = \lim_{n \to \infty} \varphi(x_n) = \lim_{n \to \infty} x_{n+1} = x.$$

THE INVERSE FUNCTION THEOREM

The inverse function theorem states, roughly speaking, that a continuously differentiable mapping $\mathbf{f}$ is invertible in a neighborhood of any point $\mathbf{x}$ at which the linear transformation $\mathbf{f}'(\mathbf{x})$ is invertible:

9.24 Theorem *Suppose $\mathbf{f}$ is a $\mathscr{C}'$-mapping of an open set $E \subset R^n$ into R^n, $\mathbf{f}'(\mathbf{a})$ is invertible for some $\mathbf{a} \in E$, and $\mathbf{b} = \mathbf{f}(\mathbf{a})$. Then*

(a) *there exist open sets U and V in R^n such that $\mathbf{a} \in U$, $\mathbf{b} \in V$, $\mathbf{f}$ is one-to-one on U, and $\mathbf{f}(U) = V$;*

(b) *if $\mathbf{g}$ is the inverse of $\mathbf{f}$ [which exists, by (a)], defined in V by*

$$\mathbf{g}(\mathbf{f}(\mathbf{x})) = \mathbf{x} \qquad (\mathbf{x} \in U),$$

then $\mathbf{g} \in \mathscr{C}'(V)$.

Writing the equation $\mathbf{y} = \mathbf{f}(\mathbf{x})$ in component form, we arrive at the following interpretation of the conclusion of the theorem: The system of n equations

$$y_i = f_i(x_1, \ldots, x_n) \qquad (1 \leq i \leq n)$$

can be solved for $x_1, \ldots, x_n$ in terms of $y_1, \ldots, y_n$, if we restrict $\mathbf{x}$ and $\mathbf{y}$ to small enough neighborhoods of $\mathbf{a}$ and $\mathbf{b}$; the solutions are unique and continuously differentiable.

Proof

(a) Put $\mathbf{f}'(\mathbf{a}) = A$, and choose λ so that

(46) $$2\lambda \|A^{-1}\| = 1.$$

Since $\mathbf{f}'$ is continuous at $\mathbf{a}$, there is an open ball $U \subset E$, with center at $\mathbf{a}$, such that

(47) $$\|\mathbf{f}'(\mathbf{x}) - A\| < \lambda \quad (\mathbf{x} \in U).$$

We associate to each $\mathbf{y} \in R^n$ a function φ, defined by

(48) $$\varphi(\mathbf{x}) = \mathbf{x} + A^{-1}(\mathbf{y} - \mathbf{f}(\mathbf{x})) \quad (\mathbf{x} \in E).$$

Note that $\mathbf{f}(\mathbf{x}) = \mathbf{y}$ if and only if $\mathbf{x}$ is a fixed point of φ.

Since $\varphi'(\mathbf{x}) = I - A^{-1}\mathbf{f}'(\mathbf{x}) = A^{-1}(A - \mathbf{f}'(\mathbf{x}))$, (46) and (47) imply that

(49) $$\|\varphi'(\mathbf{x})\| < \tfrac{1}{2} \quad (\mathbf{x} \in U).$$

Hence

(50) $$|\varphi(\mathbf{x}_1) - \varphi(\mathbf{x}_2)| \le \tfrac{1}{2}|\mathbf{x}_1 - \mathbf{x}_2| \quad (\mathbf{x}_1, \mathbf{x}_2 \in U),$$

by Theorem 9.19. It follows that φ has at most one fixed point in U, so that $\mathbf{f}(\mathbf{x}) = \mathbf{y}$ for at most one $\mathbf{x} \in U$.

Thus $\mathbf{f}$ is $1 - 1$ in U.

Next, put $V = \mathbf{f}(U)$, and pick $\mathbf{y}_0 \in V$. Then $\mathbf{y}_0 = \mathbf{f}(\mathbf{x}_0)$ for some $\mathbf{x}_0 \in U$. Let B be an open ball with center at $\mathbf{x}_0$ and radius $r > 0$, so small that its closure $\bar{B}$ lies in U. We will show that $\mathbf{y} \in V$ whenever $|\mathbf{y} - \mathbf{y}_0| < \lambda r$. This proves, of course, that V is open.

Fix $\mathbf{y}$, $|\mathbf{y} - \mathbf{y}_0| < \lambda r$. With φ as in (48),

$$|\varphi(\mathbf{x}_0) - \mathbf{x}_0| = |A^{-1}(\mathbf{y} - \mathbf{y}_0)| < \|A^{-1}\|\lambda r = \frac{r}{2}.$$

If $\mathbf{x} \in \bar{B}$, it therefore follows from (50) that

$$|\varphi(\mathbf{x}) - \mathbf{x}_0| \le |\varphi(\mathbf{x}) - \varphi(\mathbf{x}_0)| + |\varphi(\mathbf{x}_0) - \mathbf{x}_0|$$
$$< \frac{1}{2}|\mathbf{x} - \mathbf{x}_0| + \frac{r}{2} \le r;$$

hence $\varphi(\mathbf{x}) \in B$. Note that (50) holds if $\mathbf{x}_1 \in \bar{B}$, $\mathbf{x}_2 \in \bar{B}$.

Thus φ is a contraction of $\bar{B}$ into $\bar{B}$. Being a closed subset of R^n, $\bar{B}$ is complete. Theorem 9.23 implies therefore that φ has a fixed point $\mathbf{x} \in \bar{B}$. For this $\mathbf{x}$, $f(\mathbf{x}) = \mathbf{y}$. Thus $\mathbf{y} \in \mathbf{f}(\bar{B}) \subset \mathbf{f}(U) = V$.

This proves part (*a*) of the theorem.

(b) Pick $\mathbf{y} \in V$, $\mathbf{y} + \mathbf{k} \in V$. Then there exist $\mathbf{x} \in U$, $\mathbf{x} + \mathbf{h} \in U$, so that $\mathbf{y} = \mathbf{f}(\mathbf{x})$, $\mathbf{y} + \mathbf{k} = \mathbf{f}(\mathbf{x} + \mathbf{h})$. With φ as in (48),

$$\varphi(\mathbf{x} + \mathbf{h}) - \varphi(\mathbf{x}) = \mathbf{h} + A^{-1}[\mathbf{f}(\mathbf{x}) - \mathbf{f}(\mathbf{x} + \mathbf{h})] = \mathbf{h} - A^{-1}\mathbf{k}.$$

By (50), $|\mathbf{h} - A^{-1}\mathbf{k}| \le \tfrac{1}{2}|\mathbf{h}|$. Hence $|A^{-1}\mathbf{k}| \ge \tfrac{1}{2}|\mathbf{h}|$, and

(51) $$|\mathbf{h}| \le 2\|A^{-1}\|\,|\mathbf{k}| = \lambda^{-1}|\mathbf{k}|.$$

By (46), (47), and Theorem 9.8, $\mathbf{f}'(\mathbf{x})$ has an inverse, say T. Since
$$\mathbf{g}(\mathbf{y} + \mathbf{k}) - \mathbf{g}(\mathbf{y}) - T\mathbf{k} = \mathbf{h} - T\mathbf{k} = -T[\mathbf{f}(\mathbf{x} + \mathbf{h}) - \mathbf{f}(\mathbf{x}) - \mathbf{f}'(\mathbf{x})\mathbf{h}],$$
(51) implies
$$\frac{|\mathbf{g}(\mathbf{y} + \mathbf{k}) - \mathbf{g}(\mathbf{y}) - T\mathbf{k}|}{|\mathbf{k}|} \leq \frac{\|T\|}{\lambda} \cdot \frac{|\mathbf{f}(\mathbf{x} + \mathbf{h}) - \mathbf{f}(\mathbf{x}) - \mathbf{f}'(\mathbf{x})\mathbf{h}|}{|\mathbf{h}|}.$$

As $\mathbf{k} \to \mathbf{0}$, (51) shows that $\mathbf{h} \to \mathbf{0}$. The right side of the last inequality thus tends to 0. Hence the same is true of the left. We have thus proved that $\mathbf{g}'(\mathbf{y}) = T$. But T was chosen to be the inverse of $\mathbf{f}'(\mathbf{x}) = \mathbf{f}'(\mathbf{g}(\mathbf{y}))$. Thus

(52) $$\mathbf{g}'(\mathbf{y}) = \{\mathbf{f}'(\mathbf{g}(\mathbf{y}))\}^{-1} \qquad (\mathbf{y} \in V).$$

Finally, note that $\mathbf{g}$ is a continuous mapping of V onto U (since $\mathbf{g}$ is differentiable), that $\mathbf{f}'$ is a continuous mapping of U into the set Ω of all invertible elements of $L(R^n)$, and that inversion is a continuous mapping of Ω onto Ω, by Theorem 9.8. If we combine these facts with (52), we see that $\mathbf{g} \in \mathscr{C}'(V)$.

This completes the proof.

Remark. The full force of the assumption that $\mathbf{f} \in \mathscr{C}'(E)$ was only used in the last paragraph of the preceding proof. Everything else, down to Eq. (52), was derived from the existence of $\mathbf{f}'(\mathbf{x})$ for $\mathbf{x} \in E$, the invertibility of $\mathbf{f}'(\mathbf{a})$, and the continuity of $\mathbf{f}'$ at just the point $\mathbf{a}$. In this connection, we refer to the article by A. Nijenhuis in *Amer. Math. Monthly*, vol. 81, 1974, pp. 969–980.

The following is an immediate consequence of part (*a*) of the inverse function theorem.

9.25 Theorem *If $\mathbf{f}$ is a $\mathscr{C}'$-mapping of an open set $E \subset R^n$ into R^n and if $\mathbf{f}'(\mathbf{x})$ is invertible for every $\mathbf{x} \in E$, then $\mathbf{f}(W)$ is an open subset of R^n for every open set $W \subset E$.*

In other words, $\mathbf{f}$ is an *open mapping* of E into R^n.

The hypotheses made in this theorem ensure that each point $\mathbf{x} \in E$ has a neighborhood in which $\mathbf{f}$ is 1-1. This may be expressed by saying that $\mathbf{f}$ is *locally* one-to-one in E. But $\mathbf{f}$ need not be 1-1 in E under these circumstances. For an example, see Exercise 17.

THE IMPLICIT FUNCTION THEOREM

If f is a continuously differentiable real function in the plane, then the equation $f(x, y) = 0$ can be solved for y in terms of x in a neighborhood of any point

(a, b) at which $f(a, b) = 0$ and $\partial f/\partial y \neq 0$. Likewise, one can solve for x in terms of y near (a, b) if $\partial f/\partial x \neq 0$ at (a, b). For a simple example which illustrates the need for assuming $\partial f/\partial y \neq 0$, consider $f(x, y) = x^2 + y^2 - 1$.

The preceding very informal statement is the simplest case (the case $m = n = 1$ of Theorem 9.28) of the so-called "implicit function theorem." Its proof makes strong use of the fact that continuously differentiable transformations behave locally very much like their derivatives. Accordingly, we first prove Theorem 9.27, the linear version of Theorem 9.28.

9.26 Notation If $\mathbf{x} = (x_1, \ldots, x_n) \in R^n$ and $\mathbf{y} = (y_1, \ldots, y_m) \in R^m$, let us write $(\mathbf{x}, \mathbf{y})$ for the point (or vector)

$$(x_1, \ldots, x_n, y_1, \ldots, y_m) \in R^{n+m}.$$

In what follows, the first entry in $(\mathbf{x}, \mathbf{y})$ or in a similar symbol will always be a vector in R^n, the second will be a vector in R^m.

Every $A \in L(R^{n+m}, R^n)$ can be split into two linear transformations A_x and A_y, defined by

(53) $$A_x \mathbf{h} = A(\mathbf{h}, 0), \qquad A_y \mathbf{k} = A(0, \mathbf{k})$$

for any $\mathbf{h} \in R^n$, $\mathbf{k} \in R^m$. Then $A_x \in L(R^n)$, $A_y \in L(R^m, R^n)$, and

(54) $$A(\mathbf{h}, \mathbf{k}) = A_x \mathbf{h} + A_y \mathbf{k}.$$

The linear version of the implicit function theorem is now almost obvious.

9.27 Theorem *If $A \in L(R^{n+m}, R^n)$ and if A_x is invertible, then there corresponds to every $\mathbf{k} \in R^m$ a unique $\mathbf{h} \in R^n$ such that $A(\mathbf{h}, \mathbf{k}) = 0$.*

This $\mathbf{h}$ can be computed from $\mathbf{k}$ by the formula

(55) $$\mathbf{h} = -(A_x)^{-1} A_y \mathbf{k}.$$

Proof By (54), $A(\mathbf{h}, \mathbf{k}) = 0$ if and only if

$$A_x \mathbf{h} + A_y \mathbf{k} = 0,$$

which is the same as (55) when A_x is invertible.

The conclusion of Theorem 9.27 is, in other words, that the equation $A(\mathbf{h}, \mathbf{k}) = 0$ can be solved (uniquely) for $\mathbf{h}$ if $\mathbf{k}$ is given, and that the solution $\mathbf{h}$ is a linear function of $\mathbf{k}$. Those who have some acquaintance with linear algebra will recognize this as a very familiar statement about systems of linear equations.

9.28 Theorem *Let $\mathbf{f}$ be a $\mathscr{C}'$-mapping of an open set $E \subset R^{n+m}$ into R^n, such that $\mathbf{f}(\mathbf{a}, \mathbf{b}) = 0$ for some point $(\mathbf{a}, \mathbf{b}) \in E$.*

Put $A = \mathbf{f}'(\mathbf{a}, \mathbf{b})$ and assume that A_x is invertible.

Then there exist open sets $U \subset R^{n+m}$ and $W \subset R^m$, with $(\mathbf{a}, \mathbf{b}) \in U$ and $\mathbf{b} \in W$, having the following property:

To every $\mathbf{y} \in W$ corresponds a unique $\mathbf{x}$ such that

(56) $$(\mathbf{x}, \mathbf{y}) \in U \quad \text{and} \quad \mathbf{f}(\mathbf{x}, \mathbf{y}) = \mathbf{0}.$$

If this $\mathbf{x}$ is defined to be $\mathbf{g}(\mathbf{y})$, then $\mathbf{g}$ is a $\mathscr{C}'$-mapping of W into R^n, $\mathbf{g}(\mathbf{b}) = \mathbf{a}$,

(57) $$\mathbf{f}(\mathbf{g}(\mathbf{y}), \mathbf{y}) = \mathbf{0} \quad (\mathbf{y} \in W),$$

and

(58) $$\mathbf{g}'(\mathbf{b}) = -(A_x)^{-1} A_y.$$

The function g is "implicitly" defined by (57). Hence the name of the theorem.

The equation $\mathbf{f}(\mathbf{x}, \mathbf{y}) = \mathbf{0}$ can be written as a system of n equations in $n + m$ variables:

(59) $$\begin{aligned} f_1(x_1, \ldots, x_n, y_1, \ldots, y_m) &= 0 \\ &\cdots\cdots\cdots\cdots\cdots\cdots\cdots \\ f_n(x_1, \ldots, x_n, y_1, \ldots, y_m) &= 0. \end{aligned}$$

The assumption that A_x is invertible means that the n by n matrix

$$\begin{bmatrix} D_1 f_1 & \cdots & D_n f_1 \\ \cdots\cdots\cdots\cdots\cdots\cdots \\ D_1 f_n & \cdots & D_n f_n \end{bmatrix}$$

evaluated at $(\mathbf{a}, \mathbf{b})$ defines an invertible linear operator in R^n; in other words, its column vectors should be independent, or, equivalently, its determinant should be $\neq 0$. (See Theorem 9.36.) If, furthermore, (59) holds when $\mathbf{x} = \mathbf{a}$ and $\mathbf{y} = \mathbf{b}$, then the conclusion of the theorem is that (59) can be solved for $x_1, \ldots, x_n$ in terms of $y_1, \ldots, y_m$, for every $\mathbf{y}$ near $\mathbf{b}$, and that these solutions are continuously differentiable functions of $\mathbf{y}$.

Proof Define $\mathbf{F}$ by

(60) $$\mathbf{F}(\mathbf{x}, \mathbf{y}) = (\mathbf{f}(\mathbf{x}, \mathbf{y}), \mathbf{y}) \quad ((\mathbf{x}, \mathbf{y}) \in E).$$

Then $\mathbf{F}$ is a $\mathscr{C}'$-mapping of E into R^{n+m}. We claim that $\mathbf{F}'(\mathbf{a}, \mathbf{b})$ is an invertible element of $L(R^{n+m})$:

Since $\mathbf{f}(\mathbf{a}, \mathbf{b}) = \mathbf{0}$, we have

$$\mathbf{f}(\mathbf{a} + \mathbf{h}, \mathbf{b} + \mathbf{k}) = A(\mathbf{h}, \mathbf{k}) + \mathbf{r}(\mathbf{h}, \mathbf{k}),$$

where $\mathbf{r}$ is the remainder that occurs in the definition of $\mathbf{f}'(\mathbf{a}, \mathbf{b})$. Since

$$\begin{aligned} \mathbf{F}(\mathbf{a} + \mathbf{h}, \mathbf{b} + \mathbf{k}) - \mathbf{F}(\mathbf{a}, \mathbf{b}) &= (\mathbf{f}(\mathbf{a} + \mathbf{h}, \mathbf{b} + \mathbf{k}), \mathbf{k}) \\ &= (A(\mathbf{h}, \mathbf{k}), \mathbf{k}) + (\mathbf{r}(\mathbf{h}, \mathbf{k}), \mathbf{0}) \end{aligned}$$

it follows that $\mathbf{F}'(\mathbf{a}, \mathbf{b})$ is the linear operator on R^{n+m} that maps $(\mathbf{h}, \mathbf{k})$ to $(A(\mathbf{h}, \mathbf{k}), \mathbf{k})$. If this image vector is $\mathbf{0}$, then $A(\mathbf{h}, \mathbf{k}) = \mathbf{0}$ and $\mathbf{k} = \mathbf{0}$, hence $A(\mathbf{h}, \mathbf{0}) = \mathbf{0}$, and Theorem 9.27 implies that $\mathbf{h} = \mathbf{0}$. It follows that $\mathbf{F}'(\mathbf{a}, \mathbf{b})$ is 1-1; hence it is invertible (Theorem 9.5).

The inverse function theorem can therefore be applied to $\mathbf{F}$. It shows that there exist open sets U and V in R^{n+m}, with $(\mathbf{a}, \mathbf{b}) \in U$, $(\mathbf{0}, \mathbf{b}) \in V$, such that $\mathbf{F}$ is a 1-1 mapping of U onto V.

We let W be the set of all $\mathbf{y} \in R^m$ such that $(\mathbf{0}, \mathbf{y}) \in V$. Note that $\mathbf{b} \in W$.

It is clear that W is open since V is open.

If $\mathbf{y} \in W$, then $(\mathbf{0}, \mathbf{y}) = \mathbf{F}(\mathbf{x}, \mathbf{y})$ for some $(\mathbf{x}, \mathbf{y}) \in U$. By (60), $\mathbf{f}(\mathbf{x}, \mathbf{y}) = \mathbf{0}$ for this $\mathbf{x}$.

Suppose, with the same $\mathbf{y}$, that $(\mathbf{x}', \mathbf{y}) \in U$ and $\mathbf{f}(\mathbf{x}', \mathbf{y}) = \mathbf{0}$. Then

$$\mathbf{F}(\mathbf{x}', \mathbf{y}) = (\mathbf{f}(\mathbf{x}', \mathbf{y}), \mathbf{y}) = (\mathbf{f}(\mathbf{x}, \mathbf{y}), \mathbf{y}) = \mathbf{F}(\mathbf{x}, \mathbf{y}).$$

Since $\mathbf{F}$ is 1-1 in U, it follows that $\mathbf{x}' = \mathbf{x}$.

This proves the first part of the theorem.

For the second part, define $\mathbf{g}(\mathbf{y})$, for $\mathbf{y} \in W$, so that $(\mathbf{g}(\mathbf{y}), \mathbf{y}) \in U$ and (57) holds. Then

(61) $$\mathbf{F}(\mathbf{g}(\mathbf{y}), \mathbf{y}) = (\mathbf{0}, \mathbf{y}) \qquad (\mathbf{y} \in W).$$

If $\mathbf{G}$ is the mapping of V onto U that inverts $\mathbf{F}$, then $\mathbf{G} \in \mathscr{C}'$, by the inverse function theorem, and (61) gives

(62) $$(\mathbf{g}(\mathbf{y}), \mathbf{y}) = \mathbf{G}(\mathbf{0}, \mathbf{y}) \qquad (\mathbf{y} \in W).$$

Since $\mathbf{G} \in \mathscr{C}'$, (62) shows that $\mathbf{g} \in \mathscr{C}'$.

Finally, to compute $\mathbf{g}'(\mathbf{b})$, put $(\mathbf{g}(\mathbf{y}), \mathbf{y}) = \Phi(\mathbf{y})$. Then

(63) $$\Phi'(\mathbf{y})\mathbf{k} = (\mathbf{g}'(\mathbf{y})\mathbf{k}, \mathbf{k}) \qquad (\mathbf{y} \in W, \mathbf{k} \in R^m).$$

By (57), $\mathbf{f}(\Phi(\mathbf{y})) = \mathbf{0}$ in W. The chain rule shows therefore that

$$\mathbf{f}'(\Phi(\mathbf{y}))\Phi'(\mathbf{y}) = 0.$$

When $\mathbf{y} = \mathbf{b}$, then $\Phi(\mathbf{y}) = (\mathbf{a}, \mathbf{b})$, and $\mathbf{f}'(\Phi(\mathbf{y})) = A$. Thus

(64) $$A\Phi'(\mathbf{b}) = 0.$$

It now follows from (64), (63), and (54), that

$$A_x \mathbf{g}'(\mathbf{b})\mathbf{k} + A_y \mathbf{k} = A(\mathbf{g}'(\mathbf{b})\mathbf{k}, \mathbf{k}) = A\Phi'(\mathbf{b})\mathbf{k} = 0$$

for every $\mathbf{k} \in R^m$. Thus

(65) $$A_x \mathbf{g}'(\mathbf{b}) + A_y = 0.$$

This is equivalent to (58), and completes the proof.

Note. In terms of the components of **f** and **g**, (65) becomes

$$\sum_{j=1}^{n} (D_j f_i)(\mathbf{a}, \mathbf{b})(D_k g_j)(\mathbf{b}) = -(D_{n+k} f_i)(\mathbf{a}, \mathbf{b})$$

or

$$\sum_{j=1}^{n} \left(\frac{\partial f_i}{\partial x_j}\right)\left(\frac{\partial g_j}{\partial y_k}\right) = -\left(\frac{\partial f_i}{\partial y_k}\right)$$

where $1 \leq i \leq n$, $1 \leq k \leq m$.

For each k, this is a system of n linear equations in which the derivatives $\partial g_j / \partial y_k$ $(1 \leq j \leq n)$ are the unknowns.

9.29 Example Take $n = 2$, $m = 3$, and consider the mapping $\mathbf{f} = (f_1, f_2)$ of R^5 into R^2 given by

$$f_1(x_1, x_2, y_1, y_2, y_3) = 2e^{x_1} + x_2 y_1 - 4y_2 + 3$$
$$f_2(x_1, x_2, y_1, y_2, y_3) = x_2 \cos x_1 - 6x_1 + 2y_1 - y_3.$$

If $\mathbf{a} = (0, 1)$ and $\mathbf{b} = (3, 2, 7)$, then $\mathbf{f}(\mathbf{a}, \mathbf{b}) = 0$.

With respect to the standard bases, the matrix of the transformation $A = \mathbf{f}'(\mathbf{a}, \mathbf{b})$ is

$$[A] = \begin{bmatrix} 2 & 3 & 1 & -4 & 0 \\ -6 & 1 & 2 & 0 & -1 \end{bmatrix}.$$

Hence

$$[A_x] = \begin{bmatrix} 2 & 3 \\ -6 & 1 \end{bmatrix}, \quad [A_y] = \begin{bmatrix} 1 & -4 & 0 \\ 2 & 0 & -1 \end{bmatrix}.$$

We see that the column vectors of $[A_x]$ are independent. Hence A_x is invertible and the implicit function theorem asserts the existence of a $\mathscr{C}'$-mapping **g**, defined in a neighborhood of $(3, 2, 7)$, such that $\mathbf{g}(3, 2, 7) = (0, 1)$ and $\mathbf{f}(\mathbf{g}(\mathbf{y}), \mathbf{y}) = 0$.

We can use (58) to compute $\mathbf{g}'(3, 2, 7)$: Since

$$[(A_x)^{-1}] = [A_x]^{-1} = \frac{1}{20}\begin{bmatrix} 1 & -3 \\ 6 & 2 \end{bmatrix}$$

(58) gives

$$[\mathbf{g}'(3, 2, 7)] = -\frac{1}{20}\begin{bmatrix} 1 & -3 \\ 6 & 2 \end{bmatrix}\begin{bmatrix} 1 & -4 & 0 \\ 2 & 0 & -1 \end{bmatrix} = \begin{bmatrix} \frac{1}{4} & \frac{1}{5} & -\frac{3}{20} \\ -\frac{1}{2} & \frac{6}{5} & \frac{1}{10} \end{bmatrix}.$$

In terms of partial derivatives, the conclusion is that

$$D_1 g_1 = \tfrac{1}{4} \quad D_2 g_1 = \tfrac{1}{5} \quad D_3 g_1 = -\tfrac{3}{20}$$
$$D_1 g_2 = -\tfrac{1}{2} \quad D_2 g_2 = \tfrac{6}{5} \quad D_3 g_2 = \tfrac{1}{10}$$

at the point (3, 2, 7).

THE RANK THEOREM

Although this theorem is not as important as the inverse function theorem or the implicit function theorem, we include it as another interesting illustration of the general principle that the local behavior of a continuously differentiable mapping $\mathbf{F}$ near a point $\mathbf{x}$ is similar to that of the linear transformation $\mathbf{F}'(\mathbf{x})$.

Before stating it, we need a few more facts about linear transformations.

9.30 Definitions Suppose X and Y are vector spaces, and $A \in L(X, Y)$, as in Definition 9.6. The *null space* of A, $\mathcal{N}(A)$, is the set of all $\mathbf{x} \in X$ at which $A\mathbf{x} = \mathbf{0}$. It is clear that $\mathcal{N}(A)$ is a vector space in X.

Likewise, the *range* of A, $\mathcal{R}(A)$, is a vector space in Y.

The *rank of A* is defined to be the dimension of $\mathcal{R}(A)$.

For example, the invertible elements of $L(R^n)$ are precisely those whose rank is n. This follows from Theorem 9.5.

If $A \in L(X, Y)$ and A has rank 0, then $A\mathbf{x} = \mathbf{0}$ for all $x \in A$, hence $\mathcal{N}(A) = X$. In this connection, see Exercise 25.

9.31 Projections Let X be a vector space. An operator $P \in L(X)$ is said to be a *projection in X* if $P^2 = P$.

More explicitly, the requirement is that $P(P\mathbf{x}) = P\mathbf{x}$ for every $\mathbf{x} \in X$. In other words, P fixes every vector in its range $\mathcal{R}(P)$.

Here are some elementary properties of projections:

(a) *If P is a projection in X, then every $\mathbf{x} \in X$ has a unique representation of the form*

$$\mathbf{x} = \mathbf{x}_1 + \mathbf{x}_2$$

where $\mathbf{x}_1 \in \mathcal{R}(P)$, $\mathbf{x}_2 \in \mathcal{N}(P)$.

To obtain the representation, put $\mathbf{x}_1 = P\mathbf{x}$, $\mathbf{x}_2 = \mathbf{x} - \mathbf{x}_1$. Then $P\mathbf{x}_2 = P\mathbf{x} - P\mathbf{x}_1 = P\mathbf{x} - P^2\mathbf{x} = \mathbf{0}$. As regards the uniqueness, apply P to the equation $\mathbf{x} = \mathbf{x}_1 + \mathbf{x}_2$. Since $\mathbf{x}_1 \in \mathcal{R}(P)$, $P\mathbf{x}_1 = \mathbf{x}_1$; since $P\mathbf{x}_2 = \mathbf{0}$, it follows that $\mathbf{x}_1 = P\mathbf{x}$.

(b) *If X is a finite-dimensional vector space and if X_1 is a vector space in X, then there is a projection P in X with $\mathcal{R}(P) = X_1$.*

If X_1 contains only $\mathbf{0}$, this is trivial: put $P\mathbf{x} = \mathbf{0}$ for all $\mathbf{x} \in X$.

Assume dim $X_1 = k > 0$. By Theorem 9.3, X has then a basis $\{\mathbf{u}_1, \ldots, \mathbf{u}_n\}$ such that $\{\mathbf{u}_1, \ldots, \mathbf{u}_k\}$ is a basis of X_1. Define

$$P(c_1\mathbf{u}_1 + \cdots + c_n\mathbf{u}_n) = c_1\mathbf{u}_1 + \cdots + c_k\mathbf{u}_k$$

for arbitrary scalars $c_1, \ldots, c_n$.

Then $P\mathbf{x} = \mathbf{x}$ for every $\mathbf{x} \in X_1$, and $X_1 = \mathscr{R}(P)$.

Note that $\{\mathbf{u}_{k+1}, \ldots, \mathbf{u}_n\}$ is a basis of $\mathscr{N}(P)$. Note also that there are infinitely many projections in X, with range X_1, if $0 < \dim X_1 < \dim X$.

9.32 Theorem *Suppose m, n, r are nonnegative integers, $m \geq r$, $n \geq r$, $\mathbf{F}$ is a $\mathscr{C}'$-mapping of an open set $E \subset R^n$ into R^m, and $\mathbf{F}'(\mathbf{x})$ has rank r for every $\mathbf{x} \in E$.*

Fix $\mathbf{a} \in E$, put $A = \mathbf{F}'(\mathbf{a})$, let Y_1 be the range of A, and let P be a projection in R^m whose range is Y_1. Let Y_2 be the null space of P.

Then there are open sets U and V in R^n, with $\mathbf{a} \in U$, $U \subset E$, and there is a 1-1 $\mathscr{C}'$-mapping $\mathbf{H}$ of V onto U (whose inverse is also of class $\mathscr{C}'$) such that

(66) $$\mathbf{F}(\mathbf{H}(\mathbf{x})) = A\mathbf{x} + \varphi(A\mathbf{x}) \qquad (\mathbf{x} \in V)$$

where φ is a $\mathscr{C}'$-mapping of the open set $A(V) \subset Y_1$ into Y_2.

After the proof we shall give a more geometric description of the information that (66) contains.

Proof If $r = 0$, Theorem 9.19 shows that $\mathbf{F}(\mathbf{x})$ is constant in a neighborhood U of $\mathbf{a}$, and (66) holds trivially, with $V = U$, $\mathbf{H}(\mathbf{x}) = \mathbf{x}$, $\varphi(\mathbf{0}) = \mathbf{F}(\mathbf{a})$.

From now on we assume $r > 0$. Since dim $Y_1 = r$, Y_1 has a basis $\{\mathbf{y}_1, \ldots, \mathbf{y}_r\}$. Choose $\mathbf{z}_i \in R^n$ so that $A\mathbf{z}_i = \mathbf{y}_i$ ($1 \leq i \leq r$), and define a linear mapping S of Y_1 into R^n by setting

(67) $$S(c_1\mathbf{y}_1 + \cdots + c_r\mathbf{y}_r) = c_1\mathbf{z}_1 + \cdots + c_r\mathbf{z}_r$$

for all scalars $c_1, \ldots, c_r$.

Then $AS\mathbf{y}_i = A\mathbf{z}_i = \mathbf{y}_i$ for $1 \leq i \leq r$. Thus

(68) $$AS\mathbf{y} = \mathbf{y} \qquad (\mathbf{y} \in Y_1).$$

Define a mapping $\mathbf{G}$ of E into R^n by setting

(69) $$\mathbf{G}(\mathbf{x}) = \mathbf{x} + SP[\mathbf{F}(\mathbf{x}) - A\mathbf{x}] \qquad (\mathbf{x} \in E).$$

Since $\mathbf{F}'(\mathbf{a}) = A$, differentiation of (69) shows that $\mathbf{G}'(\mathbf{a}) = I$, the identity operator on R^n. By the inverse function theorem, there are open sets U and V in R^n, with $\mathbf{a} \in U$, such that $\mathbf{G}$ is a 1-1 mapping of U onto V whose inverse $\mathbf{H}$ is also of class $\mathscr{C}'$. Moreover, by shrinking U and V, if necessary, we can arrange it so that V is convex and $\mathbf{H}'(\mathbf{x})$ is invertible for every $\mathbf{x} \in V$.

Note that $ASPA = A$, since $PA = A$ and (68) holds. Therefore (69) gives

(70) $$AG(\mathbf{x}) = PF(\mathbf{x}) \qquad (\mathbf{x} \in E).$$

In particular, (70) holds for $\mathbf{x} \in U$. If we replace $\mathbf{x}$ by $\mathbf{H}(\mathbf{x})$, we obtain

(71) $$PF(\mathbf{H}(\mathbf{x})) = A\mathbf{x} \qquad (\mathbf{x} \in V).$$

Define

(72) $$\psi(\mathbf{x}) = \mathbf{F}(\mathbf{H}(\mathbf{x})) - A\mathbf{x} \qquad (\mathbf{x} \in V).$$

Since $PA = A$, (71) implies that $P\psi(\mathbf{x}) = 0$ for all $\mathbf{x} \in V$. Thus ψ is a $\mathscr{C}'$-mapping of V into Y_2.

Since V is open, it is clear that $A(V)$ is an open subset of its range $\mathscr{R}(A) = Y_1$.

To complete the proof, i.e., to go from (72) to (66), we have to show that there is a $\mathscr{C}'$-mapping φ of $A(V)$ into Y_2 which satisfies

(73) $$\varphi(A\mathbf{x}) = \psi(\mathbf{x}) \qquad (\mathbf{x} \in V).$$

As a step toward (73), we will first prove that

(74) $$\psi(\mathbf{x}_1) = \psi(\mathbf{x}_2)$$

if $\mathbf{x}_1 \in V$, $\mathbf{x}_2 \in V$, $A\mathbf{x}_1 = A\mathbf{x}_2$.

Put $\Phi(\mathbf{x}) = \mathbf{F}(\mathbf{H}(\mathbf{x}))$, for $\mathbf{x} \in V$. Since $\mathbf{H}'(\mathbf{x})$ has rank n for every $\mathbf{x} \in V$, and $\mathbf{F}'(\mathbf{x})$ has rank r for every $\mathbf{x} \in U$, it follows that

(75) $$\operatorname{rank} \Phi'(\mathbf{x}) = \operatorname{rank} \mathbf{F}'(\mathbf{H}(\mathbf{x}))\mathbf{H}'(\mathbf{x}) = r \qquad (\mathbf{x} \in V).$$

Fix $\mathbf{x} \in V$. Let M be the range of $\Phi'(\mathbf{x})$. Then $M \subset R^m$, $\dim M = r$. By (71),

(76) $$P\Phi'(\mathbf{x}) = A.$$

Thus P maps M onto $\mathscr{R}(A) = Y_1$. Since M and Y_1 have the same dimension, it follows that P (restricted to M) is 1-1.

Suppose now that $A\mathbf{h} = 0$. Then $P\Phi'(\mathbf{x})\mathbf{h} = 0$, by (76). But $\Phi'(\mathbf{x})\mathbf{h} \in M$, and P is 1-1 on M. Hence $\Phi'(\mathbf{x})\mathbf{h} = 0$. A look at (72) shows now that we have proved the following:

If $\mathbf{x} \in V$ and $A\mathbf{h} = 0$, then $\psi'(\mathbf{x})\mathbf{h} = 0$.

We can now prove (74). Suppose $\mathbf{x}_1 \in V$, $\mathbf{x}_2 \in V$, $A\mathbf{x}_1 = A\mathbf{x}_2$. Put $\mathbf{h} = \mathbf{x}_2 - \mathbf{x}_1$ and define

(77) $$\mathbf{g}(t) = \psi(\mathbf{x}_1 + t\mathbf{h}) \qquad (0 \le t \le 1).$$

The convexity of V shows that $\mathbf{x}_1 + t\mathbf{h} \in V$ for these t. Hence

(78) $$\mathbf{g}'(t) = \psi'(\mathbf{x}_1 + t\mathbf{h})\mathbf{h} = 0 \qquad (0 \le t \le 1),$$

so that $\mathbf{g}(1) = \mathbf{g}(0)$. But $\mathbf{g}(1) = \psi(\mathbf{x}_2)$ and $\mathbf{g}(0) = \psi(\mathbf{x}_1)$. This proves (74).

By (74), $\psi(\mathbf{x})$ depends only on $A\mathbf{x}$, for $\mathbf{x} \in V$. Hence (73) defines φ unambiguously in $A(V)$. It only remains to be proved that $\varphi \in \mathscr{C}'$.

Fix $\mathbf{y}_0 \in A(V)$, fix $\mathbf{x}_0 \in V$ so that $A\mathbf{x}_0 = \mathbf{y}_0$. Since V is open, $\mathbf{y}_0$ has a neighborhood W in Y_1 such that the vector

(79) $$\mathbf{x} = \mathbf{x}_0 + S(\mathbf{y} - \mathbf{y}_0)$$

lies in V for all $\mathbf{y} \in W$. By (68),

$$A\mathbf{x} = A\mathbf{x}_0 + \mathbf{y} - \mathbf{y}_0 = \mathbf{y}.$$

Thus (73) and (79) give

(80) $$\varphi(\mathbf{y}) = \psi(\mathbf{x}_0 - S\mathbf{y}_0 + S\mathbf{y}) \qquad (\mathbf{y} \in W).$$

This formula shows that $\varphi \in \mathscr{C}'$ in W, hence in $A(V)$, since $\mathbf{y}_0$ was chosen arbitrarily in $A(V)$.

The proof is now complete.

Here is what the theorem tells us about the geometry of the mapping $\mathbf{F}$.

If $\mathbf{y} \in \mathbf{F}(U)$ then $\mathbf{y} = \mathbf{F}(\mathbf{H}(\mathbf{x}))$ for some $\mathbf{x} \in V$, and (66) shows that $P\mathbf{y} = A\mathbf{x}$. Therefore

(81) $$\mathbf{y} = P\mathbf{y} + \varphi(P\mathbf{y}) \qquad (\mathbf{y} \in \mathbf{F}(U)).$$

This shows that $\mathbf{y}$ is determined by its projection $P\mathbf{y}$, and that P, restricted to $\mathbf{F}(U)$, is a 1-1 mapping of $\mathbf{F}(U)$ onto $A(V)$. Thus $\mathbf{F}(U)$ is an "r-dimensional surface" with precisely one point "over" each point of $A(V)$. We may also regard $\mathbf{F}(U)$ as the graph of φ.

If $\Phi(\mathbf{x}) = \mathbf{F}(\mathbf{H}(\mathbf{x}))$, as in the proof, then (66) shows that the level sets of Φ (these are the sets on which Φ attains a given value) are precisely the level sets of A in V. These are "flat" since they are intersections with V of translates of the vector space $\mathscr{N}(A)$. Note that dim $\mathscr{N}(A) = n - r$ (Exercise 25).

The level sets of $\mathbf{F}$ in U are the images under $\mathbf{H}$ of the flat level sets of Φ in V. They are thus "$(n - r)$-dimensional surfaces" in U.

DETERMINANTS

Determinants are numbers associated to square matrices, and hence to the operators represented by such matrices. They are 0 if and only if the corresponding operator fails to be invertible. They can therefore be used to decide whether the hypotheses of some of the preceding theorems are satisfied. They will play an even more important role in Chap. 10.

9.33 Definition If $(j_1, \ldots, j_n)$ is an ordered n-tuple of integers, define

(82) $$s(j_1, \ldots, j_n) = \prod_{p<q} \text{sgn}\,(j_q - j_p),$$

where $\text{sgn}\,x = 1$ if $x > 0$, $\text{sgn}\,x = -1$ if $x < 0$, $\text{sgn}\,x = 0$ if $x = 0$. Then $s(j_1, \ldots, j_n) = 1, -1$, or 0, and it changes sign if any two of the j's are interchanged.

Let $[A]$ be the matrix of a linear operator A on R^n, relative to the standard basis $\{e_1, \ldots, e_n\}$, with entries $a(i,j)$ in the ith row and jth column. The determinant of $[A]$ is defined to be the number

(83) $$\det [A] = \sum s(j_1, \ldots, j_n) a(1, j_1) a(2, j_2) \cdots a(n, j_n).$$

The sum in (83) extends over all ordered n-tuples of integers $(j_1, \ldots, j_n)$ with $1 \leq j_r \leq n$.

The column vectors $\mathbf{x}_j$ of $[A]$ are

(84) $$\mathbf{x}_j = \sum_{i=1}^{n} a(i,j) \mathbf{e}_i \qquad (1 \leq j \leq n).$$

It will be convenient to think of $\det [A]$ as a function of the column vectors of $[A]$. If we write

$$\det (\mathbf{x}_1, \ldots, \mathbf{x}_n) = \det [A],$$

det is now a real function on the set of all ordered n-tuples of vectors in R^n.

9.34 Theorem

(a) *If I is the identity operator on R^n, then*
$$\det [I] = \det (\mathbf{e}_1, \ldots, \mathbf{e}_n) = 1.$$

(b) *det is a linear function of each of the column vectors $\mathbf{x}_j$, if the others are held fixed.*

(c) *If $[A]_1$ is obtained from $[A]$ by interchanging two columns, then $\det [A]_1 = -\det [A]$.*

(d) *If $[A]$ has two equal columns, then $\det [A] = 0$.*

Proof If $A = I$, then $a(i,i) = 1$ and $a(i,j) = 0$ for $i \neq j$. Hence
$$\det [I] = s(1, 2, \ldots, n) = 1,$$

which proves (a). By (82), $s(j_1, \ldots, j_n) = 0$ if any two of the j's are equal. Each of the remaining $n!$ products in (83) contains exactly one factor from each column. This proves (b). Part (c) is an immediate consequence of the fact that $s(j_1, \ldots, j_n)$ changes sign if any two of the j's are interchanged, and (d) is a corollary of (c).

9.35 Theorem *If [A] and [B] are n by n matrices, then*
$$\det([B][A]) = \det[B] \det[A].$$

Proof If $\mathbf{x}_1, \ldots, \mathbf{x}_n$ are the columns of $[A]$, define

(85) $$\Delta_B(\mathbf{x}_1, \ldots, \mathbf{x}_n) = \Delta_B[A] = \det([B][A]).$$

The columns of $[B][A]$ are the vectors $B\mathbf{x}_1, \ldots, B\mathbf{x}_n$. Thus

(86) $$\Delta_B(\mathbf{x}_1, \ldots, \mathbf{x}_n) = \det(B\mathbf{x}_1, \ldots, B\mathbf{x}_n).$$

By (86) and Theorem 9.34, Δ_B also has properties 9.34 (b) to (d). By (b) and (84),

$$\Delta_B[A] = \Delta_B\left(\sum_i a(i, 1)\mathbf{e}_i, \mathbf{x}_2, \ldots, \mathbf{x}_n\right) = \sum_i a(i, 1) \Delta_B(\mathbf{e}_i, \mathbf{x}_2, \ldots, \mathbf{x}_n).$$

Repeating this process with $\mathbf{x}_2, \ldots, \mathbf{x}_n$, we obtain

(87) $$\Delta_B[A] = \sum a(i_1, 1)a(i_2, 2) \cdots a(i_n, n) \Delta_B(\mathbf{e}_{i_1}, \ldots, \mathbf{e}_{i_n}),$$

the sum being extended over all ordered *n*-tuples $(i_1, \ldots, i_n)$ with $1 \leq i_r \leq n$. By (c) and (d),

(88) $$\Delta_B(\mathbf{e}_{i_1}, \ldots, \mathbf{e}_{i_n}) = t(i_1, \ldots, i_n) \Delta_B(\mathbf{e}_1, \ldots, \mathbf{e}_n),$$

where $t = 1, 0,$ or -1, and since $[B][I] = [B]$, (85) shows that

(89) $$\Delta_B(\mathbf{e}_1, \ldots, \mathbf{e}_n) = \det[B].$$

Substituting (89) and (88) into (87), we obtain

$$\det([B][A]) = \left\{\sum a(i_1, 1) \cdots a(i_n, n)t(i_1, \ldots, i_n)\right\} \det[B],$$

for all *n* by *n* matrices $[A]$ and $[B]$. Taking $B = I$, we see that the above sum in braces is $\det[A]$. This proves the theorem.

9.36 Theorem *A linear operator A on R^n is invertible if and only if* $\det[A] \neq 0$.

Proof If A is invertible, Theorem 9.35 shows that

$$\det[A] \det[A^{-1}] = \det[AA^{-1}] = \det[I] = 1,$$

so that $\det[A] \neq 0$.

If A is not invertible, the columns $\mathbf{x}_1, \ldots, \mathbf{x}_n$ of $[A]$ are dependent (Theorem 9.5); hence there is one, say, $\mathbf{x}_k$, such that

(90) $$\mathbf{x}_k + \sum_{j \neq k} c_j \mathbf{x}_j = 0$$

for certain scalars c_j. By 9.34 (b) and (d), $\mathbf{x}_k$ can be replaced by $\mathbf{x}_k + c_j \mathbf{x}_j$ without altering the determinant, if $j \neq k$. Repeating, we see that $\mathbf{x}_k$ can be replaced by the left side of (90), i.e., by **0**, without altering the determinant. But a matrix which has **0** for one column has determinant 0. Hence det $[A] = 0$.

9.37 Remark Suppose $\{e_1, \ldots, e_n\}$ and $\{u_1, \ldots, u_n\}$ are bases in R^n. Every linear operator A on R^n determines matrices $[A]$ and $[A]_U$, with entries a_{ij} and α_{ij}, given by

$$A e_j = \sum_i a_{ij} e_i, \qquad A u_j = \sum_i \alpha_{ij} u_i.$$

If $u_j = B e_j = \Sigma b_{ij} e_i$, then $A u_j$ is equal to

$$\sum_k \alpha_{kj} B e_k = \sum_k \alpha_{kj} \sum_i b_{ik} e_i = \sum_i \left(\sum_k b_{ik} \alpha_{kj} \right) e_i,$$

and also to

$$A B e_j = A \sum_k b_{kj} e_k = \sum_i \left(\sum_k a_{ik} b_{kj} \right) e_i.$$

Thus $\Sigma b_{ik} \alpha_{kj} = \Sigma a_{ik} b_{kj}$, or

(91) $$[B][A]_U = [A][B].$$

Since B is invertible, det $[B] \neq 0$. Hence (91), combined with Theorem 9.35, shows that

(92) $$\det [A]_U = \det [A].$$

The determinant of the matrix of a linear operator does therefore not depend on the basis which is used to construct the matrix. *It is thus meaningful to speak of the determinant of a linear operator, without having any basis in mind.*

9.38 Jacobians If **f** maps an open set $E \subset R^n$ into R^n, and if **f** is differentiable at a point $x \in E$, the determinant of the linear operator $\mathbf{f}'(x)$ is called the *Jacobian of* **f** *at* **x**. In symbols,

(93) $$J_\mathbf{f}(x) = \det \mathbf{f}'(x).$$

We shall also use the notation

(94) $$\frac{\partial(y_1, \ldots, y_n)}{\partial(x_1, \ldots, x_n)}$$

for $J_\mathbf{f}(x)$, if $(y_1, \ldots, y_n) = \mathbf{f}(x_1, \ldots, x_n)$.

In terms of Jacobians, the crucial hypothesis in the inverse function theorem is that $J_\mathbf{f}(a) \neq 0$ (compare Theorem 9.36). If the implicit function theorem is stated in terms of the functions (59), the assumption made there on A amounts to

$$\frac{\partial(f_1, \ldots, f_n)}{\partial(x_1, \ldots, x_n)} \neq 0.$$

DERIVATIVES OF HIGHER ORDER

9.39 Definition Suppose f is a real function defined in an open set $E \subset R^n$, with partial derivatives $D_1 f, \ldots, D_n f$. If the functions $D_j f$ are themselves differentiable, then the *second-order partial derivatives* of f are defined by

$$D_{ij} f = D_i D_j f \qquad (i, j = 1, \ldots, n).$$

If all these functions $D_{ij} f$ are continuous in E, we say that f is of class $\mathscr{C}''$ in E, or that $f \in \mathscr{C}''(E)$.

A mapping $\mathbf{f}$ of E into R^m is said to be of class $\mathscr{C}''$ if each component of $\mathbf{f}$ is of class $\mathscr{C}''$.

It can happen that $D_{ij} f \neq D_{ji} f$ at some point, although both derivatives exist (see Exercise 27). However, we shall see below that $D_{ij} f = D_{ji} f$ whenever these derivatives are continuous.

For simplicity (and without loss of generality) we state our next two theorems for real functions of two variables. The first one is a mean value theorem.

9.40 Theorem *Suppose f is defined in an open set $E \subset R^2$, and $D_1 f$ and $D_{21} f$ exist at every point of E. Suppose $Q \subset E$ is a closed rectangle with sides parallel to the coordinate axes, having (a, b) and $(a + h, b + k)$ as opposite vertices $(h \neq 0, k \neq 0)$. Put*

$$\Delta(f, Q) = f(a + h, b + k) - f(a + h, b) - f(a, b + k) + f(a, b).$$

Then there is a point (x, y) in the interior of Q such that

(95) $$\Delta(f, Q) = hk(D_{21} f)(x, y).$$

Note the analogy between (95) and Theorem 5.10; the area of Q is hk.

Proof Put $u(t) = f(t, b + k) - f(t, b)$. Two applications of Theorem 5.10 show that there is an x between a and $a + h$, and that there is a y between b and $b + k$, such that

$$\begin{aligned} \Delta(f, Q) &= u(a + h) - u(a) \\ &= h u'(x) \\ &= h[(D_1 f)(x, b + k) - (D_1 f)(x, b)] \\ &= hk(D_{21} f)(x, y). \end{aligned}$$

9.41 Theorem *Suppose f is defined in an open set $E \subset R^2$, suppose that $D_1 f$, $D_{21} f$, and $D_2 f$ exist at every point of E, and $D_{21} f$ is continuous at some point $(a, b) \in E$.*

Then $D_{12}f$ exists at (a, b) and

(96) $$(D_{12}f)(a, b) = (D_{21}f)(a, b).$$

Corollary $D_{21}f = D_{12}f$ if $f \in \mathscr{C}''(E)$.

Proof Put $A = (D_{21}f)(a, b)$. Choose $\varepsilon > 0$. If Q is a rectangle as in Theorem 9.40, and if h and k are sufficiently small, we have
$$|A - (D_{21}f)(x, y)| < \varepsilon$$
for all $(x, y) \in Q$. Thus
$$\left|\frac{\Delta(f, Q)}{hk} - A\right| < \varepsilon,$$
by (95). Fix h, and let $k \to 0$. Since $D_2 f$ exists in E, the last inequality implies that

(97) $$\left|\frac{(D_2 f)(a + h, b) - (D_2 f)(a, b)}{h} - A\right| \le \varepsilon.$$

Since ε was arbitrary, and since (97) holds for all sufficiently small $h \ne 0$, it follows that $(D_{12}f)(a, b) = A$. This gives (96).

DIFFERENTIATION OF INTEGRALS

Suppose φ is a function of two variables which can be integrated with respect to one and which can be differentiated with respect to the other. Under what conditions will the result be the same if these two limit processes are carried out in the opposite order? To state the question more precisely: Under what conditions on φ can one prove that the equation

(98) $$\frac{d}{dt}\int_a^b \varphi(x, t)\, dx = \int_a^b \frac{\partial \varphi}{\partial t}(x, t)\, dx$$

is true? (A counter example is furnished by Exercise 28.)
It will be convenient to use the notation

(99) $$\varphi^t(x) = \varphi(x, t).$$

Thus φ^t is, for each t, a function of one variable.

9.42 Theorem *Suppose*

(a) $\varphi(x, t)$ *is defined for* $a \le x \le b$, $c \le t \le d$;
(b) α *is an increasing function on* $[a, b]$;

(c) $\varphi^t \in \mathscr{R}(\alpha)$ for every $t \in [c, d]$;
(d) $c < s < d$, and to every $\varepsilon > 0$ corresponds a $\delta > 0$ such that
$$|(D_2 \varphi)(x, t) - (D_2 \varphi)(x, s)| < \varepsilon$$
for all $x \in [a, b]$ and for all $t \in (s - \delta, s + \delta)$.

Define

(100)
$$f(t) = \int_a^b \varphi(x, t) \, d\alpha(x) \qquad (c \le t \le d).$$

Then $(D_2 \varphi)^s \in \mathscr{R}(\alpha)$, $f'(s)$ exists, and

(101)
$$f'(s) = \int_a^b (D_2 \varphi)(x, s) \, d\alpha(x).$$

Note that (c) simply asserts the existence of the integrals (100) for all $t \in [c, d]$. Note also that (d) certainly holds whenever $D_2 \varphi$ is continuous on the rectangle on which φ is defined.

Proof Consider the difference quotients
$$\psi(x, t) = \frac{\varphi(x, t) - \varphi(x, s)}{t - s}$$
for $0 < |t - s| < \delta$. By Theorem 5.10 there corresponds to each (x, t) a number u between s and t such that
$$\psi(x, t) = (D_2 \varphi)(x, u).$$
Hence (d) implies that

(102)
$$|\psi(x, t) - (D_2 \varphi)(x, s)| < \varepsilon \qquad (a \le x \le b, \ 0 < |t - s| < \delta).$$

Note that

(103)
$$\frac{f(t) - f(s)}{t - s} = \int_a^b \psi(x, t) \, d\alpha(x).$$

By (102), $\psi^t \to (D_2 \varphi)^s$, uniformly on $[a, b]$, as $t \to s$. Since each $\psi^t \in \mathscr{R}(\alpha)$, the desired conclusion follows from (103) and Theorem 7.16.

9.43 Example One can of course prove analogues of Theorem 9.42 with $(-\infty, \infty)$ in place of $[a, b]$. Instead of doing this, let us simply look at an example. Define

(104)
$$f(t) = \int_{-\infty}^{\infty} e^{-x^2} \cos(xt) \, dx$$

and

(105) $$g(t) = -\int_{-\infty}^{\infty} xe^{-x^2} \sin(xt)\, dx,$$

for $-\infty < t < \infty$. Both integrals exist (they converge absolutely) since the absolute values of the integrands are at most $\exp(-x^2)$ and $|x|\exp(-x^2)$, respectively.

Note that g is obtained from f by differentiating the integrand with respect to t. We claim that f is differentiable and that

(106) $$f'(t) = g(t) \qquad (-\infty < t < \infty).$$

To prove this, let us first examine the difference quotients of the cosine: if $\beta > 0$, then

(107) $$\frac{\cos(\alpha + \beta) - \cos\alpha}{\beta} + \sin\alpha = \frac{1}{\beta}\int_{\alpha}^{\alpha+\beta} (\sin\alpha - \sin t)\, dt.$$

Since $|\sin\alpha - \sin t| \leq |t - \alpha|$, the right side of (107) is at most $\beta/2$ in absolute value; the case $\beta < 0$ is handled similarly. Thus

(108) $$\left|\frac{\cos(\alpha + \beta) - \cos\alpha}{\beta} + \sin\alpha\right| \leq |\beta|$$

for all β (if the left side is interpreted to be 0 when $\beta = 0$).

Now fix t, and fix $h \neq 0$. Apply (108) with $\alpha = xt$, $\beta = xh$; it follows from (104) and (105) that

$$\left|\frac{f(t+h) - f(t)}{h} - g(t)\right| \leq |h|\int_{-\infty}^{\infty} x^2 e^{-x^2}\, dx.$$

When $h \to 0$, we thus obtain (106).

Let us go a step further: An integration by parts, applied to (104), shows that

(109) $$f(t) = 2\int_{-\infty}^{\infty} xe^{-x^2}\frac{\sin(xt)}{t}\, dx.$$

Thus $tf(t) = -2g(t)$, and (106) implies now that f satisfies the differential equation

(110) $$2f'(t) + tf(t) = 0.$$

If we solve this differential equation and use the fact that $f(0) = \sqrt{\pi}$ (see Sec. 8.21), we find that

(111) $$f(t) = \sqrt{\pi}\exp\left(-\frac{t^2}{4}\right).$$

The integral (104) is thus explicitly determined.

EXERCISES

1. If S is a nonempty subset of a vector space X, prove (as asserted in Sec. 9.1) that the span of S is a vector space.

2. Prove (as asserted in Sec. 9.6) that BA is linear if A and B are linear transformations. Prove also that A^{-1} is linear and invertible.

3. Assume $A \in L(X, Y)$ and $A\mathbf{x} = \mathbf{0}$ only when $\mathbf{x} = \mathbf{0}$. Prove that A is then 1-1.

4. Prove (as asserted in Sec. 9.30) that null spaces and ranges of linear transformations are vector spaces.

5. Prove that to every $A \in L(R^n, R^1)$ corresponds a unique $\mathbf{y} \in R^n$ such that $A\mathbf{x} = \mathbf{x} \cdot \mathbf{y}$. Prove also that $\|A\| = |\mathbf{y}|$.

   Hint: Under certain conditions, equality holds in the Schwarz inequality.

6. If $f(0, 0) = 0$ and
$$f(x, y) = \frac{xy}{x^2 + y^2} \quad \text{if } (x, y) \neq (0, 0),$$
prove that $(D_1 f)(x, y)$ and $(D_2 f)(x, y)$ exist at every point of R^2, although f is not continuous at $(0, 0)$.

7. Suppose that f is a real-valued function defined in an open set $E \subset R^n$, and that the partial derivatives $D_1 f, \ldots, D_n f$ are bounded in E. Prove that f is continuous in E.

   Hint: Proceed as in the proof of Theorem 9.21.

8. Suppose that f is a differentiable real function in an open set $E \subset R^n$, and that f has a local maximum at a point $\mathbf{x} \in E$. Prove that $f'(\mathbf{x}) = 0$.

9. If $\mathbf{f}$ is a differentiable mapping of a *connected* open set $E \subset R^n$ into R^m, and if $\mathbf{f}'(\mathbf{x}) = 0$ for every $\mathbf{x} \in E$, prove that $\mathbf{f}$ is constant in E.

10. If f is a real function defined in a convex open set $E \subset R^n$, such that $(D_1 f)(\mathbf{x}) = 0$ for every $\mathbf{x} \in E$, prove that $f(\mathbf{x})$ depends only on $x_2, \ldots, x_n$.

    Show that the convexity of E can be replaced by a weaker condition, but that some condition is required. For example, if $n = 2$ and E is shaped like a horseshoe, the statement may be false.

11. If f and g are differentiable real functions in R^n, prove that
$$\nabla(fg) = f \nabla g + g \nabla f$$
and that $\nabla(1/f) = -f^{-2} \nabla f$ wherever $f \neq 0$.

12. Fix two real numbers a and b, $0 < a < b$. Define a mapping $\mathbf{f} = (f_1, f_2, f_3)$ of R^2 into R^3 by
$$f_1(s, t) = (b + a \cos s) \cos t$$
$$f_2(s, t) = (b + a \cos s) \sin t$$
$$f_3(s, t) = a \sin s.$$

Describe the range K of $\mathbf{f}$. (It is a certain compact subset of R^3.)

(a) Show that there are exactly 4 points $\mathbf{p} \in K$ such that
$$(\nabla f_1)(\mathbf{f}^{-1}(\mathbf{p})) = \mathbf{0}.$$

Find these points.

(b) Determine the set of all $\mathbf{q} \in K$ such that
$$(\nabla f_3)(\mathbf{f}^{-1}(\mathbf{q})) = \mathbf{0}.$$

(c) Show that one of the points $\mathbf{p}$ found in part (a) corresponds to a local maximum of f_1, one corresponds to a local minimum, and that the other two are neither (they are so-called "saddle points").

Which of the points $\mathbf{q}$ found in part (b) correspond to maxima or minima?

(d) Let λ be an irrational real number, and define $\mathbf{g}(t) = \mathbf{f}(t, \lambda t)$. Prove that $\mathbf{g}$ is a 1-1 mapping of R^1 onto a dense subset of K. Prove that
$$|\mathbf{g}'(t)|^2 = a^2 + \lambda^2(b + a\cos t)^2.$$

13. Suppose $\mathbf{f}$ is a differentiable mapping of R^1 into R^3 such that $|\mathbf{f}(t)| = 1$ for every t. Prove that $\mathbf{f}'(t) \cdot \mathbf{f}(t) = 0$.

Interpret this result geometrically.

14. Define $f(0, 0) = 0$ and
$$f(x, y) = \frac{x^3}{x^2 + y^2} \quad \text{if } (x, y) \neq (0, 0).$$

(a) Prove that $D_1 f$ and $D_2 f$ are bounded functions in R^2. (Hence f is continuous.)

(b) Let $\mathbf{u}$ be any unit vector in R^2. Show that the directional derivative $(D_\mathbf{u} f)(0, 0)$ exists, and that its absolute value is at most 1.

(c) Let γ be a differentiable mapping of R^1 into R^2 (in other words, γ is a differentiable curve in R^2), with $\gamma(0) = (0, 0)$ and $|\gamma'(0)| > 0$. Put $g(t) = f(\gamma(t))$ and prove that g is differentiable for every $t \in R^1$.

If $\gamma \in \mathscr{C}'$, prove that $g \in \mathscr{C}'$.

(d) In spite of this, prove that f is not differentiable at $(0, 0)$.

Hint: Formula (40) fails.

15. Define $f(0, 0) = 0$, and put
$$f(x, y) = x^2 + y^2 - 2x^2 y - \frac{4x^6 y^2}{(x^4 + y^2)^2}$$

if $(x, y) \neq (0, 0)$.

(a) Prove, for all $(x, y) \in R^2$, that
$$4x^4 y^2 \leq (x^4 + y^2)^2.$$

Conclude that f is continuous.

(b) For $0 \leq \theta \leq 2\pi$, $-\infty < t < \infty$, define
$$g_\theta(t) = f(t \cos \theta, t \sin \theta).$$
Show that $g_\theta(0) = 0$, $g'_\theta(0) = 0$, $g''_\theta(0) = 2$. Each g_θ has therefore a strict local minimum at $t = 0$.

In other words, the restriction of f to each line through $(0, 0)$ has a strict local minimum at $(0, 0)$.

(c) Show that $(0, 0)$ is nevertheless not a local minimum for f, since $f(x, x^2) = -x^4$.

16. Show that the continuity of $\mathbf{f}'$ at the point $\mathbf{a}$ is needed in the inverse function theorem, even in the case $n = 1$: If
$$f(t) = t + 2t^2 \sin\left(\frac{1}{t}\right)$$
for $t \neq 0$, and $f(0) = 0$, then $f'(0) = 1$, f' is bounded in $(-1, 1)$, but f is not one-to-one in any neighborhood of 0.

17. Let $\mathbf{f} = (f_1, f_2)$ be the mapping of R^2 into R^2 given by
$$f_1(x, y) = e^x \cos y, \qquad f_2(x, y) = e^x \sin y.$$
(a) What is the range of $\mathbf{f}$?

(b) Show that the Jacobian of $\mathbf{f}$ is not zero at any point of R^2. Thus every point of R^2 has a neighborhood in which $\mathbf{f}$ is one-to-one. Nevertheless, $\mathbf{f}$ is not one-to-one on R^2.

(c) Put $\mathbf{a} = (0, \pi/3)$, $\mathbf{b} = \mathbf{f}(\mathbf{a})$, let $\mathbf{g}$ be the continuous inverse of $\mathbf{f}$, defined in a neighborhood of $\mathbf{b}$, such that $\mathbf{g}(\mathbf{b}) = \mathbf{a}$. Find an explicit formula for $\mathbf{g}$, compute $\mathbf{f}'(\mathbf{a})$ and $\mathbf{g}'(\mathbf{b})$, and verify the formula (52).

(d) What are the images under $\mathbf{f}$ of lines parallel to the coordinate axes?

18. Answer analogous questions for the mapping defined by
$$u = x^2 - y^2, \qquad v = 2xy.$$

19. Show that the system of equations
$$3x + y - z + u^2 = 0$$
$$x - y + 2z + u = 0$$
$$2x + 2y - 3z + 2u = 0$$
can be solved for x, y, u in terms of z; for x, z, u in terms of y; for y, z, u in terms of x; but not for x, y, z in terms of u.

20. Take $n = m = 1$ in the implicit function theorem, and interpret the theorem (as well as its proof) graphically.

21. Define f in R^2 by
$$f(x, y) = 2x^3 - 3x^2 + 2y^3 + 3y^2.$$
(a) Find the four points in R^2 at which the gradient of f is zero. Show that f has exactly one local maximum and one local minimum in R^2.

(b) Let S be the set of all $(x, y) \in R^2$ at which $f(x, y) = 0$. Find those points of S that have no neighborhoods in which the equation $f(x, y) = 0$ can be solved for y in terms of x (or for x in terms of y). Describe S as precisely as you can.

22. Give a similar discussion for
$$f(x, y) = 2x^3 + 6xy^2 - 3x^2 + 3y^2.$$

23. Define f in R^3 by
$$f(x, y_1, y_2) = x^2 y_1 + e^x + y_2.$$
Show that $f(0, 1, -1) = 0$, $(D_1 f)(0, 1, -1) \neq 0$, and that there exists therefore a differentiable function g in some neighborhood of $(1, -1)$ in R^2, such that $g(1, -1) = 0$ and
$$f(g(y_1, y_2), y_1, y_2) = 0.$$
Find $(D_1 g)(1, -1)$ and $(D_2 g)(1, -1)$.

24. For $(x, y) \neq (0, 0)$, define $\mathbf{f} = (f_1, f_2)$ by
$$f_1(x, y) = \frac{x^2 - y^2}{x^2 + y^2}, \qquad f_2(x, y) = \frac{xy}{x^2 + y^2}.$$
Compute the rank of $\mathbf{f}'(x, y)$, and find the range of $\mathbf{f}$.

25. Suppose $A \in L(R^n, R^m)$, let r be the rank of A.
    (a) Define S as in the proof of Theorem 9.32. Show that SA is a projection in R^n whose null space is $\mathcal{N}(A)$ and whose range is $\mathcal{R}(S)$. Hint: By (68), $SASA = SA$.
    (b) Use (a) to show that
$$\dim \mathcal{N}(A) + \dim \mathcal{R}(A) = n.$$

26. Show that the existence (and even the continuity) of $D_{12} f$ does not imply the existence of $D_1 f$. For example, let $f(x, y) = g(x)$, where g is nowhere differentiable.

27. Put $f(0, 0) = 0$, and
$$f(x, y) = \frac{xy(x^2 - y^2)}{x^2 + y^2}$$
if $(x, y) \neq (0, 0)$. Prove that
    (a) f, $D_1 f$, $D_2 f$ are continuous in R^2;
    (b) $D_{12} f$ and $D_{21} f$ exist at every point of R^2, and are continuous except at $(0, 0)$;
    (c) $(D_{12} f)(0, 0) = 1$, and $(D_{21} f)(0, 0) = -1$.

28. For $t \geq 0$, put
$$\varphi(x, t) = \begin{cases} x & (0 \leq x \leq \sqrt{t}) \\ -x + 2\sqrt{t} & (\sqrt{t} \leq x \leq 2\sqrt{t}) \\ 0 & \text{(otherwise)}, \end{cases}$$
and put $\varphi(x, t) = -\varphi(x, |t|)$ if $t < 0$.

Show that φ is continuous on R^2, and
$$(D_2\varphi)(x, 0) = 0$$
for all x. Define
$$f(t) = \int_{-1}^{1} \varphi(x, t)\, dx.$$
Show that $f(t) = t$ if $|t| < \frac{1}{4}$. Hence
$$f'(0) \neq \int_{-1}^{1} (D_2\varphi)(x, 0)\, dx.$$

29. Let E be an open set in R^n. The classes $\mathscr{C}'(E)$ and $\mathscr{C}''(E)$ are defined in the text. By induction, $\mathscr{C}^{(k)}(E)$ can be defined as follows, for all positive integers k: To say that $f \in \mathscr{C}^{(k)}(E)$ means that the partial derivatives $D_1 f, \ldots, D_n f$ belong to $\mathscr{C}^{(k-1)}(E)$.

Assume $f \in \mathscr{C}^{(k)}(E)$, and show (by repeated application of Theorem 9.41) that the kth-order derivative
$$D_{i_1 i_2 \cdots i_k} f = D_{i_1} D_{i_2} \cdots D_{i_k} f$$
is unchanged if the subscripts $i_1, \ldots, i_k$ are permuted.

For instance, if $n \geq 3$, then
$$D_{1213} f = D_{3112} f$$
for every $f \in \mathscr{C}^{(4)}$.

30. Let $f \in \mathscr{C}^{(m)}(E)$, where E is an open subset of R^n. Fix $\mathbf{a} \in E$, and suppose $\mathbf{x} \in R^n$ is so close to $\mathbf{0}$ that the points
$$\mathbf{p}(t) = \mathbf{a} + t\mathbf{x}$$
lie in E whenever $0 \leq t \leq 1$. Define
$$h(t) = f(\mathbf{p}(t))$$
for all $t \in R^1$ for which $\mathbf{p}(t) \in E$.

(a) For $1 \leq k \leq m$, show (by repeated application of the chain rule) that
$$h^{(k)}(t) = \sum (D_{i_1 \cdots i_k} f)(\mathbf{p}(t))\, x_{i_1} \cdots x_{i_k}.$$
The sum extends over all ordered k-tuples $(i_1, \ldots, i_k)$ in which each i_j is one of the integers $1, \ldots, n$.

(b) By Taylor's theorem (5.15),
$$h(1) = \sum_{k=0}^{m-1} \frac{h^{(k)}(0)}{k!} + \frac{h^{(m)}(t)}{m!}$$
for some $t \in (0, 1)$. Use this to prove Taylor's theorem in n variables by showing that the formula

$$f(\mathbf{a}+\mathbf{x}) = \sum_{k=0}^{m-1} \frac{1}{k!} \sum (D_{i_1 \ldots i_k} f)(\mathbf{a}) x_{i_1} \cdots x_{i_k} + r(\mathbf{x})$$

represents $f(\mathbf{a}+\mathbf{x})$ as the sum of its so-called "Taylor polynomial of degree $m-1$," plus a remainder that satisfies

$$\lim_{\mathbf{x} \to 0} \frac{r(\mathbf{x})}{|\mathbf{x}|^{m-1}} = 0.$$

Each of the inner sums extends over all ordered k-tuples $(i_1, \ldots, i_k)$, as in part (a); as usual, the zero-order derivative of f is simply f, so that the constant term of the Taylor polynomial of f at $\mathbf{a}$ is $f(\mathbf{a})$.

(c) Exercise 29 shows that repetition occurs in the Taylor polynomial as written in part (b). For instance, D_{113} occurs three times, as $D_{113}, D_{131}, D_{311}$. The sum of the corresponding three terms can be written in the form

$$3(D_1^2 D_3 f)(\mathbf{a}) x_1^2 x_3.$$

Prove (by calculating how often each derivative occurs) that the Taylor polynomial in (b) can be written in the form

$$\sum \frac{(D_1^{s_1} \cdots D_n^{s_n} f)(\mathbf{a})}{s_1! \cdots s_n!} x_1^{s_1} \cdots x_n^{s_n}.$$

Here the summation extends over all ordered n-tuples $(s_1, \ldots, s_n)$ such that each s_i is a nonnegative integer, and $s_1 + \cdots + s_n \leq m-1$.

31. Suppose $f \in \mathscr{C}^{(3)}$ in some neighborhood of a point $\mathbf{a} \in R^2$, the gradient of f is $\mathbf{0}$ at $\mathbf{a}$, but not all second-order derivatives of f are 0 at $\mathbf{a}$. Show how one can then determine from the Taylor polynomial of f at $\mathbf{a}$ (of degree 2) whether f has a local maximum, or a local minimum, or neither, at the point $\mathbf{a}$.

Extend this to R^n in place of R^2.

10
INTEGRATION OF DIFFERENTIAL FORMS

Integration can be studied on many levels. In Chap. 6, the theory was developed for reasonably well-behaved functions on subintervals of the real line. In Chap. 11 we shall encounter a very highly developed theory of integration that can be applied to much larger classes of functions, whose domains are more or less arbitrary sets, not necessarily subsets of R^n. The present chapter is devoted to those aspects of integration theory that are closely related to the geometry of euclidean spaces, such as the change of variables formula, line integrals, and the machinery of differential forms that is used in the statement and proof of the n-dimensional analogue of the fundamental theorem of calculus, namely Stokes' theorem.

INTEGRATION

10.1 Definition Suppose I^k is a k-cell in R^k, consisting of all
$$\mathbf{x} = (x_1, \ldots, x_k)$$
such that
(1) $$a_i \leq x_i \leq b_i \quad (i = 1, \ldots, k),$$

I^j is the j-cell in R^j defined by the first j inequalities (1), and f is a real continuous function on I^k.

Put $f = f_k$, and define f_{k-1} on I^{k-1} by

$$f_{k-1}(x_1, \ldots, x_{k-1}) = \int_{a_k}^{b_k} f_k(x_1, \ldots, x_{k-1}, x_k)\, dx_k.$$

The uniform continuity of f_k on I^k shows that f_{k-1} is continuous on I^{k-1}. Hence we can repeat this process and obtain functions f_j, continuous on I^j, such that f_{j-1} is the integral of f_j, with respect to x_j, over $[a_j, b_j]$. After k steps we arrive at a *number* f_0, which we call the *integral of f over I^k*; we write it in the form

$$(2) \qquad \int_{I^k} f(\mathbf{x})\, d\mathbf{x} \quad \text{or} \quad \int_{I^k} f.$$

A priori, this definition of the integral depends on the order in which the k integrations are carried out. However, this dependence is only apparent. To prove this, let us introduce the temporary notation $L(f)$ for the integral (2) and $L'(f)$ for the result obtained by carrying out the k integrations in some other order.

10.2 Theorem *For every $f \in \mathscr{C}(I^k)$, $L(f) = L'(f)$.*

Proof If $h(\mathbf{x}) = h_1(x_1) \cdots h_k(x_k)$, where $h_j \in \mathscr{C}([a_j, b_j])$, then

$$L(h) = \prod_{i=1}^{k} \int_{a_i}^{b_i} h_i(x_i)\, dx_i = L'(h).$$

If $\mathscr{A}$ is the set of all finite sums of such functions h, it follows that $L(g) = L'(g)$ for all $g \in \mathscr{A}$. Also, $\mathscr{A}$ is an algebra of functions on I^k to which the Stone-Weierstrass theorem applies.

Put $V = \prod_{1}^{k}(b_i - a_i)$. If $f \in \mathscr{C}(I^k)$ and $\varepsilon > 0$, there exists $g \in \mathscr{A}$ such that $\|f - g\| < \varepsilon/V$, where $\|f\|$ is defined as $\max |f(\mathbf{x})|$ ($\mathbf{x} \in I^k$). Then $|L(f - g)| < \varepsilon$, $|L'(f - g)| < \varepsilon$, and since

$$L(f) - L'(f) = L(f - g) + L'(g - f),$$

we conclude that $|L(f) - L'(f)| < 2\varepsilon$.

In this connection, Exercise 2 is relevant.

10.3 Definition The *support* of a (real or complex) function f on R^k is the closure of the set of all points $\mathbf{x} \in R^k$ at which $f(\mathbf{x}) \neq 0$. If f is a continuous function with compact support, let I^k be any k-cell which contains the support of f, and define

(3) $$\int_{R^k} f = \int_{I^k} f.$$

The integral so defined is evidently independent of the choice of I^k, provided only that I^k contains the support of f.

It is now tempting to extend the definition of the integral over R^k to functions which are limits (in some sense) of continuous functions with compact support. We do not want to discuss the conditions under which this can be done; the proper setting for this question is the Lebesgue integral. We shall merely describe one very simple example which will be used in the proof of Stokes' theorem.

10.4 Example Let Q^k be the k-simplex which consists of all points $\mathbf{x} = (x_1, \ldots, x_k)$ in R^k for which $x_1 + \cdots + x_k \leq 1$ and $x_i \geq 0$ for $i = 1, \ldots, k$. If $k = 3$, for example, Q^k is a tetrahedron, with vertices at $\mathbf{0}, \mathbf{e}_1, \mathbf{e}_2, \mathbf{e}_3$. If $f \in \mathscr{C}(Q^k)$, extend f to a function on I^k by setting $f(\mathbf{x}) = 0$ off Q^k, and define

(4) $$\int_{Q^k} f = \int_{I^k} f.$$

Here I^k is the "unit cube" defined by

$$0 \leq x_i \leq 1 \qquad (1 \leq i \leq k).$$

Since f may be discontinuous on I^k, the existence of the integral on the right of (4) needs proof. We also wish to show that this integral is independent of the order in which the k single integrations are carried out.

To do this, suppose $0 < \delta < 1$, put

(5) $$\varphi(t) = \begin{cases} 1 & (t \leq 1 - \delta) \\ \dfrac{(1-t)}{\delta} & (1 - \delta < t \leq 1) \\ 0 & (1 < t), \end{cases}$$

and define

(6) $$F(\mathbf{x}) = \varphi(x_1 + \cdots + x_k) f(\mathbf{x}) \qquad (\mathbf{x} \in I^k).$$

Then $F \in \mathscr{C}(I^k)$.

Put $\mathbf{y} = (x_1, \ldots, x_{k-1})$, $\mathbf{x} = (\mathbf{y}, x_k)$. For each $\mathbf{y} \in I^{k-1}$, the set of all x_k such that $F(\mathbf{y}, x_k) \neq f(\mathbf{y}; x_k)$ is either empty or is a segment whose length does not exceed δ. Since $0 \leq \varphi \leq 1$, it follows that

(7) $$|F_{k-1}(\mathbf{y}) - f_{k-1}(\mathbf{y})| \leq \delta \|f\| \qquad (\mathbf{y} \in I^{k-1}),$$

where $\|f\|$ has the same meaning as in the proof of Theorem 10.2, and F_{k-1}, f_{k-1} are as in Definition 10.1.

As $\delta \to 0$, (7) exhibits f_{k-1} as a uniform limit of a sequence of continuous functions. Thus $f_{k-1} \in \mathscr{C}(I^{k-1})$, and the further integrations present no problem.

This proves the existence of the integral (4). Moreover, (7) shows that

$$(8) \qquad \left| \int_{I^k} F(\mathbf{x}) \, d\mathbf{x} - \int_{I^k} f(\mathbf{x}) \, d\mathbf{x} \right| \leq \delta \|f\|.$$

Note that (8) is true, regardless of the order in which the k single integrations are carried out. Since $F \in \mathscr{C}(I^k)$, $\int F$ is unaffected by any change in this order. Hence (8) shows that the same is true of $\int f$.

This completes the proof.

Our next goal is the change of variables formula stated in Theorem 10.9. To facilitate its proof, we first discuss so-called primitive mappings, and partitions of unity. Primitive mappings will enable us to get a clearer picture of the local action of a $\mathscr{C}'$-mapping with invertible derivative, and partitions of unity are a very useful device that makes it possible to use local information in a global setting.

PRIMITIVE MAPPINGS

10.5 Definition If $\mathbf{G}$ maps an open set $E \subset R^n$ into R^n, and if there is an integer m and a real function g with domain E such that

$$(9) \qquad \mathbf{G}(\mathbf{x}) = \sum_{i \neq m} x_i \mathbf{e}_i + g(\mathbf{x}) \mathbf{e}_m \qquad (\mathbf{x} \in E),$$

then we call $\mathbf{G}$ *primitive*. A primitive mapping is thus one that changes at most one coordinate. Note that (9) can also be written in the form

$$(10) \qquad \mathbf{G}(\mathbf{x}) = \mathbf{x} + [g(\mathbf{x}) - x_m] \mathbf{e}_m.$$

If g is differentiable at some point $\mathbf{a} \in E$, so is $\mathbf{G}$. The matrix $[\alpha_{ij}]$ of the operator $\mathbf{G}'(\mathbf{a})$ has

$$(11) \qquad (D_1 g)(\mathbf{a}), \ldots, (D_m g)(\mathbf{a}), \ldots, (D_n g)(\mathbf{a})$$

as its mth row. For $j \neq m$, we have $\alpha_{jj} = 1$ and $\alpha_{ij} = 0$ if $i \neq j$. The Jacobian of $\mathbf{G}$ at $\mathbf{a}$ is thus given by

$$(12) \qquad J_{\mathbf{G}}(\mathbf{a}) = \det[\mathbf{G}'(\mathbf{a})] = (D_m g)(\mathbf{a}),$$

and we see (by Theorem 9.36) that $\mathbf{G}'(\mathbf{a})$ *is invertible if and only if* $(D_m g)(\mathbf{a}) \neq 0$.

10.6 Definition A linear operator B on R^n that interchanges some pair of members of the standard basis and leaves the others fixed will be called a *flip*.

For example, the flip B on R^4 that interchanges $\mathbf{e}_2$ and $\mathbf{e}_4$ has the form

(13) $\quad B(x_1\mathbf{e}_1 + x_2\mathbf{e}_2 + x_3\mathbf{e}_3 + x_4\mathbf{e}_4) = x_1\mathbf{e}_1 + x_2\mathbf{e}_4 + x_3\mathbf{e}_3 + x_4\mathbf{e}_2$

or, equivalently,

(14) $\quad B(x_1\mathbf{e}_1 + x_2\mathbf{e}_2 + x_3\mathbf{e}_3 + x_4\mathbf{e}_4) = x_1\mathbf{e}_1 + x_4\mathbf{e}_2 + x_3\mathbf{e}_3 + x_2\mathbf{e}_4.$

Hence B can also be thought of as interchanging two of the coordinates, rather than two basis vectors.

In the proof that follows, we shall use the projections $P_0, \ldots, P_n$ in R^n, defined by $P_0\mathbf{x} = \mathbf{0}$ and

(15) $\quad P_m\mathbf{x} = x_1\mathbf{e}_1 + \cdots + x_m\mathbf{e}_m$

for $1 \le m \le n$. Thus P_m is the projection whose range and null space are spanned by $\{\mathbf{e}_1, \ldots, \mathbf{e}_m\}$ and $\{\mathbf{e}_{m+1}, \ldots, \mathbf{e}_n\}$, respectively.

10.7 Theorem *Suppose $\mathbf{F}$ is a $\mathscr{C}'$-mapping of an open set $E \subset R^n$ into R^n, $\mathbf{0} \in E$, $\mathbf{F}(\mathbf{0}) = \mathbf{0}$, and $\mathbf{F}'(\mathbf{0})$ is invertible.*

Then there is a neighborhood of $\mathbf{0}$ in R^n in which a representation

(16) $\quad \mathbf{F}(\mathbf{x}) = B_1 \cdots B_{n-1}\mathbf{G}_n \circ \cdots \circ \mathbf{G}_1(\mathbf{x})$

is valid.

In (16), each $\mathbf{G}_i$ is a primitive $\mathscr{C}'$-mapping in some neighborhood of $\mathbf{0}$; $\mathbf{G}_i(\mathbf{0}) = \mathbf{0}$, $\mathbf{G}'_i(\mathbf{0})$ is invertible, and each B_i is either a flip or the identity operator.

Briefly, (16) represents $\mathbf{F}$ locally as a composition of primitive mappings and flips.

Proof Put $\mathbf{F} = \mathbf{F}_1$. Assume $1 \le m \le n - 1$, and make the following induction hypothesis (which evidently holds for $m = 1$):

V_m is a neighborhood of $\mathbf{0}$, $\mathbf{F}_m \in \mathscr{C}'(V_m)$, $\mathbf{F}_m(\mathbf{0}) = \mathbf{0}$, $\mathbf{F}'_m(\mathbf{0})$ is invertible, and

(17) $\quad P_{m-1}\mathbf{F}_m(\mathbf{x}) = P_{m-1}\mathbf{x} \quad (\mathbf{x} \in V_m).$

By (17), we have

(18) $\quad \mathbf{F}_m(\mathbf{x}) = P_{m-1}\mathbf{x} + \sum_{i=m}^{n} \alpha_i(\mathbf{x})\mathbf{e}_i,$

where $\alpha_m, \ldots, \alpha_n$ are real $\mathscr{C}'$-functions in V_m. Hence

(19) $\quad \mathbf{F}'_m(\mathbf{0})\mathbf{e}_m = \sum_{i=m}^{n} (D_m\alpha_i)(\mathbf{0})\mathbf{e}_i.$

Since $\mathbf{F}'_m(0)$ is invertible, the left side of (19) is not $\mathbf{0}$, and therefore there is a k such that $m \leq k \leq n$ and $(D_m \alpha_k)(0) \neq 0$.

Let B_m be the flip that interchanges m and this k (if $k = m$, B_m is the identity) and define

(20) $$\mathbf{G}_m(\mathbf{x}) = \mathbf{x} + [\alpha_k(\mathbf{x}) - x_m]\mathbf{e}_m \qquad (\mathbf{x} \in V_m).$$

Then $\mathbf{G}_m \in \mathscr{C}'(V_m)$, $\mathbf{G}_m$ is primitive, and $\mathbf{G}'_m(0)$ is invertible, since $(D_m \alpha_k)(0) \neq 0$.

The inverse function theorem shows therefore that there is an open set U_m, with $\mathbf{0} \in U_m \subset V_m$, such that $\mathbf{G}_m$ is a 1-1 mapping of U_m onto a neighborhood V_{m+1} of $\mathbf{0}$, in which $\mathbf{G}_m^{-1}$ is continuously differentiable. Define $\mathbf{F}_{m+1}$ by

(21) $$\mathbf{F}_{m+1}(\mathbf{y}) = B_m \mathbf{F}_m \circ \mathbf{G}_m^{-1}(\mathbf{y}) \qquad (\mathbf{y} \in V_{m+1}).$$

Then $\mathbf{F}_{m+1} \in \mathscr{C}'(V_{m+1})$, $\mathbf{F}_{m+1}(\mathbf{0}) = \mathbf{0}$, and $\mathbf{F}'_{m+1}(0)$ is invertible (by the chain rule). Also, for $\mathbf{x} \in U_m$,

(22) $$\begin{aligned} P_m \mathbf{F}_{m+1}(\mathbf{G}_m(\mathbf{x})) &= P_m B_m \mathbf{F}_m(\mathbf{x}) \\ &= P_m[P_{m-1}\mathbf{x} + \alpha_k(\mathbf{x})\mathbf{e}_m + \cdots] \\ &= P_{m-1}\mathbf{x} + \alpha_k(\mathbf{x})\mathbf{e}_m \\ &= P_m \mathbf{G}_m(\mathbf{x}) \end{aligned}$$

so that

(23) $$P_m \mathbf{F}_{m+1}(\mathbf{y}) = P_m \mathbf{y} \qquad (\mathbf{y} \in V_{m+1}).$$

Our induction hypothesis holds therefore with $m + 1$ in place of m.

[In (22), we first used (21), then (18) and the definition of B_m, then the definition of P_m, and finally (20).]

Since $B_m B_m = I$, (21), with $\mathbf{y} = \mathbf{G}_m(\mathbf{x})$, is equivalent to

(24) $$\mathbf{F}_m(\mathbf{x}) = B_m \mathbf{F}_{m+1}(\mathbf{G}_m(\mathbf{x})) \qquad (\mathbf{x} \in U_m).$$

If we apply this with $m = 1, \ldots, n-1$, we successively obtain

$$\begin{aligned} \mathbf{F} = \mathbf{F}_1 &= B_1 \mathbf{F}_2 \circ \mathbf{G}_1 \\ &= B_1 B_2 \mathbf{F}_3 \circ \mathbf{G}_2 \circ \mathbf{G}_1 = \cdots \\ &= B_1 \cdots B_{n-1} \mathbf{F}_n \circ \mathbf{G}_{n-1} \circ \cdots \circ \mathbf{G}_1 \end{aligned}$$

in some neighborhood of $\mathbf{0}$. By (17), $\mathbf{F}_n$ is primitive. This completes the proof.

PARTITIONS OF UNITY

10.8 Theorem *Suppose K is a compact subset of R^n, and $\{V_\alpha\}$ is an open cover of K. Then there exist functions $\psi_1, \ldots, \psi_s \in \mathscr{C}(R^n)$ such that*

(a) $0 \leq \psi_i \leq 1$ for $1 \leq i \leq s$;
(b) each ψ_i has its support in some V_α, and
(c) $\psi_1(\mathbf{x}) + \cdots + \psi_s(\mathbf{x}) = 1$ for every $\mathbf{x} \in K$.

Because of (c), $\{\psi_i\}$ is called a *partition of unity*, and (b) is sometimes expressed by saying that $\{\psi_i\}$ is *subordinate* to the cover $\{V_\alpha\}$.

Corollary *If $f \in \mathscr{C}(R^n)$ and the support of f lies in K, then*

$$(25) \qquad f = \sum_{i=1}^{s} \psi_i f.$$

Each $\psi_i f$ has its support in some V_α.

The point of (25) is that it furnishes a representation of f as a sum of continuous functions $\psi_i f$ with "small" supports.

Proof Associate with each $\mathbf{x} \in K$ an index $\alpha(\mathbf{x})$ so that $\mathbf{x} \in V_{\alpha(\mathbf{x})}$. Then there are open balls $B(\mathbf{x})$ and $W(\mathbf{x})$, centered at $\mathbf{x}$, with

$$(26) \qquad B(\mathbf{x}) \subset W(\mathbf{x}) \subset \overline{W(\mathbf{x})} \subset V_{\alpha(\mathbf{x})}.$$

Since K is compact, there are points $\mathbf{x}_1, \ldots, \mathbf{x}_s$ in K such that

$$(27) \qquad K \subset B(\mathbf{x}_1) \cup \cdots \cup B(\mathbf{x}_s).$$

By (26), there are functions $\varphi_1, \ldots, \varphi_s \in \mathscr{C}(R^n)$, such that $\varphi_i(\mathbf{x}) = 1$ on $B(\mathbf{x}_i)$, $\varphi_i(\mathbf{x}) = 0$ outside $W(\mathbf{x}_i)$, and $0 \leq \varphi_i(\mathbf{x}) \leq 1$ on R^n. Define $\psi_1 = \varphi_1$ and

$$(28) \qquad \psi_{i+1} = (1 - \varphi_1) \cdots (1 - \varphi_i)\varphi_{i+1}$$

for $i = 1, \ldots, s - 1$.

Properties (a) and (b) are clear. The relation

$$(29) \qquad \psi_1 + \cdots + \psi_i = 1 - (1 - \varphi_1) \cdots (1 - \varphi_i)$$

is trivial for $i = 1$. If (29) holds for some $i < s$, addition of (28) and (29) yields (29) with $i + 1$ in place of i. It follows that

$$(30) \qquad \sum_{i=1}^{s} \psi_i(\mathbf{x}) = 1 - \prod_{i=1}^{s} [1 - \varphi_i(\mathbf{x})] \qquad (\mathbf{x} \in R^n).$$

If $\mathbf{x} \in K$, then $\mathbf{x} \in B(\mathbf{x}_i)$ for some i, hence $\varphi_i(\mathbf{x}) = 1$, and the product in (30) is 0. This proves (c).

CHANGE OF VARIABLES

We can now describe the effect of a change of variables on a multiple integral. For simplicity, we confine ourselves here to continuous functions with compact support, although this is too restrictive for many applications. This is illustrated by Exercises 9 to 13.

10.9 Theorem *Suppose T is a 1-1 $\mathscr{C}'$-mapping of an open set $E \subset R^k$ into R^k such that $J_T(\mathbf{x}) \neq 0$ for all $\mathbf{x} \in E$. If f is a continuous function on R^k whose support is compact and lies in $T(E)$, then*

$$(31) \qquad \int_{R^k} f(\mathbf{y})\, d\mathbf{y} = \int_{R^k} f(T(\mathbf{x}))|J_T(\mathbf{x})|\, d\mathbf{x}.$$

We recall that J_T is the Jacobian of T. The assumption $J_T(\mathbf{x}) \neq 0$ implies, by the inverse function theorem, that T^{-1} is continuous on $T(E)$, and this ensures that the integrand on the right of (31) has compact support in E (Theorem 4.14).

The appearance of the *absolute value* of $J_T(\mathbf{x})$ in (31) may call for a comment. Take the case $k = 1$, and suppose T is a 1-1 $\mathscr{C}'$-mapping of R^1 onto R^1. Then $J_T(x) = T'(x)$; and if T is *increasing*, we have

$$(32) \qquad \int_{R^1} f(y)\, dy = \int_{R^1} f(T(x))T'(x)\, dx,$$

by Theorems 6.19 and 6.17, for all continuous f with compact support. But if T decreases, then $T'(x) < 0$; and if f is positive in the interior of its support, the left side of (32) is positive and the right side is negative. A correct equation is obtained if T' is replaced by $|T'|$ in (32).

The point is that the integrals we are now considering are integrals of functions over subsets of R^k, and we associate no direction or orientation with these subsets. We shall adopt a different point of view when we come to integration of differential forms over surfaces.

Proof It follows from the remarks just made that (31) is true if T is a primitive $\mathscr{C}'$-mapping (see Definition 10.5), and Theorem 10.2 shows that (31) is true if T is a linear mapping which merely interchanges two coordinates.

If the theorem is true for transformations P, Q, and if $S(\mathbf{x}) = P(Q(\mathbf{x}))$, then

$$\int f(\mathbf{z})\, d\mathbf{z} = \int f(P(\mathbf{y}))|J_P(\mathbf{y})|\, d\mathbf{y}$$

$$= \int f(P(Q(\mathbf{x})))|J_P(Q(\mathbf{x}))|\,|J_Q(\mathbf{x})|\, d\mathbf{x}$$

$$= \int f(S(\mathbf{x}))|J_S(\mathbf{x})|\, d\mathbf{x},$$

since
$$J_P(Q(\mathbf{x}))J_Q(\mathbf{x}) = \det P'(Q(\mathbf{x})) \det Q'(\mathbf{x})$$
$$= \det P'(Q(\mathbf{x}))Q'(\mathbf{x}) = \det S'(\mathbf{x}) = J_S(\mathbf{x}),$$

by the multiplication theorem for determinants and the chain rule. Thus the theorem is also true for S.

Each point $\mathbf{a} \in E$ has a neighborhood $U \subset E$ in which

(33) $\qquad T(\mathbf{x}) = T(\mathbf{a}) + B_1 \cdots B_{k-1}\mathbf{G}_k \circ \mathbf{G}_{k-1} \circ \cdots \circ \mathbf{G}_1(\mathbf{x} - \mathbf{a}),$

where $\mathbf{G}_i$ and B_i are as in Theorem 10.7. Setting $V = T(U)$, it follows that (31) holds if the support of f lies in V. Thus:

Each point $\mathbf{y} \in T(E)$ lies in an open set $V_\mathbf{y} \subset T(E)$ such that (31) holds for all continuous functions whose support lies in $V_\mathbf{y}$.

Now let f be a continuous function with compact support $K \subset T(E)$. Since $\{V_\mathbf{y}\}$ covers K, the Corollary to Theorem 10.8 shows that $f = \Sigma \psi_i f$, where each ψ_i is continuous, and each ψ_i has its support in some $V_\mathbf{y}$. Thus (31) holds for each $\psi_i f$, and hence also for their sum f.

DIFFERENTIAL FORMS

We shall now develop some of the machinery that is needed for the n-dimensional version of the fundamental theorem of calculus which is usually called *Stokes' theorem*. The original form of Stokes' theorem arose in applications of vector analysis to electromagnetism and was stated in terms of the curl of a vector field. Green's theorem and the divergence theorem are other special cases. These topics are briefly discussed at the end of the chapter.

It is a curious feature of Stokes' theorem that the only thing that is difficult about it is the elaborate structure of definitions that are needed for its statement. These definitions concern differential forms, their derivatives, boundaries, and orientation. Once these concepts are understood, the statement of the theorem is very brief and succinct, and its proof presents little difficulty.

Up to now we have considered derivatives of functions of several variables only for functions defined in *open* sets. This was done to avoid difficulties that can occur at boundary points. It will now be convenient, however, to discuss differentiable functions on *compact* sets. We therefore adopt the following convention:

To say that $\mathbf{f}$ is a $\mathscr{C}'$-mapping (or a $\mathscr{C}''$-mapping) of a compact set $D \subset R^k$ into R^n means that there is a $\mathscr{C}'$-mapping (or a $\mathscr{C}''$-mapping) $\mathbf{g}$ of an open set $W \subset R^k$ into R^n such that $D \subset W$ and such that $\mathbf{g}(\mathbf{x}) = \mathbf{f}(\mathbf{x})$ for all $\mathbf{x} \in D$.

10.10 Definition Suppose E is an open set in R^n. A *k-surface* in E is a $\mathscr{C}'$-mapping Φ from a compact set $D \subset R^k$ into E.

D is called the *parameter domain* of Φ. Points of D will be denoted by $\mathbf{u} = (u_1, \ldots, u_k)$.

We shall confine ourselves to the simple situation in which D is either a k-cell or the k-simplex Q^k described in Example 10.4. The reason for this is that we shall have to integrate over D, and we have not yet discussed integration over more complicated subsets of R^k. It will be seen that this restriction on D (which will be tacitly made from now on) entails no significant loss of generality in the resulting theory of differential forms.

We stress that k-surfaces in E are defined to be *mappings* into E, not subsets of E. This agrees with our earlier definition of curves (Definition 6.26). In fact, 1-surfaces are precisely the same as continuously differentiable curves.

10.11 Definition Suppose E is an open set in R^n. A *differential form of order* $k \geq 1$ *in* E (briefly, a *k-form in E*) is a function ω, symbolically represented by the sum

$$(34) \qquad \omega = \sum a_{i_1 \cdots i_k}(\mathbf{x})\, dx_{i_1} \wedge \cdots \wedge dx_{i_k}$$

(the indices $i_1, \ldots, i_k$ range independently from 1 to n), which assigns to each k-surface Φ in E a number $\omega(\Phi) = \int_\Phi \omega$, according to the rule

$$(35) \qquad \int_\Phi \omega = \int_D \sum a_{i_1 \cdots i_k}(\Phi(\mathbf{u})) \frac{\partial(x_{i_1}, \ldots, x_{i_k})}{\partial(u_1, \ldots, u_k)}\, d\mathbf{u},$$

where D is the parameter domain of Φ.

The functions $a_{i_1 \cdots i_k}$ are assumed to be real and continuous in E. If $\phi_1, \ldots, \phi_n$ are the components of Φ, the Jacobian in (35) is the one determined by the mapping

$$(u_1, \ldots, u_k) \to (\phi_{i_1}(\mathbf{u}), \ldots, \phi_{i_k}(\mathbf{u})).$$

Note that the right side of (35) is an integral over D, as defined in Definition 10.1 (or Example 10.4) and that (35) is the *definition* of the symbol $\int_\Phi \omega$.

A k-form ω is said to be of class $\mathscr{C}'$ or $\mathscr{C}''$ if the functions $a_{i_1 \cdots i_k}$ in (34) are all of class $\mathscr{C}'$ or $\mathscr{C}''$.

A 0-form in E is defined to be a continuous function in E.

10.12 Examples

(a) Let γ be a 1-surface (a curve of class $\mathscr{C}'$) in R^3, with parameter domain $[0, 1]$.

Write (x, y, z) in place of (x_1, x_2, x_3), and put

$$\omega = x\, dy + y\, dx.$$

Then

$$\int_\gamma \omega = \int_0^1 [\gamma_1(t)\gamma_2'(t) + \gamma_2(t)\gamma_1'(t)]\, dt = \gamma_1(1)\gamma_2(1) - \gamma_1(0)\gamma_2(0).$$

Note that in this example $\int_\gamma \omega$ depends only on the initial point $\gamma(0)$ and on the end point $\gamma(1)$ of γ. In particular, $\int_\gamma \omega = 0$ for every closed curve γ. (As we shall see later, this is true for every 1-form ω which is *exact*.)

Integrals of 1-forms are often called *line integrals*.

(b) Fix $a > 0$, $b > 0$, and define

$$\gamma(t) = (a \cos t,\, b \sin t) \qquad (0 \leq t \leq 2\pi),$$

so that γ is a closed curve in R^2. (Its range is an ellipse.) Then

$$\int_\gamma x\, dy = \int_0^{2\pi} ab \cos^2 t\, dt = \pi a b,$$

whereas

$$\int_\gamma y\, dx = -\int_0^{2\pi} ab \sin^2 t\, dt = -\pi a b.$$

Note that $\int_\gamma x\, dy$ is the area of the region bounded by γ. This is a special case of Green's theorem.

(c) Let D be the 3-cell defined by

$$0 \leq r \leq 1, \qquad 0 \leq \theta \leq \pi, \qquad 0 \leq \varphi \leq 2\pi.$$

Define $\Phi(r, \theta, \varphi) = (x, y, z)$, where

$$\begin{aligned} x &= r \sin \theta \cos \varphi \\ y &= r \sin \theta \sin \varphi \\ z &= r \cos \theta. \end{aligned}$$

Then

$$J_\Phi(r, \theta, \varphi) = \frac{\partial(x, y, z)}{\partial(r, \theta, \varphi)} = r^2 \sin \theta.$$

Hence

(36) $$\int_\Phi dx \wedge dy \wedge dz = \int_D J_\Phi = \frac{4\pi}{3}.$$

Note that Φ maps D onto the closed unit ball of R^3, that the mapping is 1-1 in the interior of D (but certain boundary points are identified by Φ), and that the integral (36) is equal to the volume of $\Phi(D)$.

10.13 Elementary properties Let $\omega, \omega_1, \omega_2$ be k-forms in E. We write $\omega_1 = \omega_2$ if and only if $\omega_1(\Phi) = \omega_2(\Phi)$ for every k-surface Φ in E. In particular, $\omega = 0$ means that $\omega(\Phi) = 0$ for every k-surface Φ in E. If c is a real number, then $c\omega$ is the k-form defined by

$$(37) \qquad \int_\Phi c\omega = c \int_\Phi \omega,$$

and $\omega = \omega_1 + \omega_2$ means that

$$(38) \qquad \int_\Phi \omega = \int_\Phi \omega_1 + \int_\Phi \omega_2$$

for every k-surface Φ in E. As a special case of (37), note that $-\omega$ is defined so that

$$(39) \qquad \int_\Phi (-\omega) = -\int_\Phi d\omega.$$

Consider a k-form

$$(40) \qquad \omega = a(\mathbf{x}) \, dx_{i_1} \wedge \cdots \wedge dx_{i_k}$$

and let $\bar\omega$ be the k-form obtained by interchanging some pair of subscripts in (40). If (35) and (39) are combined with the fact that a determinant changes sign if two of its rows are interchanged, we see that

$$(41) \qquad \bar\omega = -\omega.$$

As a special case of this, note that the *anticommutative relation*

$$(42) \qquad dx_i \wedge dx_j = -dx_j \wedge dx_i$$

holds for all i and j. In particular,

$$(43) \qquad dx_i \wedge dx_i = 0 \qquad (i = 1, \ldots, n).$$

More generally, let us return to (40), and assume that $i_r = i_s$ for some $r \ne s$. If these two subscripts are interchanged, then $\bar\omega = \omega$, hence $\omega = 0$, by (41).

In other words, if ω is given by (40), then $\omega = 0$ unless the subscripts $i_1, \ldots, i_k$ are all distinct.

If ω is as in (34), the summands with repeated subscripts can therefore be omitted without changing ω.

It follows that 0 is the only k-form in any open subset of R^n, if $k > n$.

The anticommutativity expressed by (42) is the reason for the inordinate amount of attention that has to be paid to minus signs when studying differential forms.

10.14 Basic k-forms If $i_1, \ldots, i_k$ are integers such that $1 \le i_1 < i_2 < \cdots < i_k \le n$, and if I is the ordered k-tuple $\{i_1, \ldots, i_k\}$, then we call I an *increasing k-index*, and we use the brief notation

$$dx_I = dx_{i_1} \wedge \cdots \wedge dx_{i_k}. \tag{44}$$

These forms dx_I are the so-called *basic k-forms in R^n*.

It is not hard to verify that there are precisely $n!/k!(n-k)!$ basic k-forms in R^n; we shall make no use of this, however.

Much more important is the fact that every k-form can be represented in terms of basic k-forms. To see this, note that every k-tuple $\{j_1, \ldots, j_k\}$ of distinct integers can be converted to an increasing k-index J by a finite number of interchanges of pairs; each of these amounts to a multiplication by -1, as we saw in Sec. 10.13; hence

$$dx_{j_1} \wedge \cdots \wedge dx_{j_k} = \varepsilon(j_1, \ldots, j_k)\, dx_J \tag{45}$$

where $\varepsilon(j_1, \ldots, j_k)$ is 1 or -1, depending on the number of interchanges that are needed. In fact, it is easy to see that

$$\varepsilon(j_1, \ldots, j_k) = s(j_1, \ldots, j_k) \tag{46}$$

where s is as in Definition 9.33.

For example,

$$dx_1 \wedge dx_5 \wedge dx_3 \wedge dx_2 = -dx_1 \wedge dx_2 \wedge dx_3 \wedge dx_5$$

and

$$dx_4 \wedge dx_2 \wedge dx_3 = dx_2 \wedge dx_3 \wedge dx_4.$$

If every k-tuple in (34) is converted to an increasing k-index, then we obtain the so-called *standard presentation* of ω:

$$\omega = \sum_I b_I(\mathbf{x})\, dx_I. \tag{47}$$

The summation in (47) extends over all increasing k-indices I. [Of course, every increasing k-index arises from many (from $k!$, to be precise) k-tuples. Each b_I in (47) may thus be a sum of several of the coefficients that occur in (34).]

For example,

$$x_1\, dx_2 \wedge dx_1 - x_2\, dx_3 \wedge dx_2 + x_3\, dx_2 \wedge dx_3 + dx_1 \wedge dx_2$$

is a 2-form in R^3 whose standard presentation is

$$(1 - x_1)\, dx_1 \wedge dx_2 + (x_2 + x_3)\, dx_2 \wedge dx_3.$$

The following uniqueness theorem is one of the main reasons for the introduction of the standard presentation of a k-form.

10.15 Theorem *Suppose*

(48) $$\omega = \sum_I b_I(\mathbf{x}) \, dx_I$$

is the standard presentation of a k-form ω in an open set $E \subset R^n$. If $\omega = 0$ in E, then $b_I(\mathbf{x}) = 0$ for every increasing k-index I and for every $\mathbf{x} \in E$.

Note that the analogous statement would be false for sums such as (34), since, for example,

$$dx_1 \wedge dx_2 + dx_2 \wedge dx_1 = 0.$$

Proof Assume, to reach a contradiction, that $b_J(\mathbf{v}) > 0$ for some $\mathbf{v} \in E$ and for some increasing k-index $J = \{j_1, \ldots, j_k\}$. Since b_J is continuous, there exists $h > 0$ such that $b_J(\mathbf{x}) > 0$ for all $\mathbf{x} \in R^n$ whose coordinates satisfy $|x_i - v_i| \leq h$. Let D be the k-cell in R^k such that $\mathbf{u} \in D$ if and only if $|u_r| \leq h$ for $r = 1, \ldots, k$. Define

(49) $$\Phi(\mathbf{u}) = \mathbf{v} + \sum_{r=1}^{k} u_r \mathbf{e}_{j_r} \qquad (\mathbf{u} \in D).$$

Then Φ is a k-surface in E, with parameter domain D, and $b_J(\Phi(\mathbf{u})) > 0$ for every $\mathbf{u} \in D$.

We claim that

(50) $$\int_\Phi \omega = \int_D b_J(\Phi(\mathbf{u})) \, d\mathbf{u}.$$

Since the right side of (50) is positive, it follows that $\omega(\Phi) \neq 0$. Hence (50) gives our contradiction.

To prove (50), apply (35) to the presentation (48). More specifically, compute the Jacobians that occur in (35). By (49),

$$\frac{\partial(x_{j_1}, \ldots, x_{j_k})}{\partial(u_1, \ldots, u_k)} = 1.$$

For any other increasing k-index $I \neq J$, the Jacobian is 0, since it is the determinant of a matrix with at least one row of zeros.

10.16 Products of basic k-forms *Suppose*

(51) $$I = \{i_1, \ldots, i_p\}, \qquad J = \{j_1, \ldots, j_q\}$$

where $1 \leq i_1 < \cdots < i_p \leq n$ and $1 \leq j_1 < \cdots < j_q \leq n$. The *product* of the corresponding basic forms dx_I and dx_J in R^n is a $(p+q)$-form in R^n, denoted by the symbol $dx_I \wedge dx_J$, and defined by

(52) $$dx_I \wedge dx_J = dx_{i_1} \wedge \cdots \wedge dx_{i_p} \wedge dx_{j_1} \wedge \cdots \wedge dx_{j_q}.$$

If I and J have an element in common, then the discussion in Sec. 10.13 shows that $dx_I \wedge dx_J = 0$.

If I and J have no element in common, let us write $[I, J]$ for the increasing $(p + q)$-index which is obtained by arranging the members of $I \cup J$ in increasing order. Then $dx_{[I, J]}$ is a basic $(p + q)$-form. We claim that

$$(53) \qquad dx_I \wedge dx_J = (-1)^\alpha \, dx_{[I, J]}$$

where α is the number of differences $j_t - i_s$ that are *negative*. (The number of positive differences is thus $pq - \alpha$.)

To prove (53), perform the following operations on the numbers

$$(54) \qquad i_1, \ldots, i_p; j_1, \ldots, j_q.$$

Move i_p to the right, step by step, until its right neighbor is larger than i_p. The number of steps is the number of subscripts t such that $i_p < j_t$. (Note that 0 steps are a distinct possibility.) Then do the same for $i_{p-1}, \ldots, i_1$. The total number of steps taken is α. The final arrangement reached is $[I, J]$. Each step, when applied to the right side of (52), multiplies $dx_I \wedge dx_J$ by -1. Hence (53) holds.

Note that the right side of (53) is the standard presentation of $dx_I \wedge dx_J$.

Next, let $K = (k_1, \ldots, k_r)$ be an increasing r-index in $\{1, \ldots, n\}$. We shall use (53) to prove that

$$(55) \qquad (dx_I \wedge dx_J) \wedge dx_K = dx_I \wedge (dx_J \wedge dx_K).$$

If any two of the sets I, J, K have an element in common, then each side of (55) is 0, hence they are equal.

So let us assume that I, J, K are pairwise disjoint. Let $[I, J, K]$ denote the increasing $(p + q + r)$-index obtained from their union. Associate β with the ordered pair (J, K) and γ with the ordered pair (I, K) in the way that α was associated with (I, J) in (53). The left side of (55) is then

$$(-1)^\alpha \, dx_{[I, J]} \wedge dx_K = (-1)^\alpha (-1)^{\beta + \gamma} \, dx_{[I, J, K]}$$

by two applications of (53), and the right side of (55) is

$$(-1)^\beta \, dx_I \wedge dx_{[J, K]} = (-1)^\beta (-1)^{\alpha + \gamma} \, dx_{[I, J, K]}.$$

Hence (55) is correct.

10.17 Multiplication Suppose ω and λ are p- and q-forms, respectively, in some open set $E \subset R^n$, with standard presentations

$$(56) \qquad \omega = \sum_I b_I(\mathbf{x}) \, dx_I, \qquad \lambda = \sum_J c_J(\mathbf{x}) \, dx_J$$

where I and J range over all increasing p-indices and over all increasing q-indices taken from the set $\{1, \ldots, n\}$.

Their product, denoted by the symbol $\omega \wedge \lambda$, is defined to be

(57) $$\omega \wedge \lambda = \sum_{I,J} b_I(\mathbf{x}) c_J(\mathbf{x})\, dx_I \wedge dx_J.$$

In this sum, I and J range independently over their possible values, and $dx_I \wedge dx_J$ is as in Sec. 10.16. Thus $\omega \wedge \lambda$ is a $(p+q)$-form in E.

It is quite easy to see (we leave the details as an exercise) that the distributive laws

$$(\omega_1 + \omega_2) \wedge \lambda = (\omega_1 \wedge \lambda) + (\omega_2 \wedge \lambda)$$

and

$$\omega \wedge (\lambda_1 + \lambda_2) = (\omega \wedge \lambda_1) + (\omega \wedge \lambda_2)$$

hold, with respect to the addition defined in Sec. 10.13. If these distributive laws are combined with (55), we obtain the associative law

(58) $$(\omega \wedge \lambda) \wedge \sigma = \omega \wedge (\lambda \wedge \sigma)$$

for arbitrary forms ω, λ, σ in E.

In this discussion it was tacitly assumed that $p \geq 1$ and $q \geq 1$. The product of a 0-form f with the p-form ω given by (56) is simply defined to be the p-form

$$f\omega = \omega f = \sum_I f(\mathbf{x}) b_I(\mathbf{x})\, dx_I.$$

It is customary to write $f\omega$, rather than $f \wedge \omega$, when f is a 0-form.

10.18 Differentiation We shall now define a differentiation operator d which associates a $(k+1)$-form $d\omega$ to each k-form ω of class $\mathscr{C}'$ in some open set $E \subset R^n$.

A 0-form of class $\mathscr{C}'$ in E is just a real function $f \in \mathscr{C}'(E)$, and we define

(59) $$df = \sum_{i=1}^{n} (D_i f)(\mathbf{x})\, dx_i.$$

If $\omega = \Sigma b_I(\mathbf{x})\, dx_I$ is the standard presentation of a k-form ω, and $b_I \in \mathscr{C}'(E)$ for each increasing k-index I, then we define

(60) $$d\omega = \sum_I (db_I) \wedge dx_I.$$

10.19 Example Suppose E is open in R^n, $f \in \mathscr{C}'(E)$, and γ is a continuously differentiable curve in E, with domain $[0, 1]$. By (59) and (35),

(61) $$\int_\gamma df = \int_0^1 \sum_{i=1}^{n} (D_i f)(\gamma(t)) \gamma_i'(t)\, dt.$$

By the chain rule, the last integrand is $(f \circ \gamma)'(t)$. Hence

$$\int_\gamma df = f(\gamma(1)) - f(\gamma(0)), \tag{62}$$

and we see that $\int_\gamma df$ is the same for all γ with the same initial point and the same end point, as in (a) of Example 10.12.

Comparison with Example 10.12(b) shows therefore that the 1-form $x\, dy$ is not the derivative of any 0-form f. This could also be deduced from part (b) of the following theorem, since

$$d(x\, dy) = dx \wedge dy \neq 0.$$

10.20 Theorem

(a) If ω and λ are k- and m-forms, respectively, of class $\mathscr{C}'$ in E, then

$$d(\omega \wedge \lambda) = (d\omega) \wedge \lambda + (-1)^k \omega \wedge d\lambda. \tag{63}$$

(b) If ω is of class $\mathscr{C}''$ in E, then $d^2\omega = 0$.

Here $d^2\omega$ means, of course, $d(d\omega)$.

Proof Because of (57) and (60), (a) follows if (63) is proved for the special case

$$\omega = f\, dx_I, \qquad \lambda = g\, dx_J \tag{64}$$

where $f, g \in \mathscr{C}'(E)$, dx_I is a basic k-form, and dx_J is a basic m-form. [If k or m or both are 0, simply omit dx_I or dx_J in (64); the proof that follows is unaffected by this.] Then

$$\omega \wedge \lambda = fg\, dx_I \wedge dx_J.$$

Let us assume that I and J have no element in common. [In the other case each of the three terms in (63) is 0.] Then, using (53),

$$d(\omega \wedge \lambda) = d(fg\, dx_I \wedge dx_J) = (-1)^\alpha d(fg\, dx_{[I,J]}).$$

By (59), $d(fg) = f\, dg + g\, df$. Hence (60) gives

$$d(\omega \wedge \lambda) = (-1)^\alpha (f\, dg + g\, df) \wedge dx_{[I,J]}$$
$$= (g\, df + f\, dg) \wedge dx_I \wedge dx_J.$$

Since dg is a 1-form and dx_I is a k-form, we have

$$dg \wedge dx_I = (-1)^k dx_I \wedge dg,$$

by (42). Hence
$$d(\omega \wedge \lambda) = (df \wedge dx_I) \wedge (g\, dx_J) + (-1)^k (f\, dx_I) \wedge (dg \wedge dx_J)$$
$$= (d\omega) \wedge \lambda + (-1)^k \omega \wedge d\lambda,$$
which proves (a).

Note that the associative law (58) was used freely.

Let us prove (b) first for a 0-form $f \in \mathscr{C}''$:
$$d^2 f = d\left(\sum_{j=1}^n (D_j f)(\mathbf{x})\, dx_j\right)$$
$$= \sum_{j=1}^n d(D_j f) \wedge dx_j$$
$$= \sum_{i,j=1}^n (D_{ij} f)(\mathbf{x})\, dx_i \wedge dx_j.$$

Since $D_{ij} f = D_{ji} f$ (Theorem 9.41) and $dx_i \wedge dx_j = -dx_j \wedge dx_i$, we see that $d^2 f = 0$.

If $\omega = f\, dx_I$, as in (64), then $d\omega = (df) \wedge dx_I$. By (60), $d(dx_I) = 0$. Hence (63) shows that
$$d^2\omega = (d^2 f) \wedge dx_I = 0.$$

10.21 Change of variables Suppose E is an open set in R^n, T is a $\mathscr{C}'$-mapping of E into an open set $V \subset R^m$, and ω is a k-form in V, whose standard presentation is

(65)
$$\omega = \sum_I b_I(\mathbf{y})\, dy_I.$$

(We use $\mathbf{y}$ for points of V, $\mathbf{x}$ for points of E.)

Let $t_1, \ldots, t_m$ be the components of T: If
$$\mathbf{y} = (y_1, \ldots, y_m) = T(\mathbf{x})$$
then $y_i = t_i(\mathbf{x})$. As in (59),

(66)
$$dt_i = \sum_{j=1}^n (D_j t_i)(\mathbf{x})\, dx_j \qquad (1 \leq i \leq m).$$

Thus each dt_i is a 1-form in E.

The mapping T transforms ω into a k-form ω_T in E, whose definition is

(67)
$$\omega_T = \sum_I b_I(T(\mathbf{x}))\, dt_{i_1} \wedge \cdots \wedge dt_{i_k}.$$

In each summand of (67), $I = \{i_1, \ldots, i_k\}$ is an increasing k-index.

Our next theorem shows that addition, multiplication, and differentiation of forms are defined in such a way that they commute with changes of variables.

10.22 Theorem *With E and T as in Sec. 10.21, let ω and λ be k- and m-forms in V, respectively. Then*

(a) $(\omega + \lambda)_T = \omega_T + \lambda_T$ *if* $k = m$;
(b) $(\omega \wedge \lambda)_T = \omega_T \wedge \lambda_T$;
(c) $d(\omega_T) = (d\omega)_T$ *if ω is of class $\mathscr{C}'$ and T is of class $\mathscr{C}''$.*

Proof Part (a) follows immediately from the definitions. Part (b) is almost as obvious, once we realize that

(68) $$(dy_{i_1} \wedge \cdots \wedge dy_{i_r})_T = dt_{i_1} \wedge \cdots \wedge dt_{i_r}$$

regardless of whether $\{i_1, \ldots, i_r\}$ is increasing or not; (68) holds because the same number of minus signs are needed on each side of (68) to produce increasing rearrangements.

We turn to the proof of (c). If f is a 0-form of class $\mathscr{C}'$ in V, then

$$f_T(\mathbf{x}) = f(T(\mathbf{x})), \qquad df = \sum_i (D_i f)(\mathbf{y}) \, dy_i.$$

By the chain rule, it follows that

(69) $$\begin{aligned} d(f_T) &= \sum_j (D_j f_T)(\mathbf{x}) \, dx_j \\ &= \sum_j \sum_i (D_i f)(T(\mathbf{x}))(D_j t_i)(\mathbf{x}) \, dx_j \\ &= \sum_i (D_i f)(T(\mathbf{x})) \, dt_i \\ &= (df)_T. \end{aligned}$$

If $dy_I = dy_{i_1} \wedge \cdots \wedge dy_{i_k}$, then $(dy_I)_T = dt_{i_1} \wedge \cdots \wedge dt_{i_k}$, and Theorem 10.20 shows that

(70) $$d((dy_I)_T) = 0.$$

(This is where the assumption $T \in \mathscr{C}''$ is used.)

Assume now that $\omega = f \, dy_I$. Then

$$\omega_T = f_T(\mathbf{x}) \, (dy_I)_T$$

and the preceding calculations lead to

$$\begin{aligned} d(\omega_T) &= d(f_T) \wedge (dy_I)_T = (df)_T \wedge (dy_I)_T \\ &= ((df) \wedge dy_I)_T = (d\omega)_T. \end{aligned}$$

The first equality holds by (63) and (70), the second by (69), the third by part (b), and the last by the definition of $d\omega$.

The general case of (c) follows from the special case just proved, if we apply (a). This completes the proof.

Our next objective is Theorem 10.25. This will follow directly from two other important transformation properties of differential forms, which we state first.

10.23 Theorem *Suppose T is a $\mathscr{C}'$-mapping of an open set $E \subset R^n$ into an open set $V \subset R^m$, S is a $\mathscr{C}'$-mapping of V into an open set $W \subset R^p$, and ω is a k-form in W, so that ω_S is a k-form in V and both $(\omega_S)_T$ and ω_{ST} are k-forms in E, where ST is defined by $(ST)(\mathbf{x}) = S(T(\mathbf{x}))$. Then*

(71) $$(\omega_S)_T = \omega_{ST}.$$

Proof If ω and λ are forms in W, Theorem 10.22 shows that

$$((\omega \wedge \lambda)_S)_T = (\omega_S \wedge \lambda_S)_T = (\omega_S)_T \wedge (\lambda_S)_T$$

and

$$(\omega \wedge \lambda)_{ST} = \omega_{ST} \wedge \lambda_{ST}.$$

Thus if (71) holds for ω and for λ, it follows that (71) also holds for $\omega \wedge \lambda$. Since every form can be built up from 0-forms and 1-forms by addition and multiplication, and since (71) is trivial for 0-forms, it is enough to prove (71) in the case $\omega = dz_q$, $q = 1, \ldots, p$. (We denote the points of E, V, W by $\mathbf{x}$, $\mathbf{y}$, $\mathbf{z}$, respectively.)

Let $t_1, \ldots, t_m$ be the components of T, let $s_1, \ldots, s_p$ be the components of S, and let $r_1, \ldots, r_p$ be the components of ST. If $\omega = dz_q$, then

$$\omega_S = ds_q = \sum_j (D_j s_q)(\mathbf{y}) \, dy_j,$$

so that the chain rule implies

$$\begin{aligned}(\omega_S)_T &= \sum_j (D_j s_q)(T(\mathbf{x})) \, dt_j \\ &= \sum_j (D_j s_q)(T(\mathbf{x})) \sum_i (D_i t_j)(\mathbf{x}) \, dx_i \\ &= \sum_i (D_i r_q)(\mathbf{x}) \, dx_i = dr_q = \omega_{ST}.\end{aligned}$$

10.24 Theorem *Suppose ω is a k-form in an open set $E \subset R^n$, Φ is a k-surface in E, with parameter domain $D \subset R^k$, and Δ is the k-surface in R^k, with parameter domain D, defined by $\Delta(\mathbf{u}) = \mathbf{u}(\mathbf{u} \in D)$. Then*

$$\int_\Phi \omega = \int_\Delta \omega_\Phi.$$

Proof We need only consider the case

$$\omega = a(\mathbf{x}) \, dx_{i_1} \wedge \cdots \wedge dx_{i_k}.$$

If $\phi_1, \ldots, \phi_n$ are the components of Φ, then
$$\omega_\Phi = a(\Phi(\mathbf{u}))\, d\phi_{i_1} \wedge \cdots \wedge d\phi_{i_k}.$$
The theorem will follow if we can show that
$$(72) \qquad d\phi_{i_1} \wedge \cdots \wedge d\phi_{i_k} = J(\mathbf{u})\, du_1 \wedge \cdots \wedge du_k,$$
where
$$J(\mathbf{u}) = \frac{\partial(x_{i_1}, \ldots, x_{i_k})}{\partial(u_1, \ldots, u_k)},$$
since (72) implies
$$\int_\Phi \omega = \int_D a(\Phi(\mathbf{u})) J(\mathbf{u})\, d\mathbf{u}$$
$$= \int_\Delta a(\Phi(\mathbf{u})) J(\mathbf{u})\, du_1 \wedge \cdots \wedge du_k = \int_\Delta \omega_\Phi.$$

Let $[A]$ be the k by k matrix with entries
$$\alpha(p, q) = (D_q \phi_{i_p})(\mathbf{u}) \qquad (p, q = 1, \ldots, k).$$
Then
$$d\phi_{i_p} = \sum_q \alpha(p, q)\, du_q$$
so that
$$d\phi_{i_1} \wedge \cdots \wedge d\phi_{i_k} = \sum \alpha(1, q_1) \cdots \alpha(k, q_k)\, du_{q_1} \wedge \cdots \wedge du_{q_k}.$$
In this last sum, $q_1, \ldots, q_k$ range independently over $1, \ldots, k$. The anticommutative relation (42) implies that
$$du_{q_1} \wedge \cdots \wedge du_{q_k} = s(q_1, \ldots, q_k)\, du_1 \wedge \cdots \wedge du_k,$$
where s is as in Definition 9.33; applying this definition, we see that
$$d\phi_{i_1} \wedge \cdots \wedge d\phi_{i_k} = \det[A]\, du_1 \wedge \cdots \wedge du_k;$$
and since $J(\mathbf{u}) = \det[A]$, (72) is proved.

The final result of this section combines the two preceding theorems.

10.25 Theorem *Suppose T is a $\mathscr{C}'$-mapping of an open set $E \subset R^n$ into an open set $V \subset R^m$, Φ is a k-surface in E, and ω is a k-form in V.*
Then
$$\int_{T\Phi} \omega = \int_\Phi \omega_T.$$

Proof Let D be the parameter domain of Φ (hence also of $T\Phi$) and define Δ as in Theorem 10.24.
Then
$$\int_{T\Phi} \omega = \int_\Delta \omega_{T\Phi} = \int_\Delta (\omega_T)_\Phi = \int_\Phi \omega_T.$$
The first of these equalities is Theorem 10.24, applied to $T\Phi$ in place of Φ. The second follows from Theorem 10.23. The third is Theorem 10.24, with ω_T in place of ω.

SIMPLEXES AND CHAINS

10.26 Affine simplexes A mapping $\mathbf{f}$ that carries a vector space X into a vector space Y is said to be *affine* if $\mathbf{f} - \mathbf{f}(0)$ is linear. In other words, the requirement is that

(73) $$\mathbf{f}(\mathbf{x}) = \mathbf{f}(0) + A\mathbf{x}$$

for some $A \in L(X, Y)$.

An affine mapping of R^k into R^n is thus determined if we know $\mathbf{f}(0)$ and $\mathbf{f}(\mathbf{e}_i)$ for $1 \leq i \leq k$; as usual, $\{\mathbf{e}_1, \ldots, \mathbf{e}_k\}$ is the standard basis of R^k.

We define the *standard simplex* Q^k to be the set of all $\mathbf{u} \in R^k$ of the form

(74) $$\mathbf{u} = \sum_{i=1}^k \alpha_i \mathbf{e}_i$$

such that $\alpha_i \geq 0$ for $i = 1, \ldots, k$ and $\Sigma \alpha_i \leq 1$.

Assume now that $\mathbf{p}_0, \mathbf{p}_1, \ldots, \mathbf{p}_k$ are points of R^n. The *oriented affine k-simplex*

(75) $$\sigma = [\mathbf{p}_0, \mathbf{p}_1, \ldots, \mathbf{p}_k]$$

is defined to be the k-surface in R^n with parameter domain Q^k which is given by the affine mapping

(76) $$\sigma(\alpha_1 \mathbf{e}_1 + \cdots + \alpha_k \mathbf{e}_k) = \mathbf{p}_0 + \sum_{i=1}^k \alpha_i (\mathbf{p}_i - \mathbf{p}_0).$$

Note that σ is characterized by

(77) $$\sigma(0) = \mathbf{p}_0, \quad \sigma(\mathbf{e}_i) = \mathbf{p}_i \quad (\text{for } 1 \leq i \leq k),$$

and that

(78) $$\sigma(\mathbf{u}) = \mathbf{p}_0 + A\mathbf{u} \quad (\mathbf{u} \in Q^k)$$

where $A \in L(R^k, R^n)$ and $A\mathbf{e}_i = \mathbf{p}_i - \mathbf{p}_0$ for $1 \leq i \leq k$.

We call σ *oriented* to emphasize that the ordering of the vertices $\mathbf{p}_0, \ldots, \mathbf{p}_k$ is taken into account. If

(79) $$\bar{\sigma} = [p_{i_0}, p_{i_1}, \ldots, p_{i_k}],$$

where $\{i_0, i_1, \ldots, i_k\}$ is a permutation of the ordered set $\{0, 1, \ldots, k\}$, we adopt the notation

(80) $$\bar{\sigma} = s(i_0, i_1, \ldots, i_k)\sigma,$$

where s is the function defined in Definition 9.33. Thus $\bar{\sigma} = \pm\sigma$, depending on whether $s = 1$ or $s = -1$. Strictly speaking, having adopted (75) and (76) as the definition of σ, we should not write $\bar{\sigma} = \sigma$ unless $i_0 = 0, \ldots, i_k = k$, even if $s(i_0, \ldots, i_k) = 1$; what we have here is an equivalence relation, not an equality. However, for our purposes the notation is justified by Theorem 10.27.

If $\bar{\sigma} = \varepsilon\sigma$ (using the above convention) and if $\varepsilon = 1$, we say that $\bar{\sigma}$ and σ have the *same orientation*; if $\varepsilon = -1$, $\bar{\sigma}$ and σ are said to have *opposite orientations*. Note that we have not defined what we mean by the "orientation of a simplex." What we have defined is a relation between pairs of simplexes having the same set of vertices, the relation being that of "having the same orientation."

There is, however, one situation where the orientation of a simplex can be defined in a natural way. This happens when $n = k$ and when the vectors $\mathbf{p}_i - \mathbf{p}_0$ $(1 \leq i \leq k)$ are *independent*. In that case, the linear transformation A that appears in (78) is invertible, and its determinant (which is the same as the Jacobian of σ) is not 0. Then σ is said to be *positively* (or *negatively*) oriented if $\det A$ is positive (or negative). In particular, the simplex $[\mathbf{0}, \mathbf{e}_1, \ldots, \mathbf{e}_k]$ in R^k, given by the identity mapping, has positive orientation.

So far we have assumed that $k \geq 1$. An *oriented 0-simplex* is defined to be a point with a sign attached. We write $\sigma = +\mathbf{p}_0$ or $\sigma = -\mathbf{p}_0$. If $\sigma = \varepsilon\mathbf{p}_0$ ($\varepsilon = \pm 1$) and if f is a 0-form (i.e., a real function), we define

$$\int_\sigma f = \varepsilon f(p_0).$$

10.27 Theorem *If σ is an oriented rectilinear k-simplex in an open set $E \subset R^n$ and if $\bar{\sigma} = \varepsilon\sigma$ then*

(81) $$\int_{\bar{\sigma}} \omega = \varepsilon \int_\sigma \omega$$

for every k-form ω in E.

Proof For $k = 0$, (81) follows from the preceding definition. So we assume $k \geq 1$ and assume that σ is given by (75).

Suppose $1 \leq j \leq k$, and suppose $\bar{\sigma}$ is obtained from σ by interchanging $\mathbf{p}_0$ and $\mathbf{p}_j$. Then $\varepsilon = -1$, and

$$\bar{\sigma}(\mathbf{u}) = \mathbf{p}_j + B\mathbf{u} \qquad (\mathbf{u} \in Q^k),$$

where B is the linear mapping of R^k into R^n defined by $B\mathbf{e}_j = \mathbf{p}_0 - \mathbf{p}_j$, $B\mathbf{e}_i = \mathbf{p}_i - \mathbf{p}_j$ if $i \neq j$. If we write $A\mathbf{e}_i = \mathbf{x}_i$ ($1 \leq i \leq k$), where A is given by (78), the column vectors of B (that is, the vectors $B\mathbf{e}_i$) are

$$\mathbf{x}_1 - \mathbf{x}_j, \ldots, \mathbf{x}_{j-1} - \mathbf{x}_j, -\mathbf{x}_j, \mathbf{x}_{j+1} - \mathbf{x}_j, \ldots, \mathbf{x}_k - \mathbf{x}_j.$$

If we subtract the jth column from each of the others, none of the determinants in (35) are affected, and we obtain columns $\mathbf{x}_1, \ldots, \mathbf{x}_{j-1}, -\mathbf{x}_j, \mathbf{x}_{j+1}, \ldots, \mathbf{x}_k$. These differ from those of A only in the sign of the jth column. Hence (81) holds for this case.

Suppose next that $0 < i < j \leq k$ and that $\bar{\sigma}$ is obtained from σ by interchanging $\mathbf{p}_i$ and $\mathbf{p}_j$. Then $\bar{\sigma}(\mathbf{u}) = \mathbf{p}_0 + C\mathbf{u}$, where C has the same columns as A, except that the ith and jth columns have been interchanged. This again implies that (81) holds, since $\varepsilon = -1$.

The general case follows, since every permutation of $\{0, 1, \ldots, k\}$ is a composition of the special cases we have just dealt with.

10.28 Affine chains An *affine k-chain* Γ in an open set $E \subset R^n$ is a collection of finitely many oriented affine k-simplexes $\sigma_1, \ldots, \sigma_r$ in E. These need not be distinct; a simplex may thus occur in Γ with a certain multiplicity.

If Γ is as above, and if ω is a k-form in E, we define

(82)
$$\int_\Gamma \omega = \sum_{i=1}^r \int_{\sigma_i} \omega.$$

We may view a k-surface Φ in E as a function whose domain is the collection of all k-forms in E and which assigns the number $\int_\Phi \omega$ to ω. Since real-valued functions can be added (as in Definition 4.3), this suggests the use of the notation

(83)
$$\Gamma = \sigma_1 + \cdots + \sigma_r$$

or, more compactly,

(84)
$$\Gamma = \sum_{i=1}^r \sigma_i$$

to state the fact that (82) holds for every k-form ω in E.

To avoid misunderstanding, we point out explicitly that the notations introduced by (83) and (80) have to be handled with care. The point is that every oriented affine k-simplex σ in R^n is a function in two ways, with different domains and different ranges, and that therefore two entirely different operations of addition are possible. Originally, σ was defined as an R^n-valued function with domain Q^k; accordingly, $\sigma_1 + \sigma_2$ *could* be interpreted to be the function σ that assigns the vector $\sigma_1(\mathbf{u}) + \sigma_2(\mathbf{u})$ to every $\mathbf{u} \in Q^k$; note that σ is then again an oriented affine k-simplex in R^n! This is *not* what is meant by (83).

For example, if $\sigma_2 = -\sigma_1$ as in (80) (that is to say, if σ_1 and σ_2 have the same set of vertices but are oppositely oriented) and if $\Gamma = \sigma_1 + \sigma_2$, then $\int_\Gamma \omega = 0$ for all ω, and we may express this by writing $\Gamma = 0$ or $\sigma_1 + \sigma_2 = 0$. This does not mean that $\sigma_1(\mathbf{u}) + \sigma_2(\mathbf{u})$ is the null vector of R^n.

10.29 Boundaries For $k \geq 1$, the *boundary* of the oriented affine k-simplex

$$\sigma = [\mathbf{p}_0, \mathbf{p}_1, \ldots, \mathbf{p}_k]$$

is defined to be the affine $(k-1)$-chain

(85) $$\partial \sigma = \sum_{j=0}^{k} (-1)^j [\mathbf{p}_0, \ldots, \mathbf{p}_{j-1}, \mathbf{p}_{j+1}, \ldots, \mathbf{p}_k].$$

For example, if $\sigma = [\mathbf{p}_0, \mathbf{p}_1, \mathbf{p}_2]$, then

$$\partial \sigma = [\mathbf{p}_1, \mathbf{p}_2] - [\mathbf{p}_0, \mathbf{p}_2] + [\mathbf{p}_0, \mathbf{p}_1] = [\mathbf{p}_0, \mathbf{p}_1] + [\mathbf{p}_1, \mathbf{p}_2] + [\mathbf{p}_2, \mathbf{p}_0],$$

which coincides with the usual notion of the oriented boundary of a triangle.

For $1 \leq j \leq k$, observe that the simplex $\sigma_j = [\mathbf{p}_0, \ldots, \mathbf{p}_{j-1}, \mathbf{p}_{j+1}, \ldots, \mathbf{p}_k]$ which occurs in (85) has Q^{k-1} as its parameter domain and that it is defined by

(86) $$\sigma_j(\mathbf{u}) = \mathbf{p}_0 + B\mathbf{u} \qquad (\mathbf{u} \in Q^{k-1}),$$

where B is the linear mapping from R^{k-1} to R^n determined by

$$B\mathbf{e}_i = \mathbf{p}_i - \mathbf{p}_0 \qquad (\text{if } 1 \leq i \leq j-1),$$
$$B\mathbf{e}_i = \mathbf{p}_{i+1} - \mathbf{p}_0 \qquad (\text{if } j \leq i \leq k-1).$$

The simplex

$$\sigma_0 = [\mathbf{p}_1, \mathbf{p}_2, \ldots, \mathbf{p}_k],$$

which also occurs in (85), is given by the mapping

$$\sigma_0(\mathbf{u}) = \mathbf{p}_1 + B\mathbf{u},$$

where $B\mathbf{e}_i = \mathbf{p}_{i+1} - \mathbf{p}_1$ for $1 \leq i \leq k-1$.

10.30 Differentiable simplexes and chains Let T be a $\mathscr{C}''$-mapping of an open set $E \subset R^n$ into an open set $V \subset R^m$; T need not be one-to-one. If σ is an oriented affine k-simplex in E, then the composite mapping $\Phi = T \circ \sigma$ (which we shall sometimes write in the simpler form $T\sigma$) is a k-surface in V, with parameter domain Q^k. We call Φ an *oriented k-simplex of class $\mathscr{C}''$*.

A finite collection Ψ of oriented k-simplexes $\Phi_1, \ldots, \Phi_r$ of class $\mathscr{C}''$ in V is called a *k-chain of class $\mathscr{C}''$* in V. If ω is a k-form in V, we define

$$\int_\Psi \omega = \sum_{i=1}^r \int_{\Phi_i} \omega \tag{87}$$

and use the corresponding notation $\Psi = \Sigma \Phi_i$.

If $\Gamma = \Sigma \sigma_i$ is an affine chain and if $\Phi_i = T \circ \sigma_i$, we also write $\Psi = T \circ \Gamma$, or

$$T(\sum \sigma_i) = \sum T\sigma_i. \tag{88}$$

The boundary $\partial \Phi$ of the oriented k-simplex $\Phi = T \circ \sigma$ is defined to be the $(k-1)$ chain

$$\partial \Phi = T(\partial \sigma). \tag{89}$$

In justification of (89), observe that if T is affine, then $\Phi = T \circ \sigma$ is an oriented affine k-simplex, in which case (89) is not a matter of definition, but is seen to be a *consequence* of (85). Thus (89) generalizes this special case.

It is immediate that $\partial \Phi$ is of class $\mathscr{C}''$ if this is true of Φ.

Finally, we define the boundary $\partial \Psi$ of the k-chain $\Psi = \Sigma \Phi_i$ to be the $(k-1)$ chain

$$\partial \Psi = \sum \partial \Phi_i. \tag{90}$$

10.31 Positively oriented boundaries So far we have associated boundaries to chains, not to subsets of R^n. This notion of boundary is exactly the one that is most suitable for the statement and proof of Stokes' theorem. However, in applications, especially in R^2 or R^3, it is customary and convenient to talk about "oriented boundaries" of certain sets as well. We shall now describe this briefly.

Let Q^n be the standard simplex in R^n, let σ_0 be the identity mapping with domain Q^n. As we saw in Sec. 10.26, σ_0 may be regarded as a positively oriented n-simplex in R^n. Its boundary $\partial \sigma_0$ is an affine $(n-1)$-chain. This chain is called the *positively oriented boundary of the set Q^n*.

For example, the positively oriented boundary of Q^3 is

$$[\mathbf{e}_1, \mathbf{e}_2, \mathbf{e}_3] - [0, \mathbf{e}_2, \mathbf{e}_3] + [0, \mathbf{e}_1, \mathbf{e}_3] - [0, \mathbf{e}_1, \mathbf{e}_2].$$

Now let T be a 1-1 mapping of Q^n into R^n, of class $\mathscr{C}''$, whose Jacobian is positive (at least in the interior of Q^n). Let $E = T(Q^n)$. By the inverse function theorem, E is the closure of an open subset of R^n. We define the positively oriented boundary of the set E to be the $(n-1)$-chain

$$\partial T = T(\partial \sigma_0),$$

and we may denote this $(n-1)$-chain by ∂E.

An obvious question occurs here: If $E = T_1(Q^n) = T_2(Q^n)$, and if both T_1 and T_2 have positive Jacobians, is it true that $\partial T_1 = \partial T_2$? That is to say, does the equality

$$\int_{\partial T_1} \omega = \int_{\partial T_2} \omega$$

hold for every $(n-1)$-form ω? The answer is yes, but we shall omit the proof. (To see an example, compare the end of this section with Exercise 17.)

One can go further. Let

$$\Omega = E_1 \cup \cdots \cup E_r,$$

where $E_i = T_i(Q^n)$, each T_i has the properties that T had above, and the interiors of the sets E_i are pairwise disjoint. Then the $(n-1)$-chain

$$\partial T_1 + \cdots + \partial T_r = \partial \Omega$$

is called the positively oriented boundary of Ω.

For example, the unit square I^2 in R^2 is the union of $\sigma_1(Q^2)$ and $\sigma_2(Q^2)$, where

$$\sigma_1(\mathbf{u}) = \mathbf{u}, \qquad \sigma_2(\mathbf{u}) = \mathbf{e}_1 + \mathbf{e}_2 - \mathbf{u}.$$

Both σ_1 and σ_2 have Jacobian $1 > 0$. Since

$$\sigma_1 = [\mathbf{0}, \mathbf{e}_1, \mathbf{e}_2], \qquad \sigma_2 = [\mathbf{e}_1 + \mathbf{e}_2, \mathbf{e}_2, \mathbf{e}_1]$$

we have

$$\partial \sigma_1 = [\mathbf{e}_1, \mathbf{e}_2] - [\mathbf{0}, \mathbf{e}_2] + [\mathbf{0}, \mathbf{e}_1],$$
$$\partial \sigma_2 = [\mathbf{e}_2, \mathbf{e}_1] - [\mathbf{e}_1 + \mathbf{e}_2, \mathbf{e}_1] + [\mathbf{e}_1 + \mathbf{e}_2, \mathbf{e}_2];$$

The sum of these two boundaries is

$$\partial I^2 = [\mathbf{0}, \mathbf{e}_1] + [\mathbf{e}_1, \mathbf{e}_1 + \mathbf{e}_2] + [\mathbf{e}_1 + \mathbf{e}_2, \mathbf{e}_2] + [\mathbf{e}_2, \mathbf{0}],$$

the positively oriented boundary of I^2. Note that $[\mathbf{e}_1, \mathbf{e}_2]$ canceled $[\mathbf{e}_2, \mathbf{e}_1]$.

If Φ is a 2-surface in R^m, with parameter domain I^2, then Φ (regarded as a function on 2-forms) is the same as the 2-chain

$$\Phi \circ \sigma_1 + \Phi \circ \sigma_2.$$

Thus

$$\partial \Phi = \partial(\Phi \circ \sigma_1) + \partial(\Phi \circ \sigma_2)$$
$$= \Phi(\partial \sigma_1) + \Phi(\partial \sigma_2) = \Phi(\partial I^2).$$

In other words, if the parameter domain of Φ is the square I^2, we need not refer back to the simplex Q^2, but can obtain $\partial \Phi$ directly from ∂I^2.

Other examples may be found in Exercises 17 to 19.

10.32 Example For $0 \leq u \leq \pi$, $0 \leq v \leq 2\pi$, define

$$\Sigma(u, v) = (\sin u \cos v, \sin u \sin v, \cos u).$$

Then Σ is a 2-surface in R^3, whose parameter domain is a rectangle $D \subset R^2$, and whose range is the unit sphere in R^3. Its boundary is

$$\partial \Sigma = \Sigma(\partial D) = \gamma_1 + \gamma_2 + \gamma_3 + \gamma_4$$

where

$$\gamma_1(u) = \Sigma(u, 0) = (\sin u, 0, \cos u),$$
$$\gamma_2(v) = \Sigma(\pi, v) = (0, 0, -1),$$
$$\gamma_3(u) = \Sigma(\pi - u, 2\pi) = (\sin u, 0, -\cos u),$$
$$\gamma_4(v) = \Sigma(0, 2\pi - v) = (0, 0, 1),$$

with $[0, \pi]$ and $[0, 2\pi]$ as parameter intervals for u and v, respectively.

Since γ_2 and γ_4 are constant, their derivatives are 0, hence the integral of any 1-form over γ_2 or γ_4 is 0. [See Example 1.12(a).]

Since $\gamma_3(u) = \gamma_1(\pi - u)$, direct application of (35) shows that

$$\int_{\gamma_3} \omega = - \int_{\gamma_1} \omega$$

for every 1-form ω. Thus $\int_{\partial \Sigma} \omega = 0$, and we conclude that $\partial \Sigma = 0$.

(In geographic terminology, $\partial \Sigma$ starts at the north pole N, runs to the south pole S along a meridian, pauses at S, returns to N along the same meridian, and finally pauses at N. The two passages along the meridian are in opposite directions. The corresponding two line integrals therefore cancel each other. In Exercise 32 there is also one curve which occurs twice in the boundary, but without cancellation.)

STOKES' THEOREM

10.33 Theorem *If Ψ is a k-chain of class $\mathscr{C}''$ in an open set $V \subset R^m$ and if ω is a $(k-1)$-form of class $\mathscr{C}'$ in V, then*

(91) $$\int_\Psi d\omega = \int_{\partial \Psi} \omega.$$

The case $k = m = 1$ is nothing but the fundamental theorem of calculus (with an additional differentiability assumption). The case $k = m = 2$ is Green's theorem, and $k = m = 3$ gives the so-called "divergence theorem" of Gauss. The case $k = 2$, $m = 3$ is the one originally discovered by Stokes. (Spivak's book describes some of the historical background.) These special cases will be discussed further at the end of the present chapter.

Proof It is enough to prove that

(92) $$\int_\Phi d\omega = \int_{\partial\Phi} \omega$$

for every oriented k-simplex Φ of class $\mathscr{C}''$ in V. For if (92) is proved and if $\Psi = \Sigma\Phi_i$, then (87) and (89) imply (91).

Fix such a Φ and put

(93) $$\sigma = [0, e_1, \ldots, e_k].$$

Thus σ is the oriented affine k-simplex with parameter domain Q^k which is defined by the identity mapping. Since Φ is also defined on Q^k (see Definition 10.30) and $\Phi \in \mathscr{C}''$, there is an open set $E \subset R^k$ which contains Q^k, and there is a $\mathscr{C}''$-mapping T of E into V such that $\Phi = T \circ \sigma$. By Theorems 10.25 and 10.22(c), the left side of (92) is equal to

$$\int_{T\sigma} d\omega = \int_\sigma (d\omega)_T = \int_\sigma d(\omega_T).$$

Another application of Theorem 10.25 shows, by (89), that the right side of (92) is

$$\int_{\partial(T\sigma)} \omega = \int_{T(\partial\sigma)} \omega = \int_{\partial\sigma} \omega_T.$$

Since ω_T is a $(k-1)$-form in E, we see that *in order to prove* (92) *we merely have to show that*

(94) $$\int_\sigma d\lambda = \int_{\partial\sigma} \lambda$$

for the special simplex (93) *and for every* $(k-1)$-*form* λ *of class* $\mathscr{C}'$ *in* E.

If $k = 1$, the definition of an oriented 0-simplex shows that (94) merely asserts that

(95) $$\int_0^1 f'(u)\, du = f(1) - f(0)$$

for every continuously differentiable function f on $[0, 1]$, which is true by the fundamental theorem of calculus.

From now on we assume that $k > 1$, fix an integer r $(1 \leq r \leq k)$, and choose $f \in \mathscr{C}'(E)$. It is then enough to prove (94) for the case

(96) $$\lambda = f(\mathbf{x})\, dx_1 \wedge \cdots \wedge dx_{r-1} \wedge dx_{r+1} \wedge \cdots \wedge dx_k$$

since every $(k-1)$-form is a sum of these special ones, for $r = 1, \ldots, k$.

By (85), the boundary of the simplex (93) is

$$\partial\sigma = [\mathbf{e}_1, \ldots, \mathbf{e}_k] + \sum_{i=1}^{k}(-1)^i \tau_i$$

where

$$\tau_i = [\mathbf{0}, \mathbf{e}_1, \ldots, \mathbf{e}_{i-1}, \mathbf{e}_{i+1}, \ldots, \mathbf{e}_k]$$

for $i = 1, \ldots, k$. Put

$$\tau_0 = [\mathbf{e}_r, \mathbf{e}_1, \ldots, \mathbf{e}_{r-1}, \mathbf{e}_{r+1}, \ldots, \mathbf{e}_k].$$

Note that τ_0 is obtained from $[\mathbf{e}_1, \ldots, \mathbf{e}_k]$ by $r-1$ successive interchanges of $\mathbf{e}_r$ and its left neighbors. Thus

(97) $$\partial\sigma = (-1)^{r-1}\tau_0 + \sum_{i=1}^{k}(-1)^i \tau_i.$$

Each τ_i has Q^{k-1} as parameter domain.
If $\mathbf{x} = \tau_0(\mathbf{u})$ and $\mathbf{u} \in Q^{k-1}$, then

(98) $$x_j = \begin{cases} u_j & (1 \leq j < r), \\ 1 - (u_1 + \cdots + u_{k-1}) & (j = r), \\ u_{j-1} & (r < j \leq k). \end{cases}$$

If $1 \leq i \leq k$, $\mathbf{u} \in Q^{k-1}$, and $\mathbf{x} = \tau_i(\mathbf{u})$, then

(99) $$x_j = \begin{cases} u_j & (1 \leq j < i), \\ 0 & (j = i), \\ u_{j-1} & (i < j \leq k). \end{cases}$$

For $0 \leq i \leq k$, let J_i be the Jacobian of the mapping

(100) $$(u_1, \ldots, u_{k-1}) \to (x_1, \ldots, x_{r-1}, x_{r+1}, \ldots, x_k)$$

induced by τ_i. When $i = 0$ and when $i = r$, (98) and (99) show that (100) is the identity mapping. Thus $J_0 = 1$, $J_r = 1$. For other i, the fact that $x_i = 0$ in (99) shows that J_i has a row of zeros, hence $J_i = 0$. Thus

(101) $$\int_{\tau_i} \lambda = 0 \quad (i \neq 0, i \neq r),$$

by (35) and (96). Consequently, (97) gives

(102) $$\int_{\partial\sigma} \lambda = (-1)^{r-1} \int_{\tau_0} \lambda + (-1)^r \int_{\tau_r} \lambda$$

$$= (-1)^{r-1} \int [f(\tau_0(\mathbf{u})) - f(\tau_r(\mathbf{u}))] \, d\mathbf{u}.$$

On the other hand,

$$d\lambda = (D_r f)(\mathbf{x}) dx_r \wedge dx_1 \wedge \cdots \wedge dx_{r-1} \wedge dx_{r+1} \wedge \cdots \wedge dx_k$$
$$= (-1)^{r-1}(D_r f)(\mathbf{x}) dx_1 \wedge \cdots \wedge dx_k$$

so that

(103)
$$\int_\sigma d\lambda = (-1)^{r-1} \int_{Q^k} (D_r f)(\mathbf{x}) d\mathbf{x}.$$

We evaluate (103) by first integrating with respect to x_r, over the interval

$$[0, 1 - (x_1 + \cdots + x_{r-1} + x_{r+1} + \cdots + x_k)],$$

put $(x_1, \ldots, x_{r-1}, x_{r+1}, \ldots, x_k) = (u_1, \ldots, u_{k-1})$, and see with the aid of (98) that the integral over Q^k in (103) is equal to the integral over Q^{k-1} in (102). Thus (94) holds, and the proof is complete.

CLOSED FORMS AND EXACT FORMS

10.34 Definition Let ω be a k-form in an open set $E \subset R^n$. If there is a $(k-1)$-form λ in E such that $\omega = d\lambda$, then ω is said to be *exact in E*.

If ω is of class $\mathscr{C}'$ and $d\omega = 0$, then ω is said to be *closed*.

Theorem 10.20(b) shows that every exact form of class $\mathscr{C}'$ is closed.

In certain sets E, for example in convex ones, the converse is true; this is the content of Theorem 10.39 (usually known as *Poincaré's lemma*) and Theorem 10.40. However, Examples 10.36 and 10.37 will exhibit closed forms that are not exact.

10.35 Remarks

(a) Whether a given k-form ω is or is not closed can be verified by simply differentiating the coefficients in the standard presentation of ω. For example, a 1-form

(104)
$$\omega = \sum_{i=1}^n f_i(\mathbf{x}) dx_i,$$

with $f_i \in \mathscr{C}'(E)$ for some open set $E \subset R^n$, is closed if and only if the equations

(105)
$$(D_j f_i)(\mathbf{x}) = (D_i f_j)(\mathbf{x})$$

hold for all i, j in $\{1, \ldots, n\}$ and for all $\mathbf{x} \in E$.

Note that (105) is a "pointwise" condition; it does not involve any global properties that depend on the shape of E.

On the other hand, to show that ω is exact in E, one has to prove the existence of a form λ, defined in E, such that $d\lambda = \omega$. This amounts to solving a system of partial differential equations, not just locally, but in all of E. For example, to show that (104) is exact in a set E, one has to find a function (or 0-form) $g \in \mathscr{C}'(E)$ such that

(106) $$(D_i g)(\mathbf{x}) = f_i(\mathbf{x}) \qquad (\mathbf{x} \in E, 1 \le i \le n).$$

Of course, (105) is a necessary condition for the solvability of (106).

(b) Let ω be an *exact* k-form in E. Then there is a $(k-1)$-form λ in E with $d\lambda = \omega$, and Stokes' theorem asserts that

(107) $$\int_\Psi \omega = \int_\Psi d\lambda = \int_{\partial \Psi} \lambda$$

for every k-chain Ψ of class $\mathscr{C}''$ in E.

If Ψ_1 and Ψ_2 are such chains, and if they have the same boundaries, it follows that

$$\int_{\Psi_1} \omega = \int_{\Psi_2} \omega.$$

In particular, *the integral of an exact k-form in E is 0 over every k-chain in E whose boundary is 0.*

As an important special case of this, note that integrals of exact 1-forms in E are 0 over closed (differentiable) curves in E.

(c) Let ω be a *closed* k-form in E. Then $d\omega = 0$, and Stokes' theorem asserts that

(108) $$\int_{\partial \Psi} \omega = \int_\Psi d\omega = 0$$

for every $(k+1)$-chain Ψ of class $\mathscr{C}''$ in E.

In other words, *integrals of closed k-forms in E are 0 over k-chains that are boundaries of $(k+1)$-chains in E.*

(d) Let Ψ be a $(k+1)$-chain in E and let λ be a $(k-1)$-form in E, both of class $\mathscr{C}''$. Since $d^2\lambda = 0$, two applications of Stokes' theorem show that

(109) $$\int_{\partial \partial \Psi} \lambda = \int_{\partial \Psi} d\lambda = \int_\Psi d^2\lambda = 0.$$

We conclude that $\partial^2 \Psi = 0$. In other words, *the boundary of a boundary is 0.*

See Exercise 16 for a more direct proof of this.

10.36 Example Let $E = R^2 - \{0\}$, the plane with the origin removed. The 1-form

$$\eta = \frac{x\,dy - y\,dx}{x^2 + y^2} \tag{110}$$

is *closed* in $R^2 - \{0\}$. This is easily verified by differentiation. Fix $r > 0$, and define

$$\gamma(t) = (r\cos t, r\sin t) \qquad (0 \le t \le 2\pi). \tag{111}$$

Then γ is a curve (an "oriented 1-simplex") in $R^2 - \{0\}$. Since $\gamma(0) = \gamma(2\pi)$, we have

$$\partial \gamma = 0. \tag{112}$$

Direct computation shows that

$$\int_\gamma \eta = 2\pi \ne 0. \tag{113}$$

The discussion in Remarks 10.35(*b*) and (*c*) shows that we can draw two conclusions from (113):

First, η *is not exact in* $R^2 - \{0\}$, for otherwise (112) would force the integral (113) to be 0.

Secondly, γ *is not the boundary of any 2-chain in* $R^2 - \{0\}$ (of class $\mathscr{C}''$), for otherwise the fact that η is closed would force the integral (113) to be 0.

10.37 Example Let $E = R^3 - \{0\}$, 3-space with the origin removed. Define

$$\zeta = \frac{x\,dy \wedge dz + y\,dz \wedge dx + z\,dx \wedge dy}{(x^2 + y^2 + z^2)^{3/2}} \tag{114}$$

where we have written (x, y, z) in place of (x_1, x_2, x_3). Differentiation shows that $d\zeta = 0$, so that ζ is a closed 2-form in $R^3 - \{0\}$.

Let Σ be the 2-chain in $R^3 - \{0\}$ that was constructed in Example 10.32; recall that Σ is a parametrization of the unit sphere in R^3. Using the rectangle D of Example 10.32 as parameter domain, it is easy to compute that

$$\int_\Sigma \zeta = \int_D \sin u\,du\,dv = 4\pi \ne 0. \tag{115}$$

As in the preceding example, we can now conclude that ζ is *not* exact in $R^3 - \{0\}$ (since $\partial \Sigma = 0$, as was shown in Example 10.32) and that the sphere Σ is not the boundary of any 3-chain in $R^3 - \{0\}$ (of class $\mathscr{C}''$), although $\partial \Sigma = 0$.

The following result will be used in the proof of Theorem 10.39.

10.38 Theorem *Suppose E is a convex open set in R^n, $f \in \mathscr{C}'(E)$, p is an integer, $1 \leq p \leq n$, and*

(116) $$(D_j f)(\mathbf{x}) = 0 \qquad (p < j \leq n, \mathbf{x} \in E).$$

Then there exists an $F \in \mathscr{C}'(E)$ such that

(117) $$(D_p F)(\mathbf{x}) = f(\mathbf{x}), \qquad (D_j F)(\mathbf{x}) = 0 \qquad (p < j \leq n, \mathbf{x} \in E).$$

Proof Write $\mathbf{x} = (\mathbf{x}', x_p, \mathbf{x}'')$, where
$$\mathbf{x}' = (x_1, \ldots, x_{p-1}), \quad \mathbf{x}'' = (x_{p+1}, \ldots, x_n).$$
(When $p = 1$, $\mathbf{x}'$ is absent; when $p = n$, $\mathbf{x}''$ is absent.) Let V be the set of all $(\mathbf{x}', x_p) \in R^p$ such that $(\mathbf{x}', x_p, \mathbf{x}'') \in E$ for some $\mathbf{x}''$. Being a projection of E, V is a convex open set in R^p. Since E is convex and (116) holds, $f(\mathbf{x})$ does not depend on $\mathbf{x}''$. Hence there is a function φ, with domain V, such that
$$f(\mathbf{x}) = \varphi(\mathbf{x}', x_p)$$
for all $\mathbf{x} \in E$.

If $p = 1$, V is a segment in R^1 (possibly unbounded). Pick $c \in V$ and define
$$F(\mathbf{x}) = \int_c^{x_1} \varphi(t)\, dt \qquad (\mathbf{x} \in E).$$

If $p > 1$, let U be the set of all $\mathbf{x}' \in R^{p-1}$ such that $(\mathbf{x}', x_p) \in V$ for some x_p. Then U is a convex open set in R^{p-1}, and there is a function $\alpha \in \mathscr{C}'(U)$ such that $(\mathbf{x}', \alpha(\mathbf{x}')) \in V$ for every $\mathbf{x}' \in U$; in other words, the graph of α lies in V (Exercise 29). Define
$$F(\mathbf{x}) = \int_{\alpha(\mathbf{x}')}^{x_p} \varphi(\mathbf{x}', t)\, dt \qquad (\mathbf{x} \in E).$$

In either case, F satisfies (117).

(*Note:* Recall the usual convention that $\int_a^b$ means $-\int_b^a$ if $b < a$.)

10.39 Theorem *If $E \subset R^n$ is convex and open, if $k \geq 1$, if ω is a k-form of class $\mathscr{C}'$ in E, and if $d\omega = 0$, then there is a $(k-1)$-form λ in E such that $\omega = d\lambda$.*

Briefly, closed forms are exact in convex sets.

Proof For $p = 1, \ldots, n$, let Y_p denote the set of all k-forms ω, of class $\mathscr{C}'$ in E, whose standard presentation

(118) $$\omega = \sum_I f_I(\mathbf{x})\, dx_I$$

does not involve $dx_{p+1}, \ldots, dx_n$. In other words, $I \subset \{1, \ldots, p\}$ if $f_I(\mathbf{x}) \neq 0$ for some $\mathbf{x} \in E$.

We shall proceed by induction on p.

Assume first that $\omega \in Y_1$. Then $\omega = f(\mathbf{x})\, dx_1$. Since $d\omega = 0$, $(D_j f)(\mathbf{x}) = 0$ for $1 < j \leq n$, $\mathbf{x} \in E$. By Theorem 10.38 there is an $F \in \mathscr{C}'(E)$ such that $D_1 F = f$ and $D_j F = 0$ for $1 < j \leq n$. Thus
$$dF = (D_1 F)(\mathbf{x})\, dx_1 = f(\mathbf{x})\, dx_1 = \omega.$$

Now we take $p > 1$ and make the following induction hypothesis: *Every closed k-form that belongs to Y_{p-1} is exact in E.*

Choose $\omega \in Y_p$ so that $d\omega = 0$. By (118),

(119)
$$\sum_I \sum_{j=1}^{n} (D_j f_I)(\mathbf{x})\, dx_j \wedge dx_I = d\omega = 0.$$

Consider a fixed j, with $p < j \leq n$. Each I that occurs in (118) lies in $\{1, \ldots, p\}$. If I_1, I_2 are two of these k-indices, and if $I_1 \neq I_2$, then the $(k+1)$-indices (I_1, j), (I_2, j) are distinct. Thus there is no cancellation, and we conclude from (119) that every coefficient in (118) satisfies

(120)
$$(D_j f_I)(\mathbf{x}) = 0 \qquad (\mathbf{x} \in E,\, p < j \leq n).$$

We now gather those terms in (118) that contain dx_p and rewrite ω in the form

(121)
$$\omega = \alpha + \sum_{I_0} f_I(\mathbf{x})\, dx_{I_0} \wedge dx_p,$$

where $\alpha \in Y_{p-1}$, each I_0 is an increasing $(k-1)$-index in $\{1, \ldots, p-1\}$, and $I = (I_0, p)$. By (120), Theorem 10.38 furnishes functions $F_I \in \mathscr{C}'(E)$ such that

(122)
$$D_p F_I = f_I, \qquad D_j F_I = 0 \qquad (p < j \leq n).$$

Put

(123)
$$\beta = \sum_{I_0} F_I(\mathbf{x})\, dx_{I_0}$$

and define $\gamma = \omega - (-1)^{k-1}\, d\beta$. Since β is a $(k-1)$-form, it follows that

$$\gamma = \omega - \sum_{I_0} \sum_{j=1}^{p} (D_j F_I)(\mathbf{x})\, dx_{I_0} \wedge dx_j$$
$$= \alpha - \sum_{I_0} \sum_{j=1}^{p-1} (D_j F_I)(\mathbf{x})\, dx_{I_0} \wedge dx_j,$$

which is clearly in Y_{p-1}. Since $d\omega = 0$ and $d^2\beta = 0$, we have $d\gamma = 0$. Our induction hypothesis shows therefore that $\gamma = d\mu$ for some $(k-1)$-form μ in E. If $\lambda = \mu + (-1)^{k-1}\beta$, we conclude that $\omega = d\lambda$.

By induction, this completes the proof.

10.40 Theorem *Fix k, $1 \leq k \leq n$. Let $E \subset R^n$ be an open set in which every closed k-form is exact. Let T be a 1-1 $\mathscr{C}''$-mapping of E onto an open set $U \subset R^n$ whose inverse S is also of class $\mathscr{C}''$.*

Then every closed k-form in U is exact in U.

Note that every convex open set E satisfies the present hypothesis, by Theorem 10.39. The relation between E and U may be expressed by saying that they are $\mathscr{C}''$-equivalent.

Thus every closed form is exact in any set which is $\mathscr{C}''$-equivalent to a convex open set.

Proof Let ω be a k-form in U, with $d\omega = 0$. By Theorem 10.22(c), ω_T is a k-form in E for which $d(\omega_T) = 0$. Hence $\omega_T = d\lambda$ for some $(k-1)$-form λ in E. By Theorem 10.23, and another application of Theorem 10.22(c),

$$\omega = (\omega_T)_S = (d\lambda)_S = d(\lambda_S).$$

Since λ_S is a $(k-1)$-form in U, ω is exact in U.

10.41 Remark In applications, cells (see Definition 2.17) are often more convenient parameter domains than simplexes. If our whole development had been based on cells rather than simplexes, the computation that occurs in the proof of Stokes' theorem would be even simpler. (It is done that way in Spivak's book.) The reason for preferring simplexes is that the definition of the boundary of an oriented simplex seems easier and more natural than is the case for a cell. (See Exercise 19.) Also, the partitioning of sets into simplexes (called "triangulation") plays an important role in topology, and there are strong connections between certain aspects of topology, on the one hand, and differential forms, on the other. These are hinted at in Sec. 10.35. The book by Singer and Thorpe contains a good introduction to this topic.

Since every cell can be triangulated, we may regard it as a chain. For dimension 2, this was done in Example 10.32; for dimension 3, see Exercise 18.

Poincaré's lemma (Theorem 10.39) can be proved in several ways. See, for example, page 94 in Spivak's book, or page 280 in Fleming's. Two simple proofs for certain special cases are indicated in Exercises 24 and 27.

VECTOR ANALYSIS

We conclude this chapter with a few applications of the preceding material to theorems concerning vector analysis in R^3. These are special cases of theorems about differential forms, but are usually stated in different terminology. We are thus faced with the job of translating from one language to another.

10.42 Vector fields Let $\mathbf{F} = F_1 \mathbf{e}_1 + F_2 \mathbf{e}_2 + F_3 \mathbf{e}_3$ be a continuous mapping of an open set $E \subset R^3$ into R^3. Since $\mathbf{F}$ associates a vector to each point of E, $\mathbf{F}$ is sometimes called a *vector field*, especially in physics. With every such $\mathbf{F}$ is associated a 1-form

(124) $$\lambda_{\mathbf{F}} = F_1 \, dx + F_2 \, dy + F_3 \, dz$$

and a 2-form

(125) $$\omega_{\mathbf{F}} = F_1 \, dy \wedge dz + F_2 \, dz \wedge dx + F_3 \, dx \wedge dy.$$

Here, and in the rest of this chapter, we use the customary notation (x, y, z) in place of (x_1, x_2, x_3).

It is clear, conversely, that every 1-form λ in E is $\lambda_{\mathbf{F}}$ for some vector field $\mathbf{F}$ in E, and that every 2-form ω is $\omega_{\mathbf{F}}$ for some $\mathbf{F}$. In R^3, the study of 1-forms and 2-forms is thus coextensive with the study of vector fields.

If $u \in \mathscr{C}'(E)$ is a real function, then its *gradient*

$$\nabla u = (D_1 u) \mathbf{e}_1 + (D_2 u) \mathbf{e}_2 + (D_3 u) \mathbf{e}_3$$

is an example of a vector field in E.

Suppose now that $\mathbf{F}$ is a vector field in E, of class $\mathscr{C}'$. Its *curl* $\nabla \times \mathbf{F}$ is the vector field defined in E by

$$\nabla \times \mathbf{F} = (D_2 F_3 - D_3 F_2) \mathbf{e}_1 + (D_3 F_1 - D_1 F_3) \mathbf{e}_2 + (D_1 F_2 - D_2 F_1) \mathbf{e}_3$$

and its *divergence* is the real function $\nabla \cdot \mathbf{F}$ defined in E by

$$\nabla \cdot \mathbf{F} = D_1 F_1 + D_2 F_2 + D_3 F_3.$$

These quantities have various physical interpretations. We refer to the book by O. D. Kellogg for more details.

Here are some relations between gradients, curls, and divergences.

10.43 Theorem *Suppose E is an open set in R^3, $u \in \mathscr{C}''(E)$, and $\mathbf{G}$ is a vector field in E, of class C''.*

(a) *If $\mathbf{F} = \nabla u$, then $\nabla \times \mathbf{F} = \mathbf{0}$.*
(b) *If $\mathbf{F} = \nabla \times \mathbf{G}$, then $\nabla \cdot \mathbf{F} = 0$.*

Furthermore, if E is $\mathscr{C}''$-equivalent to a convex set, then (a) and (b) have converses, in which we assume that $\mathbf{F}$ is a vector field in E, of class $\mathscr{C}'$:

(a') *If $\nabla \times \mathbf{F} = \mathbf{0}$, then $\mathbf{F} = \nabla u$ for some $u \in \mathscr{C}''(E)$.*
(b') *If $\nabla \cdot \mathbf{F} = 0$, then $\mathbf{F} = \nabla \times \mathbf{G}$ for some vector field $\mathbf{G}$ in E, of class $\mathscr{C}''$.*

Proof If we compare the definitions of ∇u, $\nabla \times \mathbf{F}$, and $\nabla \cdot \mathbf{F}$ with the differential forms $\lambda_{\mathbf{F}}$ and $\omega_{\mathbf{F}}$ given by (124) and (125), we obtain the following four statements:

$$\mathbf{F} = \nabla u \quad \text{if and only if} \quad \lambda_\mathbf{F} = du.$$
$$\nabla \times \mathbf{F} = \mathbf{0} \quad \text{if and only if} \quad d\lambda_\mathbf{F} = 0.$$
$$\mathbf{F} = \nabla \times \mathbf{G} \quad \text{if and only if} \quad \omega_\mathbf{F} = d\lambda_\mathbf{G}.$$
$$\nabla \cdot \mathbf{F} = 0 \quad \text{if and only if} \quad d\omega_\mathbf{F} = 0.$$

Now if $\mathbf{F} = \nabla u$, then $\lambda_\mathbf{F} = du$, hence $d\lambda_\mathbf{F} = d^2 u = 0$ (Theorem 10.20), which means that $\nabla \times \mathbf{F} = \mathbf{0}$. Thus (a) is proved.

As regards (a'), the hypothesis amounts to saying that $d\lambda_\mathbf{F} = 0$ in E. By Theorem 10.40, $\lambda_\mathbf{F} = du$ for some 0-form u. Hence $\mathbf{F} = \nabla u$.

The proofs of (b) and (b') follow exactly the same pattern.

10.44 Volume elements The k-form
$$dx_1 \wedge \cdots \wedge dx_k$$
is called the *volume element* in R^k. It is often denoted by dV (or by dV_k if it seems desirable to indicate the dimension explicitly), and the notation

$$(126) \qquad \int_\Phi f(\mathbf{x})\, dx_1 \wedge \cdots \wedge dx_k = \int_\Phi f\, dV$$

is used when Φ is a positively oriented k-surface in R^k and f is a continuous function on the range of Φ.

The reason for using this terminology is very simple: If D is a parameter domain in R^k, and if Φ is a 1-1 $\mathscr{C}'$-mapping of D into R^k, with positive Jacobian J_Φ, then the left side of (126) is

$$\int_D f(\Phi(\mathbf{u})) J_\Phi(\mathbf{u})\, d\mathbf{u} = \int_{\Phi(D)} f(\mathbf{x})\, d\mathbf{x},$$

by (35) and Theorem 10.9.

In particular, when $f = 1$, (126) defines the *volume* of Φ. We already saw a special case of this in (36).

The usual notation for dV_2 is dA.

10.45 Green's theorem *Suppose E is an open set in R^2, $\alpha \in \mathscr{C}'(E)$, $\beta \in \mathscr{C}'(E)$, and Ω is a closed subset of E, with positively oriented boundary $\partial \Omega$, as described in Sec. 10.31. Then*

$$(127) \qquad \int_{\partial\Omega} (\alpha\, dx + \beta\, dy) = \int_\Omega \left(\frac{\partial \beta}{\partial x} - \frac{\partial \alpha}{\partial y} \right) dA.$$

Proof Put $\lambda = \alpha\, dx + \beta\, dy$. Then
$$d\lambda = (D_2\alpha)\, dy \wedge dx + (D_1\beta)\, dx \wedge dy$$
$$= (D_1\beta - D_2\alpha)\, dA,$$

and (127) is the same as
$$\int_{\partial\Omega} \lambda = \int_\Omega d\lambda,$$

which is true by Theorem 10.33.

With $\alpha(x, y) = -y$ and $\beta(x, y) = x$, (127) becomes

(128)
$$\tfrac{1}{2}\int_{\partial\Omega}(x\, dy - y\, dx) = A(\Omega),$$

the area of Ω.

With $\alpha = 0$, $\beta = x$, a similar formula is obtained. Example 10.12(b) contains a special case of this.

10.46 Area elements in R^3 Let Φ be a 2-surface in R^3, of class $\mathscr{C}'$, with parameter domain $D \subset R^2$. Associate with each point $(u, v) \in D$ the vector

(129)
$$\mathbf{N}(u, v) = \frac{\partial(y, z)}{\partial(u, v)}\mathbf{e}_1 + \frac{\partial(z, x)}{\partial(u, v)}\mathbf{e}_2 + \frac{\partial(x, y)}{\partial(u, v)}\mathbf{e}_3.$$

The Jacobians in (129) correspond to the equation

(130)
$$(x, y, z) = \Phi(u, v).$$

If f is a continuous function on $\Phi(D)$, the *area integral* of f over Φ is defined to be

(131)
$$\int_\Phi f\, dA = \int_D f(\Phi(u, v))|\mathbf{N}(u, v)|\, du\, dv.$$

In particular, when $f = 1$ we obtain the *area* of Φ, namely,

(132)
$$A(\Phi) = \int_D |\mathbf{N}(u, v)|\, du\, dv.$$

The following discussion will show that (131) and its special case (132) are reasonable definitions. It will also describe the geometric features of the vector $\mathbf{N}$.

Write $\Phi = \varphi_1 \mathbf{e}_1 + \varphi_2 \mathbf{e}_2 + \varphi_3 \mathbf{e}_3$, fix a point $\mathbf{p}_0 = (u_0, v_0) \in D$, put $\mathbf{N} = \mathbf{N}(\mathbf{p}_0)$, put

(133)
$$\alpha_i = (D_1 \varphi_i)(\mathbf{p}_0), \qquad \beta_i = (D_2 \varphi_i)(\mathbf{p}_0) \qquad (i = 1, 2, 3)$$

and let $T \in L(R^2, R^3)$ be the linear transformation given by

(134) $$T(u, v) = \sum_{i=1}^{3} (\alpha_i u + \beta_i v) \mathbf{e}_i.$$

Note that $T = \Phi'(\mathbf{p}_0)$, in accordance with Definition 9.11.

Let us now assume that the rank of T is 2. (If it is 1 or 0, then $\mathbf{N} = \mathbf{0}$, and the tangent plane mentioned below degenerates to a line or to a point.) The range of the affine mapping

$$(u, v) \to \Phi(\mathbf{p}_0) + T(u, v)$$

is then a plane Π, called the *tangent plane* to Φ at $\mathbf{p}_0$. [One would like to call Π the tangent plane at $\Phi(\mathbf{p}_0)$, rather than at $\mathbf{p}_0$; if Φ is not one-to-one, this runs into difficulties.]

If we use (133) in (129), we obtain

(135) $$\mathbf{N} = (\alpha_2 \beta_3 - \alpha_3 \beta_2)\mathbf{e}_1 + (\alpha_3 \beta_1 - \alpha_1 \beta_3)\mathbf{e}_2 + (\alpha_1 \beta_2 - \alpha_2 \beta_1)\mathbf{e}_3,$$

and (134) shows that

(136) $$T\mathbf{e}_1 = \sum_{i=1}^{3} \alpha_i \mathbf{e}_i, \qquad T\mathbf{e}_2 = \sum_{i=1}^{3} \beta_i \mathbf{e}_i.$$

A straightforward computation now leads to

(137) $$\mathbf{N} \cdot (T\mathbf{e}_1) = 0 = \mathbf{N} \cdot (T\mathbf{e}_2).$$

Hence $\mathbf{N}$ is perpendicular to Π. It is therefore called the *normal to* Φ *at* $\mathbf{p}_0$.

A second property of $\mathbf{N}$, also verified by a direct computation based on (135) and (136), is that the determinant of the linear transformation of R^3 that takes $\{\mathbf{e}_1, \mathbf{e}_2, \mathbf{e}_3\}$ to $\{T\mathbf{e}_1, T\mathbf{e}_2, \mathbf{N}\}$ is $|\mathbf{N}|^2 > 0$ (Exercise 30). The 3-simplex

(138) $$[\mathbf{0}, T\mathbf{e}_1, T\mathbf{e}_2, \mathbf{N}]$$

is thus *positively oriented*.

The third property of $\mathbf{N}$ that we shall use is a consequence of the first two: The above-mentioned determinant, whose value is $|\mathbf{N}|^2$, is the volume of the parallelepiped with edges $[\mathbf{0}, T\mathbf{e}_1]$, $[\mathbf{0}, T\mathbf{e}_2]$, $[\mathbf{0}, \mathbf{N}]$. By (137), $[\mathbf{0}, \mathbf{N}]$ is perpendicular to the other two edges. *The area of the parallelogram with vertices*

(139) $$\mathbf{0}, T\mathbf{e}_1, T\mathbf{e}_2, T(\mathbf{e}_1 + \mathbf{e}_2)$$

is therefore $|\mathbf{N}|$.

This parallelogram is the image under T of the unit square in R^2. If E is any rectangle in R^2, it follows (by the linearity of T) that the area of the parallelogram $T(E)$ is

(140) $$A(T(E)) = |\mathbf{N}| A(E) = \int_E |\mathbf{N}(u_0, v_0)| \, du \, dv.$$

We conclude that (132) is correct when Φ is affine. To justify the definition (132) in the general case, divide D into small rectangles, pick a point (u_0, v_0) in each, and replace Φ in each rectangle by the corresponding tangent plane. The sum of the areas of the resulting parallelograms, obtained via (140), is then an approximation to $A(\Phi)$. Finally, one can justify (131) from (132) by approximating f by step functions.

10.47 Example Let $0 < a < b$ be fixed. Let K be the 3-cell determined by

$$0 \le t \le a, \quad 0 \le u \le 2\pi, \quad 0 \le v \le 2\pi.$$

The equations

(141)
$$\begin{aligned} x &= t \cos u \\ y &= (b + t \sin u) \cos v \\ z &= (b + t \sin u) \sin v \end{aligned}$$

describe a mapping Ψ of R^3 into R^3 which is 1-1 in the interior of K, such that $\Psi(K)$ is a solid torus. Its Jacobian is

$$J_\Psi = \frac{\partial(x, y, z)}{\partial(t, u, v)} = t(b + t \sin u)$$

which is positive on K, except on the face $t = 0$. If we integrate J_Ψ over K, we obtain

$$\text{vol}\,(\Psi(K)) = 2\pi^2 a^2 b$$

as the volume of our solid torus.

Now consider the 2-chain $\Phi = \partial \Psi$. (See Exercise 19.) Ψ maps the faces $u = 0$ and $u = 2\pi$ of K onto the same cylindrical strip, but with opposite orientations. Ψ maps the faces $v = 0$ and $v = 2\pi$ onto the same circular disc, but with opposite orientations. Ψ maps the face $t = 0$ onto a circle, which contributes 0 to the 2-chain $\partial \Psi$. (The relevant Jacobians are 0.) Thus Φ is simply the 2-surface obtained by setting $t = a$ in (141), with parameter domain D the square defined by $0 \le u \le 2\pi$, $0 \le v \le 2\pi$.

According to (129) and (141), the normal to Φ at $(u, v) \in D$ is thus the vector

$$\mathbf{N}(u, v) = a(b + a \sin u)\mathbf{n}(u, v)$$

where

$$\mathbf{n}(u, v) = (\cos u)\mathbf{e}_1 + (\sin u \cos v)\mathbf{e}_2 + (\sin u \sin v)\mathbf{e}_3.$$

Since $|\mathbf{n}(u, v)| = 1$, we have $|\mathbf{N}(u, v)| = a(b + a \sin u)$, and if we integrate this over D, (131) gives

$$A(\Phi) = 4\pi^2 ab$$

as the surface area of our torus.

If we think of $\mathbf{N} = \mathbf{N}(u, v)$ as a directed line segment, pointing from $\Phi(u, v)$ to $\Phi(u, v) + \mathbf{N}(u, v)$, then $\mathbf{N}$ points *outward*, that is to say, away from $\Psi(K)$. This is so because $\mathbf{J}_\Psi > 0$ when $t = a$.

For example, take $u = v = \pi/2$, $t = a$. This gives the largest value of z on $\Psi(K)$, and $\mathbf{N} = a(b + a)\mathbf{e}_3$ points "upward" for this choice of (u, v).

10.48 Integrals of 1-forms in R^3 Let γ be a $\mathscr{C}'$-curve in an open set $E \subset R^3$, with parameter interval $[0, 1]$, let $\mathbf{F}$ be a vector field in E, as in Sec. 10.42, and define $\lambda_\mathbf{F}$ by (124). The integral of $\lambda_\mathbf{F}$ over γ can be rewritten in a certain way which we now describe.

For any $u \in [0, 1]$,

$$\gamma'(u) = \gamma_1'(u)\mathbf{e}_1 + \gamma_2'(u)\mathbf{e}_2 + \gamma_3'(u)\mathbf{e}_3$$

is called the *tangent vector* to γ at u. We define $\mathbf{t} = \mathbf{t}(u)$ to be the unit vector in the direction of $\gamma'(u)$. Thus

$$\gamma'(u) = |\gamma'(u)|\mathbf{t}(u).$$

[If $\gamma'(u) = \mathbf{0}$ for some u, put $\mathbf{t}(u) = \mathbf{e}_1$; any other choice would do just as well.] By (35),

(142)
$$\int_\gamma \lambda_\mathbf{F} = \sum_{i=1}^{3} \int_0^1 F_i(\gamma(u))\gamma_i'(u)\, du$$
$$= \int_0^1 \mathbf{F}(\gamma(u)) \cdot \gamma'(u)\, du$$
$$= \int_0^1 \mathbf{F}(\gamma(u)) \cdot \mathbf{t}(u)|\gamma'(u)|\, du.$$

Theorem 6.27 makes it reasonable to call $|\gamma'(u)|\, du$ the *element of arc length along* γ. A customary notation for it is ds, and (142) is rewritten in the form

(143)
$$\int_\gamma \lambda_\mathbf{F} = \int_\gamma (\mathbf{F} \cdot \mathbf{t})\, ds.$$

Since $\mathbf{t}$ is a unit tangent vector to γ, $\mathbf{F} \cdot \mathbf{t}$ is called the *tangential component* of $\mathbf{F}$ along γ.

The right side of (143) should be regarded as just an abbreviation for the last integral in (142). The point is that **F** is defined on the range of γ, but **t** is defined on [0, 1]; thus $\mathbf{F} \cdot \mathbf{t}$ has to be properly interpreted. Of course, when γ is one-to-one, then $\mathbf{t}(u)$ can be replaced by $\mathbf{t}(\gamma(u))$, and this difficulty disappears.

10.49 Integrals of 2-forms in R^3 Let Φ be a 2-surface in an open set $E \subset R^3$, of class $\mathscr{C}'$, with parameter domain $D \subset R^2$. Let **F** be a vector field in E, and define $\omega_\mathbf{F}$ by (125). As in the preceding section, we shall obtain a different representation of the integral of $\omega_\mathbf{F}$ over Φ.

By (35) and (129),

$$\int_\Phi \omega_\mathbf{F} = \int_\Phi (F_1 \, dy \wedge dz + F_2 \, dz \wedge dx + F_3 \, dx \wedge dy)$$

$$= \int_D \left\{ (F_1 \circ \Phi) \frac{\partial(y, z)}{\partial(u, v)} + (F_2 \circ \Phi) \frac{\partial(z, x)}{\partial(u, v)} + (F_3 \circ \Phi) \frac{\partial(x, y)}{\partial(u, v)} \right\} du \, dv$$

$$= \int_D \mathbf{F}(\Phi(u, v)) \cdot \mathbf{N}(u, v) \, du \, dv.$$

Now let $\mathbf{n} = \mathbf{n}(u, v)$ be the unit vector in the direction of $\mathbf{N}(u, v)$. [If $\mathbf{N}(u, v) = \mathbf{0}$ for some $(u, v) \in D$, take $\mathbf{n}(u, v) = \mathbf{e}_1$.] Then $\mathbf{N} = |\mathbf{N}|\mathbf{n}$, and therefore the last integral becomes

$$\int_D \mathbf{F}(\Phi(u, v)) \cdot \mathbf{n}(u, v) |\mathbf{N}(u, v)| \, du \, dv.$$

By (131), we can finally write this in the form

(144) $$\int_\Phi \omega_\mathbf{F} = \int_\Phi (\mathbf{F} \cdot \mathbf{n}) \, dA.$$

With regard to the meaning of $\mathbf{F} \cdot \mathbf{n}$, the remark made at the end of Sec. 10.48 applies here as well.

We can now state the original form of Stokes' theorem.

10.50 Stokes' formula *If $\mathbf{F}$ is a vector field of class $\mathscr{C}'$ in an open set $E \subset R^3$, and if Φ is a 2-surface of class $\mathscr{C}''$ in E, then*

(145) $$\int_\Phi (\nabla \times \mathbf{F}) \cdot \mathbf{n} \, dA = \int_{\partial \Phi} (\mathbf{F} \cdot \mathbf{t}) \, ds.$$

Proof Put $\mathbf{H} = \nabla \times \mathbf{F}$. Then, as in the proof of Theorem 10.43, we have

(146) $$\omega_\mathbf{H} = d\lambda_\mathbf{F}.$$

Hence
$$\int_\Phi (\nabla \times \mathbf{F}) \cdot \mathbf{n}\, dA = \int_\Phi (\mathbf{H} \cdot \mathbf{n})\, dA = \int_\Phi \omega_\mathbf{H}$$
$$= \int_\Phi d\lambda_\mathbf{F} = \int_{\partial\Phi} \lambda_\mathbf{F} = \int_{\partial\Phi} (\mathbf{F} \cdot \mathbf{t})\, ds.$$

Here we used the definition of $\mathbf{H}$, then (144) with $\mathbf{H}$ in place of $\mathbf{F}$, then (146), then—the main step—Theorem 10.33, and finally (143), extended in the obvious way from curves to 1-chains.

10.51 The divergence theorem *If $\mathbf{F}$ is a vector field of class $\mathscr{C}'$ in an open set $E \subset R^3$, and if Ω is a closed subset of E with positively oriented boundary $\partial\Omega$ (as described in Sec. 10.31) then*

(147) $$\int_\Omega (\nabla \cdot \mathbf{F})\, dV = \int_{\partial\Omega} (\mathbf{F} \cdot \mathbf{n})\, dA.$$

Proof By (125),
$$d\omega_\mathbf{F} = (\nabla \cdot \mathbf{F})\, dx \wedge dy \wedge dz = (\nabla \cdot \mathbf{F})\, dV.$$

Hence
$$\int_\Omega (\nabla \cdot \mathbf{F})\, dV = \int_\Omega d\omega_\mathbf{F} = \int_{\partial\Omega} \omega_\mathbf{F} = \int_{\partial\Omega} (\mathbf{F} \cdot \mathbf{n})\, dA,$$

by Theorem 10.33, applied to the 2-form $\omega_\mathbf{F}$, and (144).

EXERCISES

1. Let H be a compact convex set in R^k, with nonempty interior. Let $f \in \mathscr{C}(H)$, put $f(\mathbf{x}) = 0$ in the complement of H, and define $\int_H f$ as in Definition 10.3.

    Prove that $\int_H f$ is independent of the order in which the k integrations are carried out.

    Hint: Approximate f by functions that are continuous on R^k and whose supports are in H, as was done in Example 10.4.

2. For $i = 1, 2, 3, \ldots$, let $\varphi_i \in \mathscr{C}(R^1)$ have support in $(2^{-i}, 2^{1-i})$, such that $\int \varphi_i = 1$. Put
$$f(x, y) = \sum_{i=1}^\infty [\varphi_i(x) - \varphi_{i+1}(x)]\varphi_i(y)$$

    Then f has compact support in R^2, f is continuous except at $(0, 0)$, and
$$\int dy \int f(x, y)\, dx = 0 \quad \text{but} \quad \int dx \int f(x, y)\, dy = 1.$$

    Observe that f is unbounded in every neighborhood of $(0, 0)$.

3. (a) If F is as in Theorem 10.7, put $\mathbf{A} = \mathbf{F}'(0)$, $\mathbf{F}_1(\mathbf{x}) = \mathbf{A}^{-1}\mathbf{F}(\mathbf{x})$. Then $\mathbf{F}_1'(0) = I$. Show that
$$\mathbf{F}_1(\mathbf{x}) = \mathbf{G}_n \circ \mathbf{G}_{n-1} \circ \cdots \circ \mathbf{G}_1(\mathbf{x})$$
in some neighborhood of $\mathbf{0}$, for certain primitive mappings $\mathbf{G}_1, \ldots, \mathbf{G}_n$. This gives another version of Theorem 10.7:
$$\mathbf{F}(\mathbf{x}) = \mathbf{F}'(0)\mathbf{G}_n \circ \mathbf{G}_{n-1} \circ \cdots \circ \mathbf{G}_1(\mathbf{x}).$$
(b) Prove that the mapping $(x, y) \to (y, x)$ of R^2 onto R^2 is not the composition of any two primitive mappings, in any neighborhood of the origin. (This shows that the flips B_i cannot be omitted from the statement of Theorem 10.7.)

4. For $(x, y) \in R^2$, define
$$\mathbf{F}(x, y) = (e^x \cos y - 1, e^x \sin y).$$
Prove that $\mathbf{F} = \mathbf{G}_2 \circ \mathbf{G}_1$, where
$$\mathbf{G}_1(x, y) = (e^x \cos y - 1, y)$$
$$\mathbf{G}_2(u, v) = (u, (1 + u) \tan v)$$
are primitive in some neighborhood of $(0, 0)$.

Compute the Jacobians of $\mathbf{G}_1, \mathbf{G}_2, \mathbf{F}$ at $(0, 0)$. Define
$$\mathbf{H}_2(x, y) = (x, e^x \sin y)$$
and find
$$\mathbf{H}_1(u, v) = (h(u, v), v)$$
so that $\mathbf{F} = \mathbf{H}_1 \circ \mathbf{H}_2$ is some neighborhood of $(0, 0)$.

5. Formulate and prove an analogue of Theorem 10.8, in which K is a compact subset of an arbitrary metric space. (Replace the functions φ_i that occur in the proof of Theorem 10.8 by functions of the type constructed in Exercise 22 of Chap. 4.)

6. Strengthen the conclusion of Theorem 10.8 by showing that the functions ψ_i can be made differentiable, and even infinitely differentiable. (Use Exercise 1 of Chap. 8 in the construction of the auxiliary functions φ_i.)

7. (a) Show that the simplex Q^k is the smallest convex subset of R^k that contains $0, e_1, \ldots, e_k$.
(b) Show that affine mappings take convex sets to convex sets.

8. Let H be the parallelogram in R^2 whose vertices are $(1, 1), (3, 2), (4, 5), (2, 4)$. Find the affine map T which sends $(0, 0)$ to $(1, 1)$, $(1, 0)$ to $(3, 2)$, $(0, 1)$ to $(2, 4)$. Show that $J_T = 5$. Use T to convert the integral
$$\alpha = \int_H e^{x-y} \, dx \, dy$$
to an integral over I^2 and thus compute α.

9. Define $(x, y) = T(r, \theta)$ on the rectangle
$$0 \le r \le a, \qquad 0 \le \theta \le 2\pi$$
by the equations
$$x = r \cos \theta, \qquad y = r \sin \theta.$$
Show that T maps this rectangle onto the closed disc D with center at $(0, 0)$ and radius a, that T is one-to-one in the interior of the rectangle, and that $J_T(r, \theta) = r$. If $f \in \mathscr{C}(D)$, prove the formula for integration in polar coordinates:
$$\int_D f(x, y) \, dx \, dy = \int_0^a \int_0^{2\pi} f(T(r, \theta)) r \, dr \, d\theta.$$

Hint: Let D_0 be the interior of D, minus the interval from $(0, 0)$ to $(0, a)$. As it stands, Theorem 10.9 applies to continuous functions f whose support lies in D_0. To remove this restriction, proceed as in Example 10.4.

10. Let $a \to \infty$ in Exercise 9 and prove that
$$\int_{R^2} f(x, y) \, dx \, dy = \int_0^{\infty} \int_0^{2\pi} f(T(r, \theta)) r \, dr \, d\theta,$$
for continuous functions f that decrease sufficiently rapidly as $|x| + |y| \to \infty$. (Find a more precise formulation.) Apply this to
$$f(x, y) = \exp(-x^2 - y^2)$$
to derive formula (101) of Chap. 8.

11. Define $(u, v) = T(s, t)$ on the strip
$$0 < s < \infty, \qquad 0 < t < 1$$
by setting $u = s - st$, $v = st$. Show that T is a 1-1 mapping of the strip onto the positive quadrant Q in R^2. Show that $J_T(s, t) = s$.

For $x > 0, y > 0$, integrate
$$u^{x-1} e^{-u} v^{y-1} e^{-v}$$
over Q, use Theorem 10.9 to convert the integral to one over the strip, and derive formula (96) of Chap. 8 in this way.

(For this application, Theorem 10.9 has to be extended so as to cover certain improper integrals. Provide this extension.)

12. Let I^k be the set of all $\mathbf{u} = (u_1, \ldots, u_k) \in R^k$ with $0 \le u_i \le 1$ for all i; let Q^k be the set of all $\mathbf{x} = (x_1, \ldots, x_k) \in R^k$ with $x_i \ge 0, \Sigma x_i \le 1$. ($I^k$ is the unit cube; Q^k is the standard simplex in R^k.) Define $\mathbf{x} = T(\mathbf{u})$ by
$$x_1 = u_1$$
$$x_2 = (1 - u_1) u_2$$
$$\cdots\cdots\cdots\cdots\cdots\cdots\cdots$$
$$x_k = (1 - u_1) \cdots (1 - u_{k-1}) u_k.$$

Show that

$$\sum_{i=1}^{k} x_i = 1 - \prod_{i=1}^{k}(1 - u_i).$$

Show that T maps I^k onto Q^k, that T is 1-1 in the interior of I^k, and that its inverse S is defined in the interior of Q^k by $u_1 = x_1$ and

$$u_i = \frac{x_i}{1 - x_1 - \cdots - x_{i-1}}$$

for $i = 2, \ldots, k$. Show that

$$J_T(\mathbf{u}) = (1 - u_1)^{k-1}(1 - u_2)^{k-2} \cdots (1 - u_{k-1}),$$

and

$$J_S(\mathbf{x}) = [(1 - x_1)(1 - x_1 - x_2) \cdots (1 - x_1 - \cdots - x_{k-1})]^{-1}.$$

13. Let $r_1, \ldots, r_k$ be nonnegative integers, and prove that

$$\int_{Q^k} x_1^{r_1} \cdots x_k^{r_k} \, dx = \frac{r_1! \cdots r_k!}{(k + r_1 + \cdots + r_k)!}.$$

    Hint: Use Exercise 12, Theorems 10.9 and 8.20.

    Note that the special case $r_1 = \cdots = r_k = 0$ shows that the volume of Q^k is $1/k!$.

14. Prove formula (46).

15. If ω and λ are k- and m-forms, respectively, prove that

$$\omega \wedge \lambda = (-1)^{km} \lambda \wedge \omega.$$

16. If $k \geq 2$ and $\sigma = [\mathbf{p}_0, \mathbf{p}_1, \ldots, \mathbf{p}_k]$ is an oriented affine k-simplex, prove that $\partial^2 \sigma = 0$, directly from the definition of the boundary operator ∂. Deduce from this that $\partial^2 \Psi = 0$ for every chain Ψ.

    Hint: For orientation, do it first for $k = 2$, $k = 3$. In general, if $i < j$, let σ_{ij} be the $(k - 2)$-simplex obtained by deleting $\mathbf{p}_i$ and $\mathbf{p}_j$ from σ. Show that each σ_{ij} occurs twice in $\partial^2 \sigma$, with opposite sign.

17. Put $J^2 = \tau_1 + \tau_2$, where

$$\tau_1 = [0, \mathbf{e}_1, \mathbf{e}_1 + \mathbf{e}_2], \qquad \tau_2 = -[0, \mathbf{e}_2, \mathbf{e}_2 + \mathbf{e}_1].$$

    Explain why it is reasonable to call J^2 the positively oriented unit square in R^2. Show that ∂J^2 is the sum of 4 oriented affine 1-simplexes. Find these. What is $\partial(\tau_1 - \tau_2)$?

18. Consider the oriented affine 3-simplex

$$\sigma_1 = [0, \mathbf{e}_1, \mathbf{e}_1 + \mathbf{e}_2, \mathbf{e}_1 + \mathbf{e}_2 + \mathbf{e}_3]$$

    in R^3. Show that σ_1 (regarded as a linear transformation) has determinant 1. Thus σ_1 is positively oriented.

Let $\sigma_2, \ldots, \sigma_6$ be five other oriented 3-simplexes, obtained as follows: There are five permutations (i_1, i_2, i_3) of $(1, 2, 3)$, distinct from $(1, 2, 3)$. Associate with each (i_1, i_2, i_3) the simplex

$$s(i_1, i_2, i_3)[0, e_{i_1}, e_{i_1} + e_{i_2}, e_{i_1} + e_{i_2} + e_{i_3}]$$

where s is the sign that occurs in the definition of the determinant. (This is how τ_2 was obtained from τ_1 in Exercise 17.)

Show that $\sigma_2, \ldots, \sigma_6$ are positively oriented.

Put $J^3 = \sigma_1 + \cdots + \sigma_6$. Then J^3 may be called the positively oriented unit cube in R^3.

Show that ∂J^3 is the sum of 12 oriented affine 2-simplexes. (These 12 triangles cover the surface of the unit cube I^3.)

Show that $\mathbf{x} = (x_1, x_2, x_3)$ is in the range of σ_1 if and only if $0 \le x_3 \le x_2 \le x_1 \le 1$.

Show that the ranges of $\sigma_1, \ldots, \sigma_6$ have disjoint interiors, and that their union covers I^3. (Compare with Exercise 13; note that $3! = 6$.)

19. Let J^2 and J^3 be as in Exercise 17 and 18. Define

$$B_{01}(u, v) = (0, u, v), \quad B_{11}(u, v) = (1, u, v),$$
$$B_{02}(u, v) = (u, 0, v), \quad B_{12}(u, v) = (u, 1, v),$$
$$B_{03}(u, v) = (u, v, 0), \quad B_{13}(u, v) = (u, v, 1).$$

These are affine, and map R^2 into R^3.

Put $\beta_{ri} = B_{ri}(J^2)$, for $r = 0, 1, i = 1, 2, 3$. Each β_{ri} is an affine-oriented 2-chain. (See Sec. 10.30.) Verify that

$$\partial J^3 = \sum_{i=1}^{3} (-1)^i (\beta_{0i} - \beta_{1i}),$$

in agreement with Exercise 18.

20. State conditions under which the formula

$$\int_\Phi f\, d\omega = \int_{\partial \Phi} f\omega - \int_\Phi (df) \wedge \omega$$

is valid, and show that it generalizes the formula for integration by parts.
  Hint: $d(f\omega) = (df) \wedge \omega + f\, d\omega$.

21. As in Example 10.36, consider the 1-form

$$\eta = \frac{x\, dy - y\, dx}{x^2 + y^2}$$

in $R^2 - \{0\}$.

(a) Carry out the computation that leads to formula (113), and prove that $d\eta = 0$.

(b) Let $\gamma(t) = (r \cos t, r \sin t)$, for some $r > 0$, and let Γ be a $\mathscr{C}''$-curve in $R^2 - \{0\}$, with parameter interval $[0, 2\pi]$, with $\Gamma(0) = \Gamma(2\pi)$, such that the intervals $[\gamma(t), \Gamma(t)]$ do not contain $\mathbf{0}$ for any $t \in [0, 2\pi]$. Prove that

$$\int_\Gamma \eta = 2\pi.$$

Hint: For $0 \le t \le 2\pi$, $0 \le u \le 1$, define

$$\Phi(t, u) = (1 - u)\Gamma(t) + u\gamma(t).$$

Then Φ is a 2-surface in $R^2 - \{0\}$ whose parameter domain is the indicated rectangle. Because of cancellations (as in Example 10.32),

$$\partial \Phi = \Gamma - \gamma.$$

Use Stokes' theorem to deduce that

$$\int_\Gamma \eta = \int_\gamma \eta$$

because $d\eta = 0$.

(c) Take $\Gamma(t) = (a \cos t, b \sin t)$ where $a > 0$, $b > 0$ are fixed. Use part (b) to show that

$$\int_0^{2\pi} \frac{ab}{a^2 \cos^2 t + b^2 \sin^2 t} dt = 2\pi.$$

(d) Show that

$$\eta = d\left(\arc\tan \frac{y}{x}\right)$$

in any convex open set in which $x \ne 0$, and that

$$\eta = d\left(-\arc\tan \frac{x}{y}\right)$$

in any convex open set in which $y \ne 0$.

Explain why this justifies the notation $\eta = d\theta$, in spite of the fact that η is not exact in $R^2 - \{0\}$.

(e) Show that (b) can be derived from (d).

(f) If Γ is any closed $\mathscr{C}'$-curve in $R^2 - \{0\}$, prove that

$$\frac{1}{2\pi} \int_\Gamma \eta = \text{Ind}(\Gamma).$$

(See Exercise 23 of Chap. 8 for the definition of the index of a curve.)

22. As in Example 10.37, define ζ in $R^3 - \{0\}$ by

$$\zeta = \frac{x\, dy \wedge dz + y\, dz \wedge dx + z\, dx \wedge dy}{r^3}$$

where $r = (x^2 + y^2 + z^2)^{1/2}$, let D be the rectangle given by $0 \le u \le \pi$, $0 \le v \le 2\pi$, and let Σ be the 2-surface in R^3, with parameter domain D, given by

$$x = \sin u \cos v, \qquad y = \sin u \sin v, \qquad z = \cos u.$$

(a) Prove that $d\zeta = 0$ in $R^3 - \{0\}$.

(b) Let S denote the restriction of Σ to a parameter domain $E \subset D$. Prove that

$$\int_S \zeta = \int_E \sin u\, du\, dv = A(S),$$

where A denotes area, as in Sec. 10.43. Note that this contains (115) as a special case.

(c) Suppose g, h_1, h_2, h_3, are $\mathscr{C}''$-functions on $[0, 1]$, $g > 0$. Let $(x, y, z) = \Phi(s, t)$ define a 2-surface Φ, with parameter domain I^2, by

$$x = g(t)h_1(s), \qquad y = g(t)h_2(s), \qquad z = g(t)h_3(s).$$

Prove that

$$\int_\Phi \zeta = 0,$$

directly from (35).

Note the shape of the range of Φ: For fixed s, $\Phi(s, t)$ runs over an interval on a line through $\mathbf{0}$. The range of Φ thus lies in a "cone" with vertex at the origin.

(d) Let E be a closed rectangle in D, with edges parallel to those of D. Suppose $f \in \mathscr{C}''(D), f > 0$. Let Ω be the 2-surface with parameter domain E, defined by

$$\Omega(u, v) = f(u, v)\, \Sigma(u, v).$$

Define S as in (b) and prove that

$$\int_\Omega \zeta = \int_S \zeta = A(S).$$

(Since S is the "radial projection" of Ω into the unit sphere, this result makes it reasonable to call $\int_\Omega \zeta$ the "solid angle" subtended by the range of Ω at the origin.)

Hint: Consider the 3-surface Ψ given by

$$\Psi(t, u, v) = [1 - t + tf(u, v)]\, \Sigma(u, v),$$

where $(u, v) \in E$, $0 \le t \le 1$. For fixed v, the mapping $(t, u) \to \Psi(t, u, v)$ is a 2-sur-

face Φ to which (c) can be applied to show that $\int_\Phi \zeta = 0$. The same thing holds when u is fixed. By (a) and Stokes' theorem,
$$\int_{\partial \Psi} \zeta = \int_\Psi d\zeta = 0.$$

(e) Put $\lambda = -(z/r)\eta$, where
$$\eta = \frac{x\, dy - y\, dx}{x^2 + y^2},$$
as in Exercise 21. Then λ is a 1-form in the open set $V \subset R^3$ in which $x^2 + y^2 > 0$. Show that ζ is *exact in V* by showing that
$$\zeta = d\lambda.$$

(f) Derive (d) from (e), without using (c).

Hint: To begin with, assume $0 < u < \pi$ on E. By (e),
$$\int_\Omega \zeta = \int_{\partial \Omega} \lambda \quad \text{and} \quad \int_S \zeta = \int_{\partial S} \lambda.$$

Show that the two integrals of λ are equal, by using part (d) of Exercise 21, and by noting that z/r is the same at $\Sigma(u, v)$ as at $\Omega(u, v)$.

(g) Is ζ exact in the complement of every line through the origin?

23. Fix n. Define $r_k = (x_1^2 + \cdots + x_k^2)^{1/2}$ for $1 \leq k \leq n$, let E_k be the set of all $\mathbf{x} \in R^n$ at which $r_k > 0$, and let ω_k be the $(k-1)$-form defined in E_k by
$$\omega_k = (r_k)^{-k} \sum_{i=1}^k (-1)^{i-1} x_i \, dx_1 \wedge \cdots \wedge dx_{i-1} \wedge dx_{i+1} \wedge \cdots \wedge dx_k.$$

Note that $\omega_2 = \eta$, $\omega_3 = \zeta$, in the terminology of Exercises 21 and 22. Note also that
$$E_1 \subset E_2 \subset \cdots \subset E_n = R^n - \{0\}.$$

(a) Prove that $d\omega_k = 0$ in E_k.

(b) For $k = 2, \ldots, n$, prove that ω_k is exact in E_{k-1}, by showing that
$$\omega_k = d(f_k \omega_{k-1}) = (df_k) \wedge \omega_{k-1},$$
where $f_k(\mathbf{x}) = (-1)^k g_k(x_k/r_k)$ and
$$g_k(t) = \int_{-1}^t (1 - s^2)^{(k-3)/2} \, ds \qquad (-1 < t < 1).$$

Hint: f_k satisfies the differential equations
$$\mathbf{x} \cdot (\nabla f_k)(\mathbf{x}) = 0$$
and
$$(D_k f_k)(\mathbf{x}) = \frac{(-1)^k (r_{k-1})^{k-1}}{(r_k)^k}.$$

(c) Is ω_n exact in E_n?

(d) Note that (b) is a generalization of part (e) of Exercise 22. Try to extend some of the other assertions of Exercises 21 and 22 to ω_n, for arbitrary n.

24. Let $\omega = \Sigma a_i(\mathbf{x}) \, dx_i$ be a 1-form of class $\mathscr{C}''$ in a convex open set $E \subset R^n$. Assume $d\omega = 0$ and prove that ω is exact in E, by completing the following outline:

Fix $\mathbf{p} \in E$. Define
$$f(\mathbf{x}) = \int_{[\mathbf{p},\mathbf{x}]} \omega \qquad (\mathbf{x} \in E).$$

Apply Stokes' theorem to affine-oriented 2-simplexes $[\mathbf{p}, \mathbf{x}, \mathbf{y}]$ in E. Deduce that
$$f(\mathbf{y}) - f(\mathbf{x}) = \sum_{i=1}^{n} (y_i - x_i) \int_0^1 a_i((1-t)\mathbf{x} + t\mathbf{y}) \, dt$$
for $\mathbf{x} \in E$, $\mathbf{y} \in E$. Hence $(D_i f)(\mathbf{x}) = a_i(\mathbf{x})$.

25. Assume that ω is a 1-form in an open set $E \subset R^n$ such that
$$\int_\gamma \omega = 0$$
for every closed curve γ in E, of class $\mathscr{C}'$. Prove that ω is exact in E, by imitating part of the argument sketched in Exercise 24.

26. Assume ω is a 1-form in $R^3 - \{0\}$, of class $\mathscr{C}'$ and $d\omega = 0$. Prove that ω is exact in $R^3 - \{0\}$.

Hint: Every closed continuously differentiable curve in $R^3 - \{0\}$ is the boundary of a 2-surface in $R^3 - \{0\}$. Apply Stokes' theorem and Exercise 25.

27. Let E be an open 3-cell in R^3, with edges parallel to the coordinate axes. Suppose $(a, b, c) \in E$, $f_i \in \mathscr{C}'(E)$ for $i = 1, 2, 3$,
$$\omega = f_1 \, dy \wedge dz + f_2 \, dz \wedge dx + f_3 \, dx \wedge dy,$$
and assume that $d\omega = 0$ in E. Define
$$\lambda = g_1 \, dx + g_2 \, dy$$
where
$$g_1(x, y, z) = \int_c^z f_2(x, y, s) \, ds - \int_b^y f_3(x, t, c) \, dt$$
$$g_2(x, y, z) = -\int_c^z f_1(x, y, s) \, ds,$$
for $(x, y, z) \in E$. Prove that $d\lambda = \omega$ in E.

Evaluate these integrals when $\omega = \zeta$ and thus find the form λ that occurs in part (e) of Exercise 22.

28. Fix $b > a > 0$, define

$$\Phi(r, \theta) = (r\cos\theta, r\sin\theta)$$

for $a \leq r \leq b$, $0 \leq \theta \leq 2\pi$. (The range of Φ is an annulus in R^2.) Put $\omega = x^3\,dy$, and compute both

$$\int_\Phi d\omega \quad \text{and} \quad \int_{\partial\Phi} \omega$$

to verify that they are equal.

29. Prove the existence of a function α with the properties needed in the proof of Theorem 10.38, and prove that the resulting function F is of class $\mathscr{C}'$. (Both assertions become trivial if E is an open cell or an open ball, since α can then be taken to be a constant. Refer to Theorem 9.42.)

30. If $\mathbf{N}$ is the vector given by (135), prove that

$$\det \begin{bmatrix} \alpha_1 & \beta_1 & \alpha_2\beta_3 - \alpha_3\beta_2 \\ \alpha_2 & \beta_2 & \alpha_3\beta_1 - \alpha_1\beta_3 \\ \alpha_3 & \beta_3 & \alpha_1\beta_2 - \alpha_2\beta_1 \end{bmatrix} = |\mathbf{N}|^2.$$

Also, verify Eq. (137).

31. Let $E \subset R^3$ be open, suppose $g \in \mathscr{C}''(E)$, $h \in \mathscr{C}''(E)$, and consider the vector field

$$\mathbf{F} = g\,\nabla h.$$

(*a*) Prove that

$$\nabla \cdot \mathbf{F} = g\,\nabla^2 h + (\nabla g) \cdot (\nabla h)$$

where $\nabla^2 h = \nabla \cdot (\nabla h) = \Sigma \partial^2 h/\partial x_i^2$ is the so-called "Laplacian" of h.

(*b*) If Ω is a closed subset of E with positively oriented boundary $\partial\Omega$ (as in Theorem 10.51), prove that

$$\int_\Omega [g\,\nabla^2 h + (\nabla g) \cdot (\nabla h)]\,dV = \int_{\partial\Omega} g\frac{\partial h}{\partial n}\,dA$$

where (as is customary) we have written $\partial h/\partial n$ in place of $(\nabla h) \cdot \mathbf{n}$. (Thus $\partial h/\partial n$ is the directional derivative of h in the direction of the outward normal to $\partial\Omega$, the so-called *normal derivative* of h.) Interchange g and h, subtract the resulting formula from the first one, to obtain

$$\int_\Omega (g\,\nabla^2 h - h\,\nabla^2 g)\,dV = \int_{\partial\Omega} \left(g\frac{\partial h}{\partial n} - h\frac{dg}{\partial n} \right) dA.$$

These two formulas are usually called *Green's identities*.

(*c*) Assume that h is *harmonic* in E; this means that $\nabla^2 h = 0$. Take $g = 1$ and conclude that

$$\int_{\partial\Omega} \frac{\partial h}{\partial n}\,dA = 0.$$

Take $g = h$, and conclude that $h = 0$ in Ω if $h = 0$ on $\partial\Omega$.

(d) Show that Green's identities are also valid in R^2.

32. Fix δ, $0 < \delta < 1$. Let D be the set of all $(\theta, t) \in R^2$ such that $0 \leq \theta \leq \pi$, $-\delta \leq t \leq \delta$. Let Φ be the 2-surface in R^3, with parameter domain D, given by

$$x = (1 - t \sin \theta) \cos 2\theta$$
$$y = (1 - t \sin \theta) \sin 2\theta$$
$$z = t \cos \theta$$

where $(x, y, z) = \Phi(\theta, t)$. Note that $\Phi(\pi, t) = \Phi(0, -t)$, and that Φ is one-to-one on the rest of D.

The range $M = \Phi(D)$ of Φ is known as a *Möbius band*. It is the simplest example of a nonorientable surface.

Prove the various assertions made in the following description: Put $\mathbf{p}_1 = (0, -\delta)$, $\mathbf{p}_2 = (\pi, -\delta)$, $\mathbf{p}_3 = (\pi, \delta)$, $\mathbf{p}_4 = (0, \delta)$, $\mathbf{p}_5 = \mathbf{p}_1$. Put $\gamma_i = [\mathbf{p}_i, \mathbf{p}_{i+1}]$, $i = 1, \ldots, 4$, and put $\Gamma_i = \Phi \circ \gamma_i$. Then

$$\partial \Phi = \Gamma_1 + \Gamma_2 + \Gamma_3 + \Gamma_4.$$

Put $\mathbf{a} = (1, 0, -\delta)$, $\mathbf{b} = (1, 0, \delta)$. Then

$$\Phi(\mathbf{p}_1) = \Phi(\mathbf{p}_3) = \mathbf{a}, \qquad \Phi(\mathbf{p}_2) = \Phi(\mathbf{p}_4) = \mathbf{b},$$

and $\partial \Phi$ can be described as follows.

Γ_1 spirals up from $\mathbf{a}$ to $\mathbf{b}$; its projection into the (x, y)-plane has winding number $+1$ around the origin. (See Exercise 23, Chap. 8.)

$\Gamma_2 = [\mathbf{b}, \mathbf{a}]$.

Γ_3 spirals up from $\mathbf{a}$ to $\mathbf{b}$; its projection into the (x, y) plane has winding number -1 around the origin.

$\Gamma_4 = [\mathbf{b}, \mathbf{a}]$.

Thus $\partial \Phi = \Gamma_1 + \Gamma_3 + 2\Gamma_2$.

If we go from $\mathbf{a}$ to $\mathbf{b}$ along Γ_1 and continue along the "edge" of M until we return to $\mathbf{a}$, the curve traced out is

$$\Gamma = \Gamma_1 - \Gamma_3,$$

which may also be represented on the parameter interval $[0, 2\pi]$ by the equations

$$x = (1 + \delta \sin \theta) \cos 2\theta$$
$$y = (1 + \delta \sin \theta) \sin 2\theta$$
$$z = -\delta \cos \theta.$$

It should be emphasized that $\Gamma \neq \partial \Phi$: Let η be the 1-form discussed in Exercises 21 and 22. Since $d\eta = 0$, Stokes' theorem shows that

$$\int_{\partial \Phi} \eta = 0.$$

But although Γ is the "geometric" boundary of M, we have

$$\int_\Gamma \eta = 4\pi.$$

In order to avoid this possible source of confusion, Stokes' formula (Theorem 10.50) is frequently stated only for orientable surfaces Φ.

11

THE LEBESGUE THEORY

It is the purpose of this chapter to present the fundamental concepts of the Lebesgue theory of measure and integration and to prove some of the crucial theorems in a rather general setting, without obscuring the main lines of the development by a mass of comparatively trivial detail. Therefore proofs are only sketched in some cases, and some of the easier propositions are stated without proof. However, the reader who has become familiar with the techniques used in the preceding chapters will certainly find no difficulty in supplying the missing steps.

The theory of the Lebesgue integral can be developed in several distinct ways. Only one of these methods will be discussed here. For alternative procedures we refer to the more specialized treatises on integration listed in the Bibliography.

SET FUNCTIONS

If A and B are any two sets, we write $A - B$ for the set of all elements x such that $x \in A$, $x \notin B$. The notation $A - B$ does not imply that $B \subset A$. We denote the empty set by 0, and say that A and B are disjoint if $A \cap B = 0$.

11.1 Definition A family $\mathscr{R}$ of sets is called a *ring* if $A \in \mathscr{R}$ and $B \in \mathscr{R}$ implies

(1) $$A \cup B \in \mathscr{R}, \quad A - B \in \mathscr{R}.$$

Since $A \cap B = A - (A - B)$, we also have $A \cap B \in \mathscr{R}$ if $\mathscr{R}$ is a ring.

A ring $\mathscr{R}$ is called a *σ-ring* if

(2) $$\bigcup_{n=1}^{\infty} A_n \in \mathscr{R}$$

whenever $A_n \in \mathscr{R}$ $(n = 1, 2, 3, \ldots)$. Since

$$\bigcap_{n=1}^{\infty} A_n = A_1 - \bigcup_{n=1}^{\infty} (A_1 - A_n),$$

we also have

$$\bigcap_{n=1}^{\infty} A_n \in \mathscr{R}$$

if $\mathscr{R}$ is a σ-ring.

11.2 Definition We say that ϕ is a *set function* defined on $\mathscr{R}$ if ϕ assigns to every $A \in \mathscr{R}$ a number $\phi(A)$ of the extended real number system. ϕ is *additive* if $A \cap B = 0$ implies

(3) $$\phi(A \cup B) = \phi(A) + \phi(B),$$

and ϕ is *countably additive* if $A_i \cap A_j = 0$ $(i \neq j)$ implies

(4) $$\phi\left(\bigcup_{n=1}^{\infty} A_n\right) = \sum_{n=1}^{\infty} \phi(A_n).$$

We shall always assume that the range of ϕ does not contain both $+\infty$ and $-\infty$; for if it did, the right side of (3) could become meaningless. Also, we exclude set functions whose only value is $+\infty$ or $-\infty$.

It is interesting to note that the left side of (4) is independent of the order in which the A_n's are arranged. Hence the rearrangement theorem shows that the right side of (4) converges absolutely if it converges at all; if it does not converge, the partial sums tend to $+\infty$, or to $-\infty$.

If ϕ is additive, the following properties are easily verified:

(5) $$\phi(0) = 0.$$
(6) $$\phi(A_1 \cup \cdots \cup A_n) = \phi(A_1) + \cdots + \phi(A_n)$$

if $A_i \cap A_j = 0$ whenever $i \neq j$.

(7) $$\phi(A_1 \cup A_2) + \phi(A_1 \cap A_2) = \phi(A_1) + \phi(A_2).$$

If $\phi(A) \geq 0$ for all A, and $A_1 \subset A_2$, then

(8) $$\phi(A_1) \leq \phi(A_2).$$

Because of (8), nonnegative additive set functions are often called *monotonic*.

(9) $$\phi(A - B) = \phi(A) - \phi(B)$$

if $B \subset A$, and $|(\phi B)| < +\infty$.

11.3 Theorem *Suppose ϕ is countably additive on a ring $\mathscr{R}$. Suppose $A_n \in \mathscr{R}$ $(n = 1, 2, 3, \ldots)$, $A_1 \subset A_2 \subset A_3 \subset \cdots$, $A \in \mathscr{R}$, and*

$$A = \bigcup_{n=1}^{\infty} A_n.$$

Then, as $n \to \infty$,

$$\phi(A_n) \to \phi(A).$$

Proof Put $B_1 = A_1$, and

$$B_n = A_n - A_{n-1} \qquad (n = 2, 3, \ldots).$$

Then $B_i \cap B_j = 0$ for $i \neq j$, $A_n = B_1 \cup \cdots \cup B_n$, and $A = \bigcup B_n$. Hence

$$\phi(A_n) = \sum_{i=1}^{n} \phi(B_i)$$

and

$$\phi(A) = \sum_{i=1}^{\infty} \phi(B_i).$$

CONSTRUCTION OF THE LEBESGUE MEASURE

11.4 Definition Let R^p denote p-dimensional euclidean space. By an *interval* in R^p we mean the set of points $\mathbf{x} = (x_1, \ldots, x_p)$ such that

(10) $$a_i \leq x_i \leq b_i \qquad (i = 1, \ldots, p),$$

or the set of points which is characterized by (10) with any or all of the $\leq$ signs replaced by $<$. The possibility that $a_i = b_i$ for any value of i is not ruled out; in particular, the empty set is included among the intervals.

If A is the union of a finite number of intervals, A is said to be an *elementary set*.

If I is an interval, we define
$$m(I) = \prod_{i=1}^{p}(b_i - a_i),$$
no matter whether equality is included or excluded in any of the inequalities (10).

If $A = I_1 \cup \cdots \cup I_n$, and if these intervals are pairwise disjoint, we set

(11) $$m(A) = m(I_1) + \cdots + m(I_n).$$

We let $\mathscr{E}$ denote the family of all elementary subsets of R^p.

At this point, the following properties should be verified:

(12) $\mathscr{E}$ is a ring, but not a σ-ring.
(13) If $A \in \mathscr{E}$, then A is the union of a finite number of *disjoint* intervals.
(14) If $A \in \mathscr{E}$, $m(A)$ is well defined by (11); that is, if two different decompositions of A into disjoint intervals are used, each gives rise to the same value of $m(A)$.
(15) m is additive on $\mathscr{E}$.

Note that if $p = 1, 2, 3$, then m is length, area, and volume, respectively.

11.5 Definition A nonnegative additive set function ϕ defined on $\mathscr{E}$ is said to be *regular* if the following is true: To every $A \in \mathscr{E}$ and to every $\varepsilon > 0$ there exist sets $F \in \mathscr{E}$, $G \in \mathscr{E}$ such that F is closed, G is open, $F \subset A \subset G$, and

(16) $$\phi(G) - \varepsilon \leq \phi(A) \leq \phi(F) + \varepsilon.$$

11.6 Examples

(a) *The set function m is regular.*

If A is an interval, it is trivial that the requirements of Definition 11.5 are satisfied. The general case follows from (13).

(b) Take $R^p = R^1$, and let α be a monotonically increasing function, defined for all real x. Put
$$\mu([a, b)) = \alpha(b-) - \alpha(a-),$$
$$\mu([a, b]) = \alpha(b+) - \alpha(a-),$$
$$\mu((a, b]) = \alpha(b+) - \alpha(a+),$$
$$\mu((a, b)) = \alpha(b-) - \alpha(a+).$$

Here $[a, b)$ is the set $a \leq x < b$, etc. Because of the possible discontinuities of α, these cases have to be distinguished. If μ is defined for elementary sets as in (11), μ is regular on $\mathscr{E}$. The proof is just like that of (a).

Our next objective is to show that every regular set function on $\mathscr{E}$ can be extended to a countably additive set function on a σ-ring which contains $\mathscr{E}$.

11.7 Definition Let μ be additive, regular, nonnegative, and finite on $\mathscr{E}$. Consider countable coverings of any set $E \subset R^p$ by open elementary sets A_n:

$$E \subset \bigcup_{n=1}^{\infty} A_n.$$

Define

(17)
$$\mu^*(E) = \inf \sum_{n=1}^{\infty} \mu(A_n),$$

the inf being taken over all countable coverings of E by open elementary sets. $\mu^*(E)$ is called the *outer measure* of E, corresponding to μ.

It is clear that $\mu^*(E) \geq 0$ for all E and that

(18)
$$\mu^*(E_1) \leq \mu^*(E_2)$$

if $E_1 \subset E_2$.

11.8 Theorem

(a) For every $A \in \mathscr{E}$, $\mu^*(A) = \mu(A)$.

(b) If $E = \bigcup_{1}^{\infty} E_n$, then

(19)
$$\mu^*(E) \leq \sum_{n=1}^{\infty} \mu^*(E_n).$$

Note that (a) asserts that μ^* is an extension of μ from $\mathscr{E}$ to the family of *all* subsets of R^p. The property (19) is called *subadditivity*.

Proof Choose $A \in \mathscr{E}$ and $\varepsilon > 0$.

The regularity of μ shows that A is contained in an open elementary set G such that $\mu(G) \leq \mu(A) + \varepsilon$. Since $\mu^*(A) \leq \mu(G)$ and since ε was arbitrary, we have

(20)
$$\mu^*(A) \leq \mu(A).$$

The definition of μ^* shows that there is a sequence $\{A_n\}$ of open elementary sets whose union contains A, such that

$$\sum_{n=1}^{\infty} \mu(A_n) \leq \mu^*(A) + \varepsilon.$$

The regularity of μ shows that A contains a closed elementary set F such that $\mu(F) \geq \mu(A) - \varepsilon$; and since F is compact, we have
$$F \subset A_1 \cup \cdots \cup A_N$$
for some N. Hence
$$\mu(A) \leq \mu(F) + \varepsilon \leq \mu(A_1 \cup \cdots \cup A_N) + \varepsilon \leq \sum_1^N \mu(A_n) + \varepsilon \leq \mu^*(A) + 2\varepsilon.$$

In conjunction with (20), this proves (a).

Next, suppose $E = \bigcup E_n$, and assume that $\mu^*(E_n) < +\infty$ for all n. Given $\varepsilon > 0$, there are coverings $\{A_{nk}\}$, $k = 1, 2, 3, \ldots$, of E_n by open elementary sets such that

(21)
$$\sum_{k=1}^\infty \mu(A_{nk}) \leq \mu^*(E_n) + 2^{-n}\varepsilon.$$

Then
$$\mu^*(E) \leq \sum_{n=1}^\infty \sum_{k=1}^\infty \mu(A_{nk}) \leq \sum_{n=1}^\infty \mu^*(E_n) + \varepsilon,$$

and (19) follows. In the excluded case, i.e., if $\mu^*(E_n) = +\infty$ for some n, (19) is of course trivial.

11.9 Definition For any $A \subset R^p$, $B \subset R^p$, we define

(22) $$S(A, B) = (A - B) \cup (B - A),$$
(23) $$d(A, B) = \mu^*(S(A, B)).$$

We write $A_n \to A$ if
$$\lim_{n \to \infty} d(A, A_n) = 0.$$

If there is a sequence $\{A_n\}$ of elementary sets such that $A_n \to A$, we say that A is *finitely μ-measurable* and write $A \in \mathfrak{M}_F(\mu)$.

If A is the union of a countable collection of finitely μ-measurable sets, we say that A is *μ-measurable* and write $A \in \mathfrak{M}(\mu)$.

$S(A, B)$ is the so-called "symmetric difference" of A and B. We shall see that $d(A, B)$ is essentially a distance function.

The following theorem will enable us to obtain the desired extension of μ.

11.10 Theorem *$\mathfrak{M}(\mu)$ is a σ-ring, and μ^* is countably additive on $\mathfrak{M}(\mu)$.*

Before we turn to the proof of this theorem, we develop some of the properties of $S(A, B)$ and $d(A, B)$. We have

(24) $$S(A, B) = S(B, A), \quad S(A, A) = 0.$$
(25) $$S(A, B) \subset S(A, C) \cup S(C, B).$$
(26) $$\left.\begin{array}{l} S(A_1 \cup A_2, B_1 \cup B_2) \\ S(A_1 \cap A_2, B_1 \cap B_2) \\ S(A_1 - A_2, B_1 - B_2) \end{array}\right\} \subset S(A_1, B_1) \cup S(A_2, B_2).$$

(24) is clear, and (25) follows from
$$(A - B) \subset (A - C) \cup (C - B), \quad (B - A) \subset (C - A) \cup (B - C).$$
The first formula of (26) is obtained from
$$(A_1 \cup A_2) - (B_1 \cup B_2) \subset (A_1 - B_1) \cup (A_2 - B_2).$$
Next, writing E^c for the complement of E, we have
$$S(A_1 \cap A_2, B_1 \cap B_2) = S(A_1^c \cup A_2^c, B_1^c \cup B_2^c)$$
$$\subset S(A_1^c, B_1^c) \cup S(A_2^c, B_2^c) = S(A_1, B_1) \cup S(A_2, B_2);$$
and the last formula of (26) is obtained if we note that
$$A_1 - A_2 = A_1 \cap A_2^c.$$

By (23), (19), and (18), these properties of $S(A, B)$ imply
(27) $$d(A, B) = d(B, A), \quad d(A, A) = 0,$$
(28) $$d(A, B) \leq d(A, C) + d(C, B),$$
(29) $$\left.\begin{array}{l} d(A_1 \cup A_2, B_1 \cup B_2) \\ d(A_1 \cap A_2, B_1 \cap B_2) \\ d(A_1 - A_2, B_1 - B_2) \end{array}\right\} \leq d(A_1, B_1) + d(A_2, B_2).$$

The relations (27) and (28) show that $d(A, B)$ satisfies the requirements of Definition 2.15, except that $d(A, B) = 0$ does not imply $A = B$. For instance, if $\mu = m$, A is countable, and B is empty, we have
$$d(A, B) = m^*(A) = 0;$$
to see this, cover the nth point of A by an interval I_n such that
$$m(I_n) < 2^{-n}\varepsilon.$$

But if we define two sets A and B to be equivalent, provided
$$d(A, B) = 0,$$
we divide the subsets of R^p into equivalence classes, and $d(A, B)$ makes the set of these equivalence classes into a metric space. $\mathfrak{M}_F(\mu)$ is then obtained as the closure of $\mathscr{E}$. This interpretation is not essential for the proof, but it explains the underlying idea.

We need one more property of $d(A, B)$, namely,

(30) $$|\mu^*(A) - \mu^*(B)| \le d(A, B),$$

if at least one of $\mu^*(A)$, $\mu^*(B)$ is finite. For suppose $0 \le \mu^*(B) \le \mu^*(A)$. Then (28) shows that

$$d(A, 0) \le d(A, B) + d(B, 0),$$

that is,

$$\mu^*(A) \le d(A, B) + \mu^*(B).$$

Since $\mu^*(B)$ is finite, it follows that

$$\mu^*(A) - \mu^*(B) \le d(A, B).$$

Proof of Theorem 11.10 Suppose $A \in \mathfrak{M}_F(\mu)$, $B \in \mathfrak{M}_F(\mu)$. Choose $\{A_n\}$, $\{B_n\}$ such that $A_n \in \mathscr{E}$, $B_n \in \mathscr{E}$, $A_n \to A$, $B_n \to B$. Then (29) and (30) show that

(31) $$A_n \cup B_n \to A \cup B,$$
(32) $$A_n \cap B_n \to A \cap B,$$
(33) $$A_n - B_n \to A - B,$$
(34) $$\mu^*(A_n) \to \mu^*(A),$$

and $\mu^*(A) < +\infty$ since $d(A_n, A) \to 0$. By (31) and (33), $\mathfrak{M}_F(\mu)$ is a ring. By (7),

$$\mu(A_n) + \mu(B_n) = \mu(A_n \cup B_n) + \mu(A_n \cap B_n).$$

Letting $n \to \infty$, we obtain, by (34) and Theorem 11.8(a),

$$\mu^*(A) + \mu^*(B) = \mu^*(A \cup B) + \mu^*(A \cap B).$$

If $A \cap B = 0$, then $\mu^*(A \cap B) = 0$.

It follows that μ^* is additive on $\mathfrak{M}_F(\mu)$.

Now let $A \in \mathfrak{M}(\mu)$. Then A can be represented as the union of a countable collection of *disjoint* sets of $\mathfrak{M}_F(\mu)$. For if $A = \bigcup A_n'$ with $A_n' \in \mathfrak{M}_F(\mu)$, write $A_1 = A_1'$, and

$$A_n = (A_1' \cup \cdots \cup A_n') - (A_1' \cup \cdots \cup A_{n-1}') \qquad (n = 2, 3, 4, \ldots).$$

Then

(35) $$A = \bigcup_{n=1}^{\infty} A_n$$

is the required representation. By (19)

(36) $$\mu^*(A) \le \sum_{n=1}^{\infty} \mu^*(A_n).$$

On the other hand, $A \supset A_1 \cup \cdots \cup A_n$; and by the additivity of μ^* on $\mathfrak{M}_F(\mu)$ we obtain

(37) $$\mu^*(A) \geq \mu^*(A_1 \cup \cdots \cup A_n) = \mu^*(A_1) + \cdots + \mu^*(A_n).$$

Equations (36) and (37) imply

(38) $$\mu^*(A) = \sum_{n=1}^{\infty} \mu^*(A_n).$$

Suppose $\mu^*(A)$ is finite. Put $B_n = A_1 \cup \cdots \cup A_n$. Then (38) shows that

$$d(A, B_n) = \mu^*\left(\bigcup_{i=n+1}^{\infty} A_i\right) = \sum_{i=n+1}^{\infty} \mu^*(A_i) \to 0$$

as $n \to \infty$. Hence $B_n \to A$; and since $B_n \in \mathfrak{M}_F(\mu)$, it is easily seen that $A \in \mathfrak{M}_F(\mu)$.

We have thus shown that $A \in \mathfrak{M}_F(\mu)$ if $A \in \mathfrak{M}(\mu)$ and $\mu^*(A) < +\infty$. It is now clear that μ^* is countably additive on $\mathfrak{M}(\mu)$. For if

$$A = \bigcup A_n,$$

where $\{A_n\}$ is a sequence of disjoint sets of $\mathfrak{M}(\mu)$, we have shown that (38) holds if $\mu^*(A_n) < +\infty$ for every n, and in the other case (38) is trivial.

Finally, we have to show that $\mathfrak{M}(\mu)$ is a σ-ring. If $A_n \in \mathfrak{M}(\mu)$, $n = 1, 2, 3, \ldots$, it is clear that $\bigcup A_n \in \mathfrak{M}(\mu)$ (Theorem 2.12). Suppose $A \in \mathfrak{M}(\mu)$, $B \in \mathfrak{M}(\mu)$, and

$$A = \bigcup_{n=1}^{\infty} A_n, \qquad B = \bigcup_{n=1}^{\infty} B_n,$$

where $A_n, B_n \in \mathfrak{M}_F(\mu)$. Then the identity

$$A_n \cap B = \bigcup_{i=1}^{\infty} (A_n \cap B_i)$$

shows that $A_n \cap B \in \mathfrak{M}(\mu)$; and since

$$\mu^*(A_n \cap B) \leq \mu^*(A_n) < +\infty,$$

$A_n \cap B \in \mathfrak{M}_F(\mu)$. Hence $A_n - B \in \mathfrak{M}_F(\mu)$, and $A - B \in \mathfrak{M}(\mu)$ since $A - B = \bigcup_{n=1}^{\infty} (A_n - B)$.

We now replace $\mu^*(A)$ by $\mu(A)$ if $A \in \mathfrak{M}(\mu)$. Thus μ, originally only defined on $\mathscr{E}$, is extended to a countably additive set function on the σ-ring $\mathfrak{M}(\mu)$. This extended set function is called a *measure*. The special case $\mu = m$ is called the *Lebesgue measure* on R^p.

11.11 Remarks

(a) If A is open, then $A \in \mathfrak{M}(\mu)$. For every open set in R^p is the union of a countable collection of open intervals. To see this, it is sufficient to construct a countable base whose members are open intervals.

By taking complements, it follows that every closed set is in $\mathfrak{M}(\mu)$.

(b) If $A \in \mathfrak{M}(\mu)$ and $\varepsilon > 0$, there exist sets F and G such that

$$F \subset A \subset G,$$

F is closed, G is open, and

(39) $$\mu(G - A) < \varepsilon, \quad \mu(A - F) < \varepsilon.$$

The first inequality holds since μ^* was defined by means of coverings by *open* elementary sets. The second inequality then follows by taking complements.

(c) We say that E is a *Borel set* if E can be obtained by a countable number of operations, starting from open sets, each operation consisting in taking unions, intersections, or complements. The collection $\mathscr{B}$ of all Borel sets in R^p is a σ-ring; in fact, it is the smallest σ-ring which contains all open sets. By Remark (a), $E \in \mathfrak{M}(\mu)$ if $E \in \mathscr{B}$.

(d) If $A \in \mathfrak{M}(\mu)$, there exist Borel sets F and G such that $F \subset A \subset G$, and

(40) $$\mu(G - A) = \mu(A - F) = 0.$$

This follows from (b) if we take $\varepsilon = 1/n$ and let $n \to \infty$.

Since $A = F \cup (A - F)$, we see that every $A \in \mathfrak{M}(\mu)$ is the union of a Borel set and a set of measure zero.

The Borel sets are μ-measurable for every μ. But the sets of measure zero [that is, the sets E for which $\mu^*(E) = 0$] may be different for different μ's.

(e) For every μ, the sets of measure zero form a σ-ring.

(f) In case of the Lebesgue measure, every countable set has measure zero. But there are uncountable (in fact, perfect) sets of measure zero. The Cantor set may be taken as an example: Using the notation of Sec. 2.44, it is easily seen that

$$m(E_n) = (\tfrac{2}{3})^n \quad (n = 1, 2, 3, \ldots);$$

and since $P = \bigcap E_n$, $P \subset E_n$ for every n, so that $m(P) = 0$.

MEASURE SPACES

11.12 Definition Suppose X is a set, not necessarily a subset of a euclidean space, or indeed of any metric space. X is said to be a *measure space* if there exists a σ-ring $\mathfrak{M}$ of subsets of X (which are called measurable sets) and a nonnegative countably additive set function μ (which is called a measure), defined on $\mathfrak{M}$.

If, in addition, $X \in \mathfrak{M}$, then X is said to be a *measurable space*.

For instance, we can take $X = R^p$, $\mathfrak{M}$ the collection of all Lebesgue-measurable subsets of R^p, and μ Lebesgue measure.

Or, let X be the set of all positive integers, $\mathfrak{M}$ the collection of all subsets of X, and $\mu(E)$ the number of elements of E.

Another example is provided by probability theory, where events may be considered as sets, and the probability of the occurrence of events is an additive (or countably additive) set function.

In the following sections we shall always deal with measurable spaces. It should be emphasized that the integration theory which we shall soon discuss would not become simpler in any respect if we sacrificed the generality we have now attained and restricted ourselves to Lebesgue measure, say, on an interval of the real line. In fact, the essential features of the theory are brought out with much greater clarity in the more general situation, where it is seen that everything depends only on the countable additivity of μ on a σ-ring.

It will be convenient to introduce the notation

$$(41) \qquad \{x | P\}$$

for the set of all elements x which have the property P.

MEASURABLE FUNCTIONS

11.13 Definition Let f be a function defined on the measurable space X, with values in the extended real number system. The function f is said to be *measurable* if the set

$$(42) \qquad \{x | f(x) > a\}$$

is measurable for every real a.

11.14 Example If $X = R^p$ and $\mathfrak{M} = \mathfrak{M}(\mu)$ as defined in Definition 11.9, every continuous f is measurable, since then (42) is an open set.

11.15 Theorem *Each of the following four conditions implies the other three:*

(43) $\{x | f(x) > a\}$ *is measurable for every real a.*

(44) $\{x | f(x) \geq a\}$ *is measurable for every real a.*

(45) $\{x | f(x) < a\}$ *is measurable for every real a.*

(46) $\{x | f(x) \leq a\}$ *is measurable for every real a.*

Proof The relations

$$\{x | f(x) \geq a\} = \bigcap_{n=1}^{\infty} \left\{ x \Big| f(x) > a - \frac{1}{n} \right\},$$

$$\{x | f(x) < a\} = X - \{x | f(x) \geq a\},$$

$$\{x | f(x) \leq a\} = \bigcap_{n=1}^{\infty} \left\{ x \Big| f(x) < a + \frac{1}{n} \right\},$$

$$\{x | f(x) > a\} = X - \{x | f(x) \leq a\}$$

show successively that (43) implies (44), (44) implies (45), (45) implies (46), and (46) implies (43).

Hence any of these conditions may be used instead of (42) to define measurability.

11.16 Theorem *If f is measurable, then $|f|$ is measurable.*

Proof

$$\{x | |f(x)| < a\} = \{x | f(x) < a\} \cap \{x | f(x) > -a\}.$$

11.17 Theorem *Let $\{f_n\}$ be a sequence of measurable functions. For $x \in X$, put*

$$g(x) = \sup f_n(x) \quad (n = 1, 2, 3, \ldots),$$

$$h(x) = \limsup_{n \to \infty} f_n(x).$$

Then g and h are measurable.

The same is of course true of the inf and lim inf.

Proof

$$\{x | g(x) > a\} = \bigcup_{n=1}^{\infty} \{x | f_n(x) > a\},$$

$$h(x) = \inf g_m(x),$$

where $g_m(x) = \sup f_n(x) \ (n \geq m)$.

Corollaries

(a) *If f and g are measurable, then* $\max(f, g)$ *and* $\min(f, g)$ *are measurable. If*

(47) $$f^+ = \max(f, 0), \qquad f^- = -\min(f, 0),$$

it follows, in particular, that f^+ and f^- are measurable.

(b) *The limit of a convergent sequence of measurable functions is measurable.*

11.18 Theorem *Let f and g be measurable real-valued functions defined on X, let F be real and continuous on R^2, and put*

$$h(x) = F(f(x), g(x)) \qquad (x \in X).$$

Then h is measurable.

In particular, $f + g$ and fg are measurable.

Proof Let

$$G_a = \{(u, v) \mid F(u, v) > a\}.$$

Then G_a is an open subset of R^2, and we can write

$$G_a = \bigcup_{n=1}^{\infty} I_n,$$

where $\{I_n\}$ is a sequence of open intervals:

$$I_n = \{(u, v) \mid a_n < u < b_n, c_n < v < d_n\}.$$

Since

$$\{x \mid a_n < f(x) < b_n\} = \{x \mid f(x) > a_n\} \cap \{x \mid f(x) < b_n\}$$

is measurable, it follows that the set

$$\{x \mid (f(x), g(x)) \in I_n\} = \{x \mid a_n < f(x) < b_n\} \cap \{x \mid c_n < g(x) < d_n\}$$

is measurable. Hence the same is true of

$$\{x \mid h(x) > a\} = \{x \mid (f(x), g(x)) \in G_a\}$$
$$= \bigcup_{n=1}^{\infty} \{x \mid (f(x), g(x)) \in I_n\}.$$

Summing up, we may say that all ordinary operations of analysis, including limit operations, when applied to measurable functions, lead to measurable functions; in other words, all functions that are ordinarily met with are measurable.

That this is, however, only a rough statement is shown by the following example (based on Lebesgue measure, on the real line): If $h(x) = f(g(x))$, where

f is measurable and g is continuous, then h is not necessarily measurable. (For the details, we refer to McShane, page 241.)

The reader may have noticed that measure has not been mentioned in our discussion of measurable functions. In fact, the class of measurable functions on X depends only on the σ-ring $\mathfrak{M}$ (using the notation of Definition 11.12). For instance, we may speak of *Borel-measurable functions* on R^p, that is, of function f for which

$$\{x \mid f(x) > a\}$$

is always a Borel set, without reference to any particular measure.

SIMPLE FUNCTIONS

11.19 Definition Let s be a real-valued function defined on X. If the range of s is finite, we say that s is a *simple function*.

Let $E \subset X$, and put

(48) $$K_E(x) = \begin{cases} 1 & (x \in E), \\ 0 & (x \notin E). \end{cases}$$

K_E is called the *characteristic function* of E.

Suppose the range of s consists of the distinct numbers $c_1, \ldots, c_n$. Let

$$E_i = \{x \mid s(x) = c_i\} \qquad (i = 1, \ldots, n).$$

Then

(49) $$s = \sum_{n=1}^{n} c_i K_{E_i},$$

that is, every simple function is a finite linear combination of characteristic functions. It is clear that s is measurable if and only if the sets $E_1, \ldots, E_n$ are measurable.

It is of interest that every function can be approximated by simple functions:

11.20 Theorem *Let f be a real function on X. There exists a sequence $\{s_n\}$ of simple functions such that $s_n(x) \to f(x)$ as $n \to \infty$, for every $x \in X$. If f is measurable, $\{s_n\}$ may be chosen to be a sequence of measurable functions. If $f \geq 0$, $\{s_n\}$ may be chosen to be a monotonically increasing sequence.*

Proof If $f \geq 0$, define

$$E_{ni} = \left\{ x \mid \frac{i-1}{2^n} \leq f(x) < \frac{i}{2^n} \right\}, \qquad F_n = \{x \mid f(x) \geq n\}$$

for $n = 1, 2, 3, \ldots, i = 1, 2, \ldots, n2^n$. Put

(50) $$s_n = \sum_{i=1}^{n2^n} \frac{i-1}{2^n} K_{E_{ni}} + nK_{F_n}.$$

In the general case, let $f = f^+ - f^-$, and apply the preceding construction to f^+ and to f^-.

It may be noted that the sequence $\{s_n\}$ given by (50) converges uniformly to f if f is bounded.

INTEGRATION

We shall define integration on a measurable space X, in which $\mathfrak{M}$ is the σ-ring of measurable sets, and μ is the measure. The reader who wishes to visualize a more concrete situation may think of X as the real line, or an interval, and of μ as the Lebesgue measure m.

11.21 Definition Suppose

(51) $$s(x) = \sum_{i=1}^{n} c_i K_{E_i}(x) \qquad (x \in X, c_i > 0)$$

is measurable, and suppose $E \in \mathfrak{M}$. We define

(52) $$I_E(s) = \sum_{i=1}^{n} c_i \mu(E \cap E_i).$$

If f is measurable and nonnegative, we define

(53) $$\int_E f \, d\mu = \sup I_E(s),$$

where the sup is taken over all measurable simple functions s such that $0 \leq s \leq f$.

The left member of (53) is called the *Lebesgue integral* of f, with respect to the measure μ, over the set E. It should be noted that the integral may have the value $+\infty$.

It is easily verified that

(54) $$\int_E s \, d\mu = I_E(s)$$

for every nonnegative simple measurable function s.

11.22 Definition Let f be measurable, and consider the two integrals

(55) $$\int_E f^+ \, d\mu, \qquad \int_E f^- \, d\mu,$$

where f^+ and f^- are defined as in (47).

If at least one of the integrals (55) is finite, we define

(56) $$\int_E f\, d\mu = \int_E f^+\, d\mu - \int_E f^-\, d\mu.$$

If both integrals in (55) are finite, then (56) is finite, and we say that f is *integrable* (or *summable*) on E in the Lebesgue sense, with respect to μ; we write $f \in \mathscr{L}(\mu)$ on E. If $\mu = m$, the usual notation is: $f \in \mathscr{L}$ on E.

This terminology may be a little confusing: If (56) is $+\infty$ or $-\infty$, then the integral of f over E is defined, although f is not integrable in the above sense of the word; f is integrable on E only if its integral over E is finite.

We shall be mainly interested in integrable functions, although in some cases it is desirable to deal with the more general situation.

11.23 Remarks The following properties are evident:

(a) If f is measurable and bounded on E, and if $\mu(E) < +\infty$, then $f \in \mathscr{L}(\mu)$ on E.

(b) If $a \le f(x) \le b$ for $x \in E$, and $\mu(E) < +\infty$, then
$$a\mu(E) \le \int_E f\, d\mu \le b\mu(E).$$

(c) If f and $g \in \mathscr{L}(\mu)$ on E, and if $f(x) \le g(x)$ for $x \in E$, then
$$\int_E f\, d\mu \le \int_E g\, d\mu.$$

(d) If $f \in \mathscr{L}(\mu)$ on E, then $cf \in \mathscr{L}(\mu)$ on E, for every finite constant c, and
$$\int_E cf\, d\mu = c\int_E f\, d\mu.$$

(e) If $\mu(E) = 0$, and f is measurable, then
$$\int_E f\, d\mu = 0.$$

(f) If $f \in \mathscr{L}(\mu)$ on E, $A \in \mathfrak{M}$, and $A \subset E$, then $f \in \mathscr{L}(\mu)$ on A.

11.24 Theorem

(a) Suppose f is measurable and nonnegative on X. For $A \in \mathfrak{M}$, define

(57) $$\phi(A) = \int_A f\, d\mu.$$

Then ϕ is countably additive on $\mathfrak{M}$.

(b) *The same conclusion holds if $f \in \mathscr{L}(\mu)$ on X.*

Proof It is clear that (b) follows from (a) if we write $f = f^+ - f^-$ and apply (a) to f^+ and to f^-.

To prove (a), we have to show that

(58) $$\phi(A) = \sum_{n=1}^{\infty} \phi(A_n)$$

if $A_n \in \mathfrak{M}$ ($n = 1, 2, 3, \ldots$), $A_i \cap A_j = 0$ for $i \neq j$, and $A = \bigcup_1^{\infty} A_n$.

If f is a characteristic function, then the countable additivity of ϕ is precisely the same as the countable additivity of μ, since

$$\int_A K_E \, d\mu = \mu(A \cap E).$$

If f is simple, then f is of the form (51), and the conclusion again holds.

In the general case, we have, for every measurable simple function s such that $0 \leq s \leq f$,

$$\int_A s \, d\mu = \sum_{n=1}^{\infty} \int_{A_n} s \, d\mu \leq \sum_{n=1}^{\infty} \phi(A_n).$$

Therefore, by (53),

(59) $$\phi(A) \leq \sum_{n=1}^{\infty} \phi(A_n).$$

Now if $\phi(A_n) = +\infty$ for some n, (58) is trivial, since $\phi(A) \geq \phi(A_n)$. Suppose $\phi(A_n) < +\infty$ for every n.

Given $\varepsilon > 0$, we can choose a measurable function s such that $0 \leq s \leq f$, and such that

(60) $$\int_{A_1} s \, d\mu \geq \int_{A_1} f \, d\mu - \varepsilon, \qquad \int_{A_2} s \, d\mu \geq \int_{A_2} f \, d\mu - \varepsilon.$$

Hence

$$\phi(A_1 \cup A_2) \geq \int_{A_1 \cup A_2} s \, d\mu = \int_{A_1} s \, d\mu + \int_{A_2} s \, d\mu \geq \phi(A_1) + \phi(A_2) - 2\varepsilon,$$

so that

$$\phi(A_1 \cup A_2) \geq \phi(A_1) + \phi(A_2).$$

It follows that we have, for every n,

(61) $$\phi(A_1 \cup \cdots \cup A_n) \geq \phi(A_1) + \cdots + \phi(A_n).$$

Since $A \supset A_1 \cup \cdots \cup A_n$, (61) implies

(62) $$\phi(A) \geq \sum_{n=1}^{\infty} \phi(A_n),$$

and (58) follows from (59) and (62).

Corollary *If $A \in \mathfrak{M}$, $B \in \mathfrak{M}$, $B \subset A$, and $\mu(A - B) = 0$, then*

$$\int_A f \, d\mu = \int_B f \, d\mu.$$

Since $A = B \cup (A - B)$, this follows from Remark 11.23(e).

11.25 Remarks The preceding corollary shows that sets of measure zero are negligible in integration.

Let us write $f \sim g$ on E if the set

$$\{x \mid f(x) \neq g(x)\} \cap E$$

has measure zero.

Then $f \sim f$; $f \sim g$ implies $g \sim f$; and $f \sim g$, $g \sim h$ implies $f \sim h$. That is, the relation $\sim$ is an equivalence relation.

If $f \sim g$ on E, we clearly have

$$\int_A f \, d\mu = \int_A g \, d\mu,$$

provided the integrals exist, for every measurable subset A of E.

If a property P holds for every $x \in E - A$, and if $\mu(A) = 0$, it is customary to say that P holds for almost all $x \in E$, or that P holds almost everywhere on E. (This concept of "almost everywhere" depends of course on the particular measure under consideration. In the literature, unless something is said to the contrary, it usually refers to Lebesgue measure.)

If $f \in \mathscr{L}(\mu)$ on E, it is clear that $f(x)$ must be finite almost everywhere on E. In most cases we therefore do not lose any generality if we assume the given functions to be finite-valued from the outset.

11.26 Theorem *If $f \in \mathscr{L}(\mu)$ on E, then $|f| \in \mathscr{L}(\mu)$ on E, and*

(63) $$\left| \int_E f \, d\mu \right| \leq \int_E |f| \, d\mu.$$

Proof Write $E = A \cup B$, where $f(x) \geq 0$ on A and $f(x) < 0$ on B. By Theorem 11.24,

$$\int_E |f|\, d\mu = \int_A |f|\, d\mu + \int_B |f|\, d\mu = \int_A f^+\, d\mu + \int_B f^-\, d\mu < +\infty,$$

so that $|f| \in \mathscr{L}(\mu)$. Since $f \leq |f|$ and $-f \leq |f|$, we see that

$$\int_E f\, d\mu \leq \int_E |f|\, d\mu, \qquad -\int_E f\, d\mu \leq \int_E |f|\, d\mu,$$

and (63) follows.

Since the integrability of f implies that of $|f|$, the Lebesgue integral is often called an absolutely convergent integral. It is of course possible to define nonabsolutely convergent integrals, and in the treatment of some problems it is essential to do so. But these integrals lack some of the most useful properties of the Lebesgue integral and play a somewhat less important role in analysis.

11.27 Theorem *Suppose f is measurable on E, $|f| \leq g$, and $g \in \mathscr{L}(\mu)$ on E. Then $f \in \mathscr{L}(\mu)$ on E.*

Proof We have $f^+ \leq g$ and $f^- \leq g$.

11.28 Lebesgue's monotone convergence theorem *Suppose $E \in \mathfrak{M}$. Let $\{f_n\}$ be a sequence of measurable functions such that*

(64) $$0 \leq f_1(x) \leq f_2(x) \leq \cdots \qquad (x \in E).$$

Let f be defined by

(65) $$f_n(x) \to f(x) \qquad (x \in E)$$

as $n \to \infty$. Then

(66) $$\int_E f_n\, d\mu \to \int_E f\, d\mu \qquad (n \to \infty).$$

Proof By (64) it is clear that, as $n \to \infty$,

(67) $$\int_E f_n\, d\mu \to \alpha$$

for some α; and since $\int f_n \leq \int f$, we have

(68) $$\alpha \leq \int_E f\, d\mu.$$

Choose c such that $0 < c < 1$, and let s be a simple measurable function such that $0 \le s \le f$. Put

$$E_n = \{x \mid f_n(x) \ge cs(x)\} \qquad (n = 1, 2, 3, \ldots).$$

By (64), $E_1 \subset E_2 \subset E_3 \subset \cdots$; and by (65),

(69) $$E = \bigcup_{n=1}^{\infty} E_n.$$

For every n,

(70) $$\int_E f_n \, d\mu \ge \int_{E_n} f_n \, d\mu \ge c \int_{E_n} s \, d\mu.$$

We let $n \to \infty$ in (70). Since the integral is a countably additive set function (Theorem 11.24), (69) shows that we may apply Theorem 11.3 to the last integral in (70), and we obtain

(71) $$\alpha \ge c \int_E s \, d\mu.$$

Letting $c \to 1$, we see that

$$\alpha \ge \int_E s \, d\mu,$$

and (53) implies

(72) $$\alpha \ge \int_E f \, d\mu.$$

The theorem follows from (67), (68), and (72).

11.29 Theorem *Suppose $f = f_1 + f_2$, where $f_i \in \mathscr{L}(\mu)$ on E ($i = 1, 2$). Then $f \in \mathscr{L}(\mu)$ on E, and*

(73) $$\int_E f \, d\mu = \int_E f_1 \, d\mu + \int_E f_2 \, d\mu.$$

Proof First, suppose $f_1 \ge 0, f_2 \ge 0$. If f_1 and f_2 are simple, (73) follows trivially from (52) and (54). Otherwise, choose monotonically increasing sequences $\{s'_n\}, \{s''_n\}$ of nonnegative measurable simple functions which converge to f_1, f_2. Theorem 11.20 shows that this is possible. Put $s_n = s'_n + s''_n$. Then

$$\int_E s_n \, d\mu = \int_E s'_n \, d\mu + \int_E s''_n \, d\mu,$$

and (73) follows if we let $n \to \infty$ and appeal to Theorem 11.28.

Next, suppose $f_1 \geq 0, f_2 \leq 0$. Put
$$A = \{x | f(x) \geq 0\}, \qquad B = \{x | f(x) < 0\}.$$
Then f, f_1, and $-f_2$ are nonnegative on A. Hence

(74) $$\int_A f_1 d\mu = \int_A f \, d\mu + \int_A (-f_2) \, d\mu = \int_A f \, d\mu - \int_A f_2 \, d\mu.$$

Similarly, $-f, f_1$, and $-f_2$ are nonnegative on B, so that
$$\int_B (-f_2) \, d\mu = \int_B f_1 \, d\mu + \int_B (-f) \, d\mu,$$
or

(75) $$\int_B f_1 \, d\mu = \int_B f \, d\mu - \int_B f_2 \, d\mu,$$

and (73) follows if we add (74) and (75).

In the general case, E can be decomposed into four sets E_i on each of which $f_1(x)$ and $f_2(x)$ are of constant sign. The two cases we have proved so far imply
$$\int_{E_i} f \, d\mu = \int_{E_i} f_1 \, d\mu + \int_{E_i} f_2 \, d\mu \qquad (i = 1, 2, 3, 4),$$
and (73) follows by adding these four equations.

We are now in a position to reformulate Theorem 11.28 for series.

11.30 Theorem *Suppose $E \in \mathfrak{M}$. If $\{f_n\}$ is a sequence of nonnegative measurable functions and*

(76) $$f(x) = \sum_{n=1}^{\infty} f_n(x) \qquad (x \in E),$$

then
$$\int_E f \, d\mu = \sum_{n=1}^{\infty} \int_E f_n \, d\mu.$$

Proof The partial sums of (76) form a monotonically increasing sequence.

11.31 Fatou's theorem *Suppose $E \in \mathfrak{M}$. If $\{f_n\}$ is a sequence of nonnegative measurable functions and*
$$f(x) = \liminf_{n \to \infty} f_n(x) \qquad (x \in E),$$
then

(77) $$\int_E f \, d\mu \leq \liminf_{n \to \infty} \int_E f_n \, d\mu.$$

Strict inequality may hold in (77). An example is given in Exercise 5.

Proof For $n = 1, 2, 3, \ldots$ and $x \in E$, put
$$g_n(x) = \inf f_i(x) \qquad (i \geq n).$$
Then g_n is measurable on E, and

(78) $$0 \leq g_1(x) \leq g_2(x) \leq \cdots,$$

(79) $$g_n(x) \leq f_n(x),$$

(80) $$g_n(x) \to f(x) \qquad (n \to \infty).$$

By (78), (80), and Theorem 11.28,

(81) $$\int_E g_n \, d\mu \to \int_E f \, d\mu,$$

so that (77) follows from (79) and (81).

11.32 Lebesgue's dominated convergence theorem *Suppose $E \in \mathfrak{M}$. Let $\{f_n\}$ be a sequence of measurable functions such that*

(82) $$f_n(x) \to f(x) \qquad (x \in E)$$

as $n \to \infty$. If there exists a function $g \in \mathscr{L}(\mu)$ on E, such that

(83) $$|f_n(x)| \leq g(x) \qquad (n = 1, 2, 3, \ldots, x \in E),$$

then

(84) $$\lim_{n \to \infty} \int_E f_n \, d\mu = \int_E f \, d\mu.$$

Because of (83), $\{f_n\}$ is said to be *dominated* by g, and we talk about *dominated convergence*. By Remark 11.25, the conclusion is the same if (82) holds almost everywhere on E.

Proof First, (83) and Theorem 11.27 imply that $f_n \in \mathscr{L}(\mu)$ and $f \in \mathscr{L}(\mu)$ on E.

Since $f_n + g \geq 0$, Fatou's theorem shows that
$$\int_E (f + g) \, d\mu \leq \liminf_{n \to \infty} \int_E (f_n + g) \, d\mu,$$
or

(85) $$\int_E f \, d\mu \leq \liminf_{n \to \infty} \int_E f_n \, d\mu.$$

Since $g - f_n \geq 0$, we see similarly that
$$\int_E (g - f)\, d\mu \leq \liminf_{n \to \infty} \int_E (g - f_n)\, d\mu,$$
so that
$$-\int_E f\, d\mu \leq \liminf_{n \to \infty} \left[-\int_E f_n\, d\mu \right],$$
which is the same as
(86) $$\int_E f\, d\mu \geq \limsup_{n \to \infty} \int_E f_n\, d\mu.$$

The existence of the limit in (84) and the equality asserted by (84) now follow from (85) and (86).

Corollary *If $\mu(E) < +\infty$, $\{f_n\}$ is uniformly bounded on E, and $f_n(x) \to f(x)$ on E, then (84) holds.*

A uniformly bounded convergent sequence is often said to be boundedly convergent.

COMPARISON WITH THE RIEMANN INTEGRAL

Our next theorem will show that every function which is Riemann-integrable on an interval is also Lebesgue-integrable, and that Riemann-integrable functions are subject to rather stringent continuity conditions. Quite apart from the fact that the Lebesgue theory therefore enables us to integrate a much larger class of functions, its greatest advantage lies perhaps in the ease with which many limit operations can be handled; from this point of view, Lebesgue's convergence theorems may well be regarded as the core of the Lebesgue theory.

One of the difficulties which is encountered in the Riemann theory is that limits of Riemann-integrable functions (or even continuous functions) may fail to be Riemann-integrable. This difficulty is now almost eliminated, since limits of measurable functions are always measurable.

Let the measure space X be the interval $[a, b]$ of the real line, with $\mu = m$ (the Lebesgue measure), and $\mathfrak{M}$ the family of Lebesgue-measurable subsets of $[a, b]$. Instead of
$$\int_X f\, dm$$
it is customary to use the familiar notation
$$\int_a^b f\, dx$$

for the Lebesgue integral of f over $[a, b]$. To distinguish Riemann integrals from Lebesgue integrals, we shall now denote the former by

$$\mathscr{R} \int_a^b f \, dx.$$

11.33 Theorem

(a) If $f \in \mathscr{R}$ on $[a, b]$, then $f \in \mathscr{L}$ on $[a, b]$, and

(87)
$$\int_a^b f \, dx = \mathscr{R} \int_a^b f \, dx.$$

(b) Suppose f is bounded on $[a, b]$. Then $f \in \mathscr{R}$ on $[a, b]$ if and only if f is continuous almost everywhere on $[a, b]$.

Proof Suppose f is bounded. By Definition 6.1 and Theorem 6.4 there is a sequence $\{P_k\}$ of partitions of $[a, b]$, such that P_{k+1} is a refinement of P_k, such that the distance between adjacent points of P_k is less than $1/k$, and such that

(88)
$$\lim_{k \to \infty} L(P_k, f) = \mathscr{R} \underline{\int} f \, dx, \quad \lim_{k \to \infty} U(P_k, f) = \mathscr{R} \overline{\int} f \, dx.$$

(In this proof, all integrals are taken over $[a, b]$.)

If $P_k = \{x_0, x_1, \ldots, x_n\}$, with $x_0 = a$, $x_n = b$, define

$$U_k(a) = L_k(a) = f(a);$$

put $U_k(x) = M_i$ and $L_k(x) = m_i$ for $x_{i-1} < x \leq x_i$, $1 \leq i \leq n$, using the notation introduced in Definition 6.1. Then

(89)
$$L(P_k, f) = \int L_k \, dx, \quad U(P_k, f) = \int U_k \, dx,$$

and

(90)
$$L_1(x) \leq L_2(x) \leq \cdots \leq f(x) \leq \cdots \leq U_2(x) \leq U_1(x)$$

for all $x \in [a, b]$, since P_{k+1} refines P_k. By (90), there exist

(91)
$$L(x) = \lim_{k \to \infty} L_k(x), \quad U(x) = \lim_{k \to \infty} U_k(x).$$

Observe that L and U are bounded measurable functions on $[a, b]$, that

(92)
$$L(x) \leq f(x) \leq U(x) \quad (a \leq x \leq b),$$

and that

$$\int L \, dx = \underline{\mathscr{R}} \int f \, dx, \qquad \int U \, dx = \overline{\mathscr{R}} \int f \, dx, \tag{93}$$

by (88), (90), and the monotone convergence theorem.

So far, nothing has been assumed about f except that f is a bounded real function on $[a, b]$.

To complete the proof, note that $f \in \mathscr{R}$ if and only if its upper and lower Riemann integrals are equal, hence if and only if

$$\int L \, dx = \int U \, dx; \tag{94}$$

since $L \leq U$, (94) happens if and only if $L(x) = U(x)$ for almost all $x \in [a, b]$ (Exercise 1).

In that case, (92) implies that

$$L(x) = f(x) = U(x) \tag{95}$$

almost everywhere on $[a, b]$, so that f is measurable, and (87) follows from (93) and (95).

Furthermore, if x belongs to no P_k, it is quite easy to see that $U(x) = L(x)$ if and only if f is continuous at x. Since the union of the sets P_k is countable, its measure is 0, and we conclude that f is continuous almost everywhere on $[a, b]$ if and only if $L(x) = U(x)$ almost everywhere, hence (as we saw above) if and only if $f \in \mathscr{R}$.

This completes the proof.

The familiar connection between integration and differentiation is to a large degree carried over into the Lebesgue theory. If $f \in \mathscr{L}$ on $[a, b]$, and

$$F(x) = \int_a^x f \, dt \qquad (a \leq x \leq b), \tag{96}$$

then $F'(x) = f(x)$ almost everywhere on $[a, b]$.

Conversely, if F is differentiable at every point of $[a, b]$ ("almost everywhere" is not good enough here!) and if $F' \in \mathscr{L}$ on $[a, b]$, then

$$F(x) - F(a) = \int_a^x F'(t) \qquad (a \leq x \leq b).$$

For the proofs of these two theorems, we refer the reader to any of the works on integration cited in the Bibliography.

INTEGRATION OF COMPLEX FUNCTIONS

Suppose f is a complex-valued function defined on a measure space X, and $f = u + iv$, where u and v are real. We say that f is measurable if and only if both u and v are measurable.

It is easy to verify that sums and products of complex measurable functions are again measurable. Since

$$|f| = (u^2 + v^2)^{1/2},$$

Theorem 11.18 shows that $|f|$ is measurable for every complex measurable f.

Suppose μ is a measure on X, E is a measurable subset of X, and f is a complex function on X. We say that $f \in \mathscr{L}(\mu)$ on E provided that f is measurable and

(97) $$\int_E |f|\, d\mu < +\infty,$$

and we define

$$\int_E f\, d\mu = \int_E u\, d\mu + i \int_E v\, d\mu$$

if (97) holds. Since $|u| \leq |f|$, $|v| \leq |f|$, and $|f| \leq |u| + |v|$, it is clear that (97) holds if and only if $u \in \mathscr{L}(\mu)$ and $v \in \mathscr{L}(\mu)$ on E.

Theorems 11.23(a), (d), (e), (f), 11.24(b), 11.26, 11.27, 11.29, and 11.32 can now be extended to Lebesgue integrals of complex functions. The proofs are quite straightforward. That of Theorem 11.26 is the only one that offers anything of interest:

If $f \in \mathscr{L}(\mu)$ on E, there is a complex number c, $|c| = 1$, such that

$$c \int_E f\, d\mu \geq 0.$$

Put $g = cf = u + iv$, u and v real. Then

$$\left| \int_E f\, d\mu \right| = c \int_E f\, d\mu = \int_E g\, d\mu = \int_E u\, d\mu \leq \int_E |f|\, d\mu.$$

The third of the above equalities holds since the preceding ones show that $\int g\, d\mu$ is real.

FUNCTIONS OF CLASS $\mathscr{L}^2$

As an application of the Lebesgue theory, we shall now extend the Parseval theorem (which we proved only for Riemann-integrable functions in Chap. 8) and prove the Riesz-Fischer theorem for orthonormal sets of functions.

11.34 Definition Let X be a measurable space. We say that a complex function $f \in \mathscr{L}^2(\mu)$ on X if f is measurable and if

$$\int_X |f|^2 \, d\mu < +\infty.$$

If μ is Lebesgue measure, we say $f \in \mathscr{L}^2$. For $f \in \mathscr{L}^2(\mu)$ (we shall omit the phrase "on X" from now on) we define

$$\|f\| = \left\{ \int_X |f|^2 \, d\mu \right\}^{1/2}$$

and call $\|f\|$ the $\mathscr{L}^2(\mu)$ norm of f.

11.35 Theorem *Suppose $f \in \mathscr{L}^2(\mu)$ and $g \in \mathscr{L}^2(\mu)$. Then $fg \in \mathscr{L}(\mu)$, and*

$$\int_X |fg| \, d\mu \leq \|f\| \, \|g\|. \tag{98}$$

This is the Schwarz inequality, which we have already encountered for series and for Riemann integrals. It follows from the inequality

$$0 \leq \int_X (|f| + \lambda |g|)^2 \, d\mu = \|f\|^2 + 2\lambda \int_X |fg| \, d\mu + \lambda^2 \|g\|^2,$$

which holds for every real λ.

11.36 Theorem *If $f \in \mathscr{L}^2(\mu)$ and $g \in \mathscr{L}^2(\mu)$, then $f + g \in \mathscr{L}^2(\mu)$, and*

$$\|f + g\| \leq \|f\| + \|g\|.$$

Proof The Schwarz inequality shows that

$$\|f + g\|^2 = \int |f|^2 + \int f\bar{g} + \int \bar{f}g + \int |g|^2$$
$$\leq \|f\|^2 + 2\|f\| \, \|g\| + \|g\|^2$$
$$= (\|f\| + \|g\|)^2.$$

11.37 Remark If we define the distance between two functions f and g in $\mathscr{L}^2(\mu)$ to be $\|f - g\|$, we see that the conditions of Definition 2.15 are satisfied, except for the fact that $\|f - g\| = 0$ does not imply that $f(x) = g(x)$ for all x, but only for almost all x. Thus, if we identify functions which differ only on a set of measure zero, $\mathscr{L}^2(\mu)$ is a metric space.

We now consider $\mathscr{L}^2$ on an interval of the real line, with respect to Lebesgue measure.

11.38 Theorem *The continuous functions form a dense subset of $\mathscr{L}^2$ on $[a, b]$.*

More explicitly, this means that for any $f \in \mathscr{L}^2$ on $[a, b]$, and any $\varepsilon > 0$, there is a function g, continuous on $[a, b]$, such that

$$\|f - g\| = \left\{ \int_a^b |f - g|^2 \, dx \right\}^{1/2} < \varepsilon.$$

Proof We shall say that f is approximated in $\mathscr{L}^2$ by a sequence $\{g_n\}$ if $\|f - g_n\| \to 0$ as $n \to \infty$.

Let A be a closed subset of $[a, b]$, and K_A its characteristic function. Put

$$t(x) = \inf |x - y| \qquad (y \in A)$$

and

$$g_n(x) = \frac{1}{1 + nt(x)} \qquad (n = 1, 2, 3, \ldots).$$

Then g_n is continuous on $[a, b]$, $g_n(x) = 1$ on A, and $g_n(x) \to 0$ on B, where $B = [a, b] - A$. Hence

$$\|g_n - K_A\| = \left\{ \int_B g_n^2 \, dx \right\}^{1/2} \to 0$$

by Theorem 11.32. Thus characteristic functions of closed sets can be approximated in $\mathscr{L}^2$ by continuous functions.

By (39) the same is true for the characteristic function of any measurable set, and hence also for simple measurable functions.

If $f \geq 0$ and $f \in \mathscr{L}^2$, let $\{s_n\}$ be a monotonically increasing sequence of simple nonnegative measurable functions such that $s_n(x) \to f(x)$. Since $|f - s_n|^2 \leq f^2$, Theorem 11.32 shows that $\|f - s_n\| \to 0$.

The general case follows.

11.39 Definition We say that a sequence of complex functions $\{\phi_n\}$ is an *orthonormal* set of functions on a measurable space X if

$$\int_X \phi_n \bar{\phi}_m \, d\mu = \begin{cases} 0 & (n \neq m), \\ 1 & (n = m). \end{cases}$$

In particular, we must have $\phi_n \in \mathscr{L}^2(\mu)$. If $f \in \mathscr{L}^2(\mu)$ and if

$$c_n = \int_X f \bar{\phi}_n \, d\mu \qquad (n = 1, 2, 3, \ldots),$$

we write

$$f \sim \sum_{n=1}^{\infty} c_n \phi_n,$$

as in Definition 8.10.

The definition of a trigonometric Fourier series is extended in the same way to $\mathscr{L}^2$ (or even to $\mathscr{L}$) on $[-\pi, \pi]$. Theorems 8.11 and 8.12 (the Bessel inequality) hold for any $f \in \mathscr{L}^2(\mu)$. The proofs are the same, word for word.

We can now prove the Parseval theorem.

11.40 Theorem *Suppose*

$$(99) \qquad f(x) \sim \sum_{-\infty}^{\infty} c_n e^{inx},$$

where $f \in \mathscr{L}^2$ on $[-\pi, \pi]$. Let s_n be the nth partial sum of (99). Then

$$(100) \qquad \lim_{n \to \infty} \|f - s_n\| = 0,$$

$$(101) \qquad \sum_{-\infty}^{\infty} |c_n|^2 = \frac{1}{2\pi} \int_{-\pi}^{\pi} |f|^2 \, dx.$$

Proof Let $\varepsilon > 0$ be given. By Theorem 11.38, there is a continuous function g such that

$$\|f - g\| < \frac{\varepsilon}{2}.$$

Moreover, it is easy to see that we can arrange it so that $g(\pi) = g(-\pi)$. Then g can be extended to a periodic continuous function. By Theorem 8.16, there is a trigonometric polynomial T, of degree N, say, such that

$$\|g - T\| < \frac{\varepsilon}{2}.$$

Hence, by Theorem 8.11 (extended to $\mathscr{L}^2$), $n \geq N$ implies

$$\|s_n - f\| \leq \|T - f\| < \varepsilon,$$

and (100) follows. Equation (101) is deduced from (100) as in the proof of Theorem 8.16.

Corollary *If $f \in \mathscr{L}^2$ on $[-\pi, \pi]$, and if*

$$\int_{-\pi}^{\pi} f(x) e^{-inx} \, dx = 0 \qquad (n = 0, \pm 1, \pm 2, \ldots),$$

then $\|f\| = 0$.

Thus if two functions in $\mathscr{L}^2$ have the same Fourier series, they differ at most on a set of measure zero.

11.41 Definition Let f and $f_n \in \mathscr{L}^2(\mu)$ $(n = 1, 2, 3, \ldots)$. We say that $\{f_n\}$ converges to f in $\mathscr{L}^2(\mu)$ if $\|f_n - f\| \to 0$. We say that $\{f_n\}$ is a Cauchy sequence in $\mathscr{L}^2(\mu)$ if for every $\varepsilon > 0$ there is an integer N such that $n \geq N$, $m \geq N$ implies $\|f_n - f_m\| \leq \varepsilon$.

11.42 Theorem *If $\{f_n\}$ is a Cauchy sequence in $\mathscr{L}^2(\mu)$, then there exists a function $f \in \mathscr{L}^2(\mu)$ such that $\{f_n\}$ converges to f in $\mathscr{L}^2(\mu)$.*

This says, in other words, that $\mathscr{L}^2(\mu)$ is a *complete* metric space.

Proof Since $\{f_n\}$ is a Cauchy sequence, we can find a sequence $\{n_k\}$, $k = 1, 2, 3, \ldots$, such that

$$\|f_{n_k} - f_{n_{k+1}}\| < \frac{1}{2^k} \qquad (k = 1, 2, 3, \ldots).$$

Choose a function $g \in \mathscr{L}^2(\mu)$. By the Schwarz inequality,

$$\int_X |g(f_{n_k} - f_{n_{k+1}})| \, d\mu \leq \frac{\|g\|}{2^k}.$$

Hence

(102) $$\sum_{k=1}^{\infty} \int_X |g(f_{n_k} - f_{n_{k+1}})| \, d\mu \leq \|g\|.$$

By Theorem 11.30, we may interchange the summation and integration in (102). It follows that

(103) $$|g(x)| \sum_{k=1}^{\infty} |f_{n_k}(x) - f_{n_{k+1}}(x)| < +\infty$$

almost everywhere on X. Therefore

(104) $$\sum_{k=1}^{\infty} |f_{n_{k+1}}(x) - f_{n_k}(x)| < +\infty$$

almost everywhere on X. For if the series in (104) were divergent on a set E of positive measure, we could take $g(x)$ to be nonzero on a subset of E of positive measure, thus obtaining a contradiction to (103).

Since the kth partial sum of the series

$$\sum_{k=1}^{\infty} (f_{n_{k+1}}(x) - f_{n_k}(x)),$$

which converges almost everywhere on X, is

$$f_{n_{k+1}}(x) - f_{n_1}(x),$$

we see that the equation
$$f(x) = \lim_{k \to \infty} f_{n_k}(x)$$
defines $f(x)$ for almost all $x \in X$, and it does not matter how we define $f(x)$ at the remaining points of X.

We shall now show that this function f has the desired properties. Let $\varepsilon > 0$ be given, and choose N as indicated in Definition 11.41. If $n_k > N$, Fatou's theorem shows that
$$\|f - f_{n_k}\| \le \liminf_{i \to \infty} \|f_{n_i} - f_{n_k}\| \le \varepsilon.$$

Thus $f - f_{n_k} \in \mathscr{L}^2(\mu)$, and since $f = (f - f_{n_k}) + f_{n_k}$, we see that $f \in \mathscr{L}^2(\mu)$. Also, since ε is arbitrary,
$$\lim_{k \to \infty} \|f - f_{n_k}\| = 0.$$

Finally, the inequality

(105) $$\|f - f_n\| \le \|f - f_{n_k}\| + \|f_{n_k} - f_n\|$$

shows that $\{f_n\}$ converges to f in $\mathscr{L}^2(\mu)$; for if we take n and n_k large enough, each of the two terms on the right of (105) can be made arbitrarily small.

11.43 The Riesz-Fischer theorem Let $\{\phi_n\}$ be orthonormal on X. Suppose $\Sigma |c_n|^2$ converges, and put $s_n = c_1\phi_1 + \cdots + c_n\phi_n$. Then there exists a function $f \in \mathscr{L}^2(\mu)$ such that $\{s_n\}$ converges to f in $\mathscr{L}^2(\mu)$, and such that
$$f \sim \sum_{n=1}^{\infty} c_n \phi_n.$$

Proof For $n > m$,
$$\|s_n - s_m\|^2 = |c_{m+1}|^2 + \cdots + |c_n|^2,$$
so that $\{s_n\}$ is a Cauchy sequence in $\mathscr{L}^2(\mu)$. By Theorem 11.42, there is a function $f \in \mathscr{L}^2(\mu)$ such that
$$\lim_{n \to \infty} \|f - s_n\| = 0.$$

Now, for $n > k$,
$$\int_X f \bar{\phi}_k \, d\mu - c_k = \int_X f \bar{\phi}_k \, d\mu - \int_X s_n \bar{\phi}_k \, d\mu,$$

so that
$$\left| \int_X f\bar\phi_k \, d\mu - c_k \right| \leq \|f - s_n\| \cdot \|\phi_k\| + \|f - s_n\|.$$

Letting $n \to \infty$, we see that
$$c_k = \int_X f\bar\phi_k \, d\mu \qquad (k = 1, 2, 3, \ldots),$$
and the proof is complete.

11.44 Definition An orthonormal set $\{\phi_n\}$ is said to be *complete* if, for $f \in \mathscr{L}^2(\mu)$, the equations
$$\int_X f\bar\phi_n \, d\mu = 0 \qquad (n = 1, 2, 3, \ldots)$$
imply that $\|f\| = 0$.

In the Corollary to Theorem 11.40 we deduced the completeness of the trigonometric system from the Parseval equation (101). Conversely, the Parseval equation holds for every complete orthonormal set:

11.45 Theorem *Let $\{\phi_n\}$ be a complete orthonormal set. If $f \in \mathscr{L}^2(\mu)$ and if*

(106)
$$f \sim \sum_{n=1}^{\infty} c_n \phi_n,$$

then

(107)
$$\int_X |f|^2 \, d\mu = \sum_{n=1}^{\infty} |c_n|^2.$$

Proof By the Bessel inequality, $\Sigma |c_n|^2$ converges. Putting
$$s_n = c_1 \phi_1 + \cdots + c_n \phi_n,$$
the Riesz-Fischer theorem shows that there is a function $g \in \mathscr{L}^2(\mu)$ such that

(108)
$$g \sim \sum_{n=1}^{\infty} c_n \phi_n,$$

and such that $\|g - s_n\| \to 0$. Hence $\|s_n\| \to \|g\|$. Since
$$\|s_n\|^2 = |c_1|^2 + \cdots + |c_n|^2,$$
we have

(109)
$$\int_X |g|^2 \, d\mu = \sum_{n=1}^{\infty} |c_n|^2.$$

Now (106), (108), and the completeness of $\{\phi_n\}$ show that $\|f - g\| = 0$, so that (109) implies (107).

Combining Theorems 11.43 and 11.45, we arrive at the very interesting conclusion that every complete orthonormal set induces a 1-1 correspondence between the functions $f \in \mathscr{L}^2(\mu)$ (identifying those which are equal almost everywhere) on the one hand and the sequences $\{c_n\}$ for which $\Sigma |c_n|^2$ converges, on the other. The representation

$$f \sim \sum_{n=1}^{\infty} c_n \phi_n,$$

together with the Parseval equation, shows that $\mathscr{L}^2(\mu)$ may be regarded as an infinite-dimensional euclidean space (the so-called "Hilbert space"), in which the point f has coordinates c_n, and the functions ϕ_n are the coordinate vectors.

EXERCISES

1. If $f \geq 0$ and $\int_E f\, d\mu = 0$, prove that $f(x) = 0$ almost everywhere on E. *Hint:* Let E_n be the subset of E on which $f(x) > 1/n$. Write $A = \bigcup E_n$. Then $\mu(A) = 0$ if and only if $\mu(E_n) = 0$ for every n.

2. If $\int_A f\, d\mu = 0$ for every measurable subset A of a measurable set E, then $f(x) = 0$ almost everywhere on E.

3. If $\{f_n\}$ is a sequence of measurable functions, prove that the set of points x at which $\{f_n(x)\}$ converges is measurable.

4. If $f \in \mathscr{L}(\mu)$ on E and g is bounded and measurable on E, then $fg \in \mathscr{L}(\mu)$ on E.

5. Put
$$g(x) = \begin{cases} 0 & (0 \leq x \leq \tfrac{1}{2}), \\ 1 & (\tfrac{1}{2} < x \leq 1), \end{cases}$$
$$f_{2k}(x) = g(x) \qquad (0 \leq x \leq 1),$$
$$f_{2k+1}(x) = g(1-x) \qquad (0 \leq x \leq 1).$$

Show that
$$\liminf_{n \to \infty} f_n(x) = 0 \qquad (0 \leq x \leq 1),$$

but
$$\int_0^1 f_n(x)\, dx = \tfrac{1}{2}.$$

[Compare with (77).]

6. Let
$$f_n(x) = \begin{cases} \frac{1}{n} & (|x| \leq n), \\ 0 & (|x| > n). \end{cases}$$

Then $f_n(x) \to 0$ uniformly on R^1, but
$$\int_{-\infty}^{\infty} f_n \, dx = 2 \quad (n = 1, 2, 3, \ldots).$$

(We write $\int_{-\infty}^{\infty}$ in place of $\int_{R^1}$.) Thus uniform convergence does not imply dominated convergence in the sense of Theorem 11.32. However, on sets of finite measure, uniformly convergent sequences of bounded functions do satisfy Theorem 11.32.

7. Find a necessary and sufficient condition that $f \in \mathscr{R}(\alpha)$ on $[a, b]$. *Hint:* Consider Example 11.6(b) and Theorem 11.33.

8. If $f \in \mathscr{R}$ on $[a, b]$ and if $F(x) = \int_a^x f(t)\,dt$, prove that $F'(x) = f(x)$ almost everywhere on $[a, b]$.

9. Prove that the function F given by (96) is continuous on $[a, b]$.

10. If $\mu(X) < +\infty$ and $f \in \mathscr{L}^2(\mu)$ on X, prove that $f \in \mathscr{L}(\mu)$ on X. If
$$\mu(X) = +\infty,$$
this is false. For instance, if
$$f(x) = \frac{1}{1 + |x|},$$
then $f \in \mathscr{L}^2$ on R^1, but $f \notin \mathscr{L}$ on R^1.

11. If $f, g \in \mathscr{L}(\mu)$ on X, define the distance between f and g by
$$\int_X |f - g|\, d\mu.$$
Prove that $\mathscr{L}(\mu)$ is a complete metric space.

12. Suppose
   (a) $|f(x, y)| \leq 1$ if $0 \leq x \leq 1, 0 \leq y \leq 1$,
   (b) for fixed x, $f(x, y)$ is a continuous function of y,
   (c) for fixed y, $f(x, y)$ is a continuous function of x.
   Put
   $$g(x) = \int_0^1 f(x, y)\, dy \quad (0 \leq x \leq 1).$$
   Is g continuous?

13. Consider the functions
$$f_n(x) = \sin nx \quad (n = 1, 2, 3, \ldots, -\pi \leq x \leq \pi)$$

as points of $\mathscr{L}^2$. Prove that the set of these points is closed and bounded, but not compact.

14. Prove that a complex function f is measurable if and only if $f^{-1}(V)$ is measurable for every open set V in the plane.

15. Let $\mathscr{R}$ be the ring of all elementary subsets of $(0, 1]$. If $0 < a \leq b \leq 1$, define
$$\phi([a, b]) = \phi([a, b)) = \phi((a, b]) = \phi((a, b)) = b - a,$$
but define
$$\phi((0, b)) = \phi((0, b]) = 1 + b$$
if $0 < b \leq 1$. Show that this gives an additive set function ϕ on $\mathscr{R}$, which is not regular and which cannot be extended to a countably additive set function on a σ-ring.

16. Suppose $\{n_k\}$ is an increasing sequence of positive integers and E is the set of all $x \in (-\pi, \pi)$ at which $\{\sin n_k x\}$ converges. Prove that $m(E) = 0$. *Hint:* For every $A \subset E$,
$$\int_A \sin n_k x \, dx \to 0,$$
and
$$2 \int_A (\sin n_k x)^2 \, dx = \int_A (1 - \cos 2n_k x) \, dx \to m(A) \qquad \text{as } k \to \infty.$$

17. Suppose $E \subset (-\pi, \pi)$, $m(E) > 0$, $\delta > 0$. Use the Bessel inequality to prove that there are at most finitely many integers n such that $\sin nx \geq \delta$ for all $x \in E$.

18. Suppose $f \in \mathscr{L}^2(\mu)$, $g \in \mathscr{L}^2(\mu)$. Prove that
$$\left| \int f\bar{g} \, d\mu \right|^2 = \int |f|^2 \, d\mu \int |g|^2 \, d\mu$$
if and only if there is a constant c such that $g(x) = cf(x)$ almost everywhere. (Compare Theorem 11.35.)

BIBLIOGRAPHY

ARTIN, E.: "The Gamma Function," Holt, Rinehart and Winston, Inc., New York, 1964.

BOAS, R. P.: "A Primer of Real Functions," Carus Mathematical Monograph No. 13, John Wiley & Sons, Inc., New York, 1960.

BUCK, R. C. (ed.): "Studies in Modern Analysis," Prentice-Hall, Inc., Englewood Cliffs, N.J., 1962.

——: "Advanced Calculus," 2d ed., McGraw-Hill Book Company, New York, 1965.

BURKILL, J. C.: "The Lebesgue Integral," Cambridge University Press, New York, 1951.

DIEUDONNÉ, J.: "Foundations of Modern Analysis," Academic Press, Inc., New York, 1960.

FLEMING, W. H.: "Functions of Several Variables," Addison-Wesley Publishing Company, Inc., Reading, Mass., 1965.

GRAVES, L. M.: "The Theory of Functions of Real Variables," 2d ed., McGraw-Hill Book Company, New York, 1956.

HALMOS, P. R.: "Measure Theory," D. Van Nostrand Company, Inc., Princeton, N.J., 1950.

———: "Finite-dimensional Vector Spaces," 2d ed., D. Van Nostrand Company, Inc., Princeton, N.J., 1958.
HARDY, G. H.: "Pure Mathematics," 9th ed., Cambridge University Press, New York, 1947.
——— and ROGOSINSKI, W.: "Fourier Series," 2d ed., Cambridge University Press, New York, 1950.
HERSTEIN, I. N.: "Topics in Algebra," Blaisdell Publishing Company, New York, 1964.
HEWITT, E., and STROMBERG, K.: "Real and Abstract Analysis," Springer Publishing Co., Inc., New York, 1965.
KELLOGG, O. D.: "Foundations of Potential Theory," Frederick Ungar Publishing Co., New York, 1940.
KNOPP, K.: "Theory and Application of Infinite Series," Blackie & Son, Ltd., Glasgow, 1928.
LANDAU, E. G. H.: "Foundations of Analysis," Chelsea Publishing Company, New York, 1951.
MCSHANE, E. J.: "Integration," Princeton University Press, Princeton, N.J., 1944.
NIVEN, I. M.: "Irrational Numbers," Carus Mathematical Monograph No. 11, John Wiley & Sons, Inc., New York, 1956.
ROYDEN, H. L.: "Real Analysis," The Macmillan Company, New York, 1963.
RUDIN, W.: "Real and Complex Analysis," 2d ed., McGraw-Hill Book Company, New York, 1974.
SIMMONS, G. F.: "Topology and Modern Analysis," McGraw-Hill Book Company, New York, 1963.
SINGER, I. M., and THORPE, J. A.: "Lecture Notes on Elementary Topology and Geometry," Scott, Foresman and Company, Glenview, Ill., 1967.
SMITH, K. T.: "Primer of Modern Analysis," Bogden and Quigley, Tarrytown-on-Hudson, N.Y., 1971.
SPIVAK, M.: "Calculus on Manifolds," W. A. Benjamin, Inc., New York, 1965.
THURSTON, H. A.: "The Number System," Blackie & Son, Ltd., London-Glasgow, 1956.

LIST OF SPECIAL SYMBOLS

The symbols listed below are followed by a brief statement of their meaning and by the number of the page on which they are defined.

$\in$	belongs to	3
$\notin$	does not belong to	3
$\subset, \supset$	inclusion signs	3
Q	rational field	3
$<, \leq, >, \geq$	inequality signs	3
sup	least upper bound	4
inf	greatest lower bound	4
R	real field	8
$+\infty, -\infty, \infty$	infinities	11, 27
$\bar{z}$	complex conjugate	14
Re (z)	real part	14
Im (z)	imaginary part	14
$\|z\|$	absolute value	14
$\sum$	summation sign	15, 59
R^k	euclidean k-space	16
$\mathbf{0}$	null vector	16
$\mathbf{x} \cdot \mathbf{y}$	inner product	16
$\|\mathbf{x}\|$	norm of vector $\mathbf{x}$	16
$\{x_n\}$	sequence	26
$\bigcup, \cup$	union	27
$\bigcap, \cap$	intersection	27
(a, b)	segment	31
$[a, b]$	interval	31
E^c	complement of E	32
E'	limit points of E	35
$\bar{E}$	closure of E	35
lim	limit	47
$\to$	converges to	47, 98
lim sup	upper limit	56
lim inf	lower limit	56
$g \circ f$	composition	86
$f(x+)$	right-hand limit	94
$f(x-)$	left-hand limit	94
$f', \mathbf{f}'(x)$	derivatives	103, 112
$U(P,f), U(P,f,\alpha), L(P,f), L(P,f,\alpha)$ Riemann sums		121, 122

$\mathscr{R}, \mathscr{R}(\alpha)$ classes of Riemann (Stieltjes) integrable functions 121, 122
$\mathscr{C}(X)$ space of continuous functions 150
$\| \ \|$ norm 140, 150, 326
exp exponential function 179
D_N Dirichlet kernel 189
$\Gamma(x)$ gamma function 192
$\{\mathbf{e}_1, \ldots, \mathbf{e}_n\}$ standard basis 205
$L(X), L(X, Y)$ spaces of linear transformations................207
$[A]$ matrix 210
$D_j f$ partial derivative 215
∇f gradient....................217
$\mathscr{C}', \mathscr{C}''$ classes of differentiable functions 219, 235
det $[A]$ determinant.............232
$J_f(\mathbf{x})$ Jacobian 234
$\dfrac{\partial(y_1, \ldots, y_n)}{\partial(x_1, \ldots, x_n)}$ Jacobian 234

I^k k-cell 245
Q^k k-simplex 247
dx_I basic k-form 257
$\wedge$ multiplication symbol 254
d differentiation operator 260
ω_T transform of ω 262
∂ boundary operator 269
$\nabla \times \mathbf{F}$ curl 281
$\nabla \cdot \mathbf{F}$ divergence 281
$\mathscr{E}$ ring of elementary sets 303
m Lebesgue measure 303, 308
μ measure 303, 308
$\mathfrak{M}_F, \mathfrak{M}$ families of measurable sets 305
$\{x|P\}$ set with property P 310
f^+, f^- positive (negative) part of f 312
K_E characteristic function 313
$\mathscr{L}, \mathscr{L}(\mu), \mathscr{L}^2, \mathscr{L}^2(\mu)$ classes of Lebesgue-integrable functions 315, 326

INDEX

Abel, N. H., 75, 174
Absolute convergence, 71
   of integral, 138
Absolute value, 14
Addition (*see* Sum)
Addition formula, 178
Additivity, 301
Affine chain, 268
Affine mapping, 266
Affine simplex, 266
Algebra, 161
   self-adjoint, 165
   uniformly closed, 161
Algebraic numbers, 43
Almost everywhere, 317
Alternating series, 71
Analytic function, 172
Anticommutative law, 256
Arc, 136
Area element, 283
Arithmetic means, 80, 199
Artin, E., 192, 195
Associative law, 5, 28, 259
Axioms, 5

Baire's theorem, 46, 82
Ball, 31
Base, 45
Basic form, 257
Basis, 205
Bellman, R., 198
Bessel inequality, 188, 328
Beta function, 193
Binomial series, 201
Bohr-Mollerup theorem, 193
Borel-measurable function, 313

Borel set, 309
Boundary, 269
Bounded convergence, 322
Bounded function, 89
Bounded sequence, 48
Bounded set, 32
Brouwer's theorem, 203
Buck, R. C., 195

Cantor, G., 21, 30, 186
Cantor set, 41, 81, 138, 168, 309
Cardinal number, 25
Cauchy criterion, 54, 59, 147
Cauchy sequence, 21, 52, 82, 329
Cauchy's condensation test, 61
Cell, 31
$\mathscr{C}''$-equivalence, 280
Chain, 268
   affine, 268
   differentiable, 270
Chain rule, 105, 214
Change of variables, 132, 252, 262
Characteristic function, 313
Circle of convergence, 69
Closed curve, 136
Closed form, 275
Closed set, 32
Closure, 35
   uniform, 151, 161
Collection, 27
Column matrix, 217
Column vector, 210
Common refinement, 123
Commutative law, 5, 28
Compact metric space, 36
Compact set, 36

Comparison test, 60
Complement, 32
Complete metric space, 54, 82, 151, 329
Complete orthonormal set, 331
Completion, 82
Complex field, 12, 184
Complex number, 12
Complex plane, 17
Component of a function, 87, 215
Composition, 86, 105, 127, 207
Condensation point, 45
Conjugate, 14
Connected set, 42
Constant function, 85
Continuity, 85
   uniform, 90
Continuous functions, space of, 150
Continuous mapping, 85
Continuously differentiable curve, 136
Continuously differentiable mapping, 219
Contraction, 220
Convergence, 47
   absolute, 71
   bounded, 322
   dominated, 321
   of integral, 138
   pointwise, 144
   radius of, 69, 79
   of sequences, 47
   of series, 59
   uniform, 147
Convex function, 101
Convex set, 31

Coordinate function, 88
Coordinates, 16, 205
Countable additivity, 301
Countable base, 45
Countable set, 25
Cover, 36
Cunningham, F., 167
Curl, 281
Curve, 136
  closed, 136
  continuously differentiable, 136
  rectifiable, 136
  space-filling, 168
Cut, 17

Davis, P. J., 192
Decimals, 11
Dedekind, R., 21
Dense subset, 9, 32
Dependent set, 205
Derivative, 104
  directional, 218
  of a form, 260
  of higher order, 110
  of an integral, 133, 236, 324
  integration of, 134, 324
  partial, 215
  of power series, 173
  total, 213
  of a transformation, 214
  of a vector-valued function, 112
Determinant, 232
  of an operator, 234
  product of, 233
Diagonal process, 30, 157
Diameter, 52
Differentiable function, 104, 212
Differential, 213
Differential equation, 119, 170
Differential form (see Form)
Differentiation (see Derivative)
Dimension, 205
Directional derivative, 218
Dirichlet's kernel, 189
Discontinuities, 94
Disjoint sets, 27
Distance, 30
Distributive law, 6, 20, 28
Divergence, 281
Divergence theorem, 253, 272, 288
Divergent sequence, 47
Divergent series, 59
Domain, 24
Dominated convergence theorem, 155, 167, 321
Double sequence, 144

e, 63
Eberlein, W. F., 184
Elementary set, 303
Empty set, 3
Equicontinuity, 156

Equivalence relation, 25
Euclidean space, 16, 30
Euler's constant, 197
Exact form, 275
Existence theorem, 170
Exponential function, 178
Extended real number system, 11
Extension, 99

Family, 27
Fatou's theorem, 320
Fejér's kernel, 199
Fejér's theorem, 199
Field axioms, 5
Fine, N. J., 100
Finite set, 25
Fixed point, 117
  theorems, 117, 203, 220
Fleming, W. H., 280
Flip, 249
Form, 254
  basic, 257
  of class $\mathscr{C}', \mathscr{C}''$, 254
  closed, 275
  derivative of, 260
  exact, 275
  product of, 258, 260
  sum of, 256
Fourier, J. B., 186
Fourier coefficients, 186, 187
Fourier series, 186, 187, 328
Function, 24
  absolute value, 88
  analytic, 172
  Borel-measurable, 313
  bounded, 89
  characteristic, 313
  component of, 87
  constant, 85
  continuous, 85
    from left, 97
    from right, 97
  continuously differentiable, 219
  convex, 101
  decreasing, 95
  differentiable, 104, 212
  exponential, 178
  harmonic, 297
  increasing, 95
  inverse, 90
  Lebesgue-integrable, 315
  limit, 144
  linear, 206
  logarithmic, 180
  measurable, 310
  monotonic, 95
  nowhere differentiable continuous, 154
  one-to-one, 25
  orthogonal, 187
  periodic, 183
  product of, 85
  rational, 88
  Riemann-integrable, 121

Function:
  simple, 313
  sum of, 85
  summable, 315
  trigonometric, 182
  uniformly continuous, 90
  uniformly differentiable, 115
  vector-valued, 85
Fundamental theorem of calculus, 134, 324

Gamma function, 192
Geometric series, 61
Gradient, 217, 281
Graph, 99
Greatest lower bound, 4
Green's identities, 297
Green's theorem, 253, 255, 272, 282

Half-open interval, 31
Harmonic function, 297
Havin, V. P., 113
Heine-Borel theorem, 39
Helly's selection theorem, 167
Herstein, I. N., 65
Hewitt, E., 21
Higher-order derivative, 110
Hilbert space, 332
Hölder's inequality, 139

i, 13
Identity operator, 232
Image, 24
Imaginary part, 14
Implicit function theorem, 224
Improper integral, 139
Increasing index, 257
Increasing sequence, 55
Independent set, 205
Index of a curve, 201
Infimum, 4
Infinite series, 59
Infinite set, 25
Infinity, 11
Initial-value problem, 119, 170
Inner product, 16
Integrable functions, spaces of, 315, 326
Integral:
  countable additivity of, 316
  differentiation of, 133, 236, 324
  Lebesgue, 314
  lower, 121, 122
  Riemann, 121
  Stieltjes, 122
  upper, 121, 122
Integral test, 139
Integration:
  of derivative, 134, 324
  by parts, 134, 139, 141
Interior, 43

Interior point, 32
Intermediate value, 93, 100, 108
Intersection, 27
Interval, 31, 302
Into, 24
Inverse function, 90
Inverse function theorem, 221
Inverse image, 24
Inverse of linear operator, 207
Inverse mapping, 90
Invertible transformation, 207
Irrational number, 1, 10, 65
Isolated point, 32
Isometry, 82, 170
Isomorphism, 21

Jacobian, 234

Kellogg, O. D., 281
Kestelman, H., 167
Knopp, K., 21, 63

Landau, E. G. H., 21
Laplacian, 297
Least upper bound, 4
  property, 4, 18
Lebesgue, H.|L., 186
Lebesgue-integrable function, 315
Lebesgue integral, 314
Lebesgue measure, 308
Lebesgue's theorem, 155, 167, 318, 321
Left-hand limit, 94
Leibnitz, G. W., 71
Length, 136
L'Hospital's rule, 109, 113
Limit, 47, 83, 144
  left-hand, 94
  lower, 56
  pointwise, 144
  right-hand, 94
  subsequential, 51
  upper, 56
Limit function, 144
Limit point, 32
Line, 17
Line integral, 255
Linear combination, 204
Linear function, 206
Linear mapping, 206
Linear operator, 207
Linear transformation, 206
Local maximum, 107
Localization theorem, 190
Locally one-to-one mapping, 223
Logarithm, 22, 180
Logarithmic function, 180
Lower bound, 3
Lower integral, 121, 122
Lower limit, 56

McShane, E. J., 313

Mapping, 24
  affine, 266
  continuous, 85
  continuously differentiable, 219
  linear, 206
  open, 100, 223
  primitive, 248
  uniformly continuous, 90
  (*See also* Function)
Matrix, 210
  product, 211
Maximum, 90
Mean square approximation, 187
Mean value theorem, 108, 235
Measurable function, 310
Measurable set, 305, 310
Measurable space, 310
Measure, 308
  outer, 304
Measure space, 310
Measure zero, set of, 309, 317
Mertens, F., 74
Metric space, 30
Minimum, 90
Möbius band, 298
Monotone convergence theorem, 318
Monotonic function, 95, 302
Monotonic sequence, 55
Multiplication (*see* Product)

Negative number, 7
Negative orientation, 267
Neighborhood, 32
Newton's method, 118
Nijenhuis, A., 223
Niven, I., 65, 198
Nonnegative number, 60
Norm, 16, 140, 150, 326
  of operator, 208
Normal derivative, 297
Normal space, 101
Normal vector, 284
Nowhere differentiable function, 154
Null space, 228
Null vector, 16
Number:
  algebraic, 43
  cardinal, 25
  complex, 12
  decimal, 11
  finite, 12
  irrational, 1, 10, 65
  negative, 7
  nonnegative, 60
  positive, 7, 8
  rational, 1
  real, 8

One-to-one correspondence, 25
Onto, 24
Open cover, 36

Open mapping, 100, 223
Open set, 32
Order, 3, 17
  lexicographic, 22
Ordered field, 7, 20
  k-tuple, 16
  pair, 12
  set, 3, 18, 22
Oriented simplex, 266
Origin, 16
Orthogonal set of functions, 187
Orthonormal set, 187, 327, 331
Outer measure, 304

Parameter domain, 254
Parameter interval, 136
Parseval's theorem, 191, 198, 328, 331
Partial derivative, 215
Partial sum, 59, 186
Partition, 120
  of unity, 251
Perfect set, 32
Periodic function, 183, 190
π, 183
Plane, 17
Poincaré's lemma, 275, 280
Pointwise bounded sequence, 155
Pointwise convergence, 144
Polynomial, 88
  trigonometric, 185
Positive orientation, 267
Power series, 69, 172
Primes, 197
Primitive mapping, 248
Product, 5
  Cauchy, 73
  of complex numbers, 12
  of determinants, 233
  of field elements, 5
  of forms, 258, 260
  of functions, 85
  inner, 16
  of matrices, 211
  of real numbers, 19, 20
  scalar, 16
  of series, 73
  of transformations, 207
Projection, 228
Proper subset, 3

Radius, 31, 32
  of convergence, 69, 79
Range, 24, 207
Rank, 228
Rank theorem, 229
Ratio test, 66
Rational function, 88
Rational number, 1
Real field, 8
Real line, 17
Real number, 8
Real part, 14

Rearrangement, 75
Rectifiable curve, 136
Refinement, 123
Reflexive property, 25
Regular set function, 303
Relatively open set, 35
Remainder, 211, 244
Restriction, 99
Riemann, B., 76, 186
Riemann integral, 121
Riemann-Stieltjes integral, 122
Riesz-Fischer theorem, 330
Right-hand limit, 94
Ring, 301
Robison, G. B., 184
Root, 10
Root test, 65
Row matrix, 217

Saddle point, 240
Scalar product, 16
Schoenberg, I. J., 168
Schwarz inequality, 15, 139, 326
Segment, 31
Self-adjoint algebra, 165
Separable space, 45
Separated sets, 42
Separation of points, 162
Sequence, 26
   bounded, 48
   Cauchy, 52, 82, 329
   convergent, 47
   divergent, 47
   double, 144
   of functions, 143
   increasing, 55
   monotonic, 55
   pointwise bounded, 155
   pointwise convergent, 144
   uniformly bounded, 155
   uniformly convergent, 157
Series, 59
   absolutely convergent, 71
   alternating, 71
   convergent, 59
   divergent, 59
   geometric, 61
   nonabsolutely convergent, 72
   power, 69, 172
   product of, 73
   trigonometric, 186
   uniformly convergent, 157
Set, 3
   at most countable, 25
   Borel, 309
   bounded, 32
   bounded above, 3
   Cantor, 41, 81, 138, 168, 309
   closed, 32
   compact, 36
   complete orthonormal, 331
   connected, 42
   convex, 31
   countable, 25

Set,
   dense, 9, 32
   elementary, 303
   empty, 3
   finite, 25
   independent, 205
   infinite, 25
   measurable, 305, 310
   nonempty, 3
   open, 32
   ordered, 3
   perfect, 32, 41
   relatively open, 35
   uncountable, 25, 30, 41
Set function, 301
σ-ring, 301
Simple discontinuity, 94
Simple function, 313
Simplex, 247
   affine, 266
   differentiable, 269
   oriented, 266
Singer, I. M., 280
Solid angle, 294
Space:
   compact metric, 36
   complete metric, 54
   connected, 42
   of continuous functions, 150
   euclidean, 16
   Hilbert, 332
   of integrable functions, 315, 326
   measurable, 310
   measure, 310
   metric, 30
   normal, 101
   separable, 45
Span, 204
Sphere, 272, 277, 294
Spivak, M., 272, 280
Square root, 2, 81, 118
Standard basis, 205
Standard presentation, 257
Standard simplex, 266
Stark, E. L., 199
Step function, 129
Stieltjes integral, 122
Stirling's formula, 194, 200
Stokes' theorem, 253, 272, 287
Stone-Weierstrass theorem, 162, 190, 246
Stromberg, K., 21
Subadditivity, 304
Subcover, 36
Subfield, 8, 13
Subsequence, 51
Subsequential limit, 51
Subset, 3
   dense, 9, 32
   proper, 3
Sum, 5
   of complex numbers, 12
   of field elements, 5
   of forms, 256
   of functions, 85

Sum,
   of linear transformations, 207
   of oriented simplexes, 268
   of real numbers, 18
   of series, 59
   of vectors, 16
Summation by parts, 70
Support, 246
Supremum, 4
Supremum norm, 150
Surface, 254
Symmetric difference, 305

Tangent plane, 284
Tangent vector, 286
Tangential component, 286
Taylor polynomial, 244
Taylor's theorem, 110, 116, 176, 243
Thorpe, J. A., 280
Thurston, H. A., 21
Torus, 239–240, 285
Total derivative, 213
Transformation (*see* Function; Mapping)
Transitivity, 25
Triangle inequality, 14, 16, 30, 140
Trigonometric functions, 182
Trigonometric polynomial, 185
Trigonometric series, 186

Uncountable set, 25, 30, 41
Uniform boundedness, 155
Uniform closure, 151
Uniform continuity, 90
Uniform convergence, 147
Uniformly closed algebra, 161
Uniformly continuous mapping, 90
Union, 27
Uniqueness theorem, 119, 258
Unit cube, 247
Unit vector, 217
Upper bound, 3
Upper integral, 121, 122
Upper limit, 56

Value, 24
Variable of integration, 122
Vector, 16
Vector field, 281
Vector space, 16, 204
Vector-valued function, 85
   derivative of, 112
Volume, 255, 282

Weierstrass test, 148
Weierstrass theorem, 40, 159
Winding number, 201

Zero set, 98, 117
Zeta function, 141

推 荐 阅 读

线性代数高级教程：矩阵理论及应用

作者：Stephan Ramon Garcia 等 ISBN：978-7-111-64004-2 定价：99.00元

矩阵分析（原书第2版）

作者：Roger A. Horn 等 ISBN：978-7-111-47754-9 定价：119.00元

代数（原书第2版）

作者：Michael Artin ISBN：978-7-111-48212-3 定价：79.00元

概率与计算：算法与数据分析中的随机化和概率技术（原书第2版）

作者：Michael Mitzenmacher 等 ISBN：978-7-111-64411-8 定价：99.00元

推荐阅读

具体数学：计算机科学基础（英文版·原书第2版）典藏版

作者：[美] 葛立恒（Ronald L. Graham）等著 ISBN: 978-7-111-64195-7 定价：139.00元

实分析（原书第4版）

作者：[美] H. L. 罗伊登（H. L. Royden）P. M. 等著 ISBN: 978-7-111-63084-5 定价：129.00元